AF305132

THE ALGORITHM OF WAR

Code Compute Conquer

THE ALGORITHM OF WAR

Code Compute Conquer

INSIGHTS FOR INDIAN ARMY'S NEXT LEAP

Colonel Maneesh Parthsarthy, VSM

Estd 1870

United Service Institution of India
New Delhi

PENTAGON PRESS LLP

First published in 2026 by
PENTAGON PRESS LLP
206, Peacock Lane, Shahpur Jat
New Delhi-110049, India
Contact: 011-26490600

Typeset in AGaramond, 11 Point
Printed at
Quiinkit Infotech Pvt. Ltd., Gurugram

ISBN 978-81-685612-3-6

Disclaimer: The views expressed in this book are those of the author and do not necessarily reflect those of the United Service Institution of India, New Delhi, or the Government of India.

www.pentagonpress.in

Dedicated To

My family, whose unwavering support and sacrifices have been the bedrock of my journey. Your love and understanding have sustained me through the challenges of military life and the pursuit of knowledge.

My parent unit, 2ⁿᵈ Battalion The 8ᵗʰ Gorkha Rifles, where I learned the true meaning of camaraderie, discipline, and honour. The values instilled in me there have been a guiding light throughout my career.

The 51 Special Action Group, the unit I had the privilege to command. It was here that I witnessed first-hand, the transformative power of cutting-edge technology in counter-terrorism operations. The bravery and innovation of my comrades inspired me to explore the intersection of technology and warfare.

The Infantry and the Indian Army, the guardians of our nation's sovereignty. Your dedication to duty and commitment to excellence are the foundation upon which this book is built. May this work contribute to the ongoing evolution of our armed forces, ensuring they remain at the forefront of technological advancement and strategic prowess.

With deepest gratitude and respect.

—Maneesh

CONTENTS

FOREWORD

It is with great pleasure and a deep sense of strategic significance that I write this foreword to *The Algorithm of War: Code, Compute, Conquer – Insights for the Indian Army's Next Leap*. As a retired infantry general, I have witnessed firsthand the profound impact of technology on the character and conduct of war. Warfare has always been a crucible of innovation, but in today's rapidly evolving battlespace, staying ahead of the technological curve is no longer merely advantageous – it is an operational imperative.

This book, meticulously researched and thoughtfully structured by Colonel Maneesh Parthsarthy, VSM, enters this discourse at precisely the right moment. It examines the disruptive technologies that are poised to redefine the future of conflict. From artificial intelligence, robotics, and cyber systems to biotechnology, hypersonics, and quantum technologies, the author offers a comprehensive appraisal of emerging capabilities and their potential application across the combat arms and space domain of the Indian Army.

The depth, breadth, and clarity of research in this work are truly commendable. Colonel Parthsarthy does not merely catalogue technologies that are already reshaping the modern battlefield; he ventures boldly into those that are still in developmental or prototype stages. This forward-looking approach is critical for a military that must anticipate the contours of tomorrow's conflict rather than merely react to them. His structured progression – from a historical survey of technology in warfare to an analysis of emerging technologies and their specific implications for each combat arm – demonstrates rigour, vision, and admirable methodical precision.

What distinguishes this book is its India-centric orientation. The concluding chapter, with its recommendations for the Indian Army, is particularly notable. It offers actionable insights into how these technologies can be integrated into force structures, training systems, and operational doctrines. The emphasis on leveraging DRDO, ISRO, private industry, startups, and MSMEs reflects a pragmatic roadmap for innovation and self-reliance. Equally compelling is the examination of global best practices and the evocative articulation of the concept of "the soldier as scientist and the scientist as soldier" – a reminder that the future force must be anchored in a seamless civil–military technological partnership.

Colonel Parthsarthy has addressed a theme that is both critical and urgent for the Indian Army. This book is not an abstract academic exercise; it is a call to action. It compels us to rethink assumptions, embrace disruption, and harness the full spectrum of emerging technologies to ensure that the Indian Army remains future-ready, resilient, and operationally superior in an era of complex threats.

I am confident that *The Algorithm of War: Code, Compute, Conquer* will become an invaluable reference for military leaders, policymakers, researchers, and all those engaged in shaping India's defence preparedness. It is a testament to the author's dedication, scholarship, and strategic vision. I wholeheartedly commend Colonel Maneesh Parthsarthy, VSM, for this timely and significant contribution to the intellectual growth of the Indian armed forces.

Maj Gen B K Sharma, AVSM, SM (Retd)**
Director General
United Service Institution of India

PREFACE

In an era where the line between science fiction and reality grows ever thinner, the nature of warfare is undergoing a profound and rapid transformation. The battlefield of the future will not be shaped solely by the bravery of soldiers or the might of traditional weaponry but by the invisible forces of algorithms, artificial intelligence, and cutting-edge technologies that are only beginning to emerge. As someone who has spent years studying the interplay between military strategy and technological innovation, I have seen how each leap forward – whether the wheel, gunpowder, or the atomic bomb – has redefined the art of war. Today, however, the pace of change is unlike anything in history, and the implications for the Indian Army, a force tasked with defending a nation of unparalleled diversity and strategic complexity, are both urgent and profound.

This book, *The Algorithm of War: Code, Compute, Conquer – Insights for the Indian Army's Next Leap*, is my attempt to chart this uncharted territory. It is not merely a catalogue of technological marvels but a strategic guide designed to help the Indian Army harness these advancements to maintain its edge in an increasingly unpredictable world. My aim is to bridge the gap between the abstract potential of emerging technologies and their practical application on the battlefield, offering a vision of how India's military can leap forward into the future.

A Journey of Exploration

The genesis of this book lies in a question that has haunted me for years: How will the disruptive technologies of today and tomorrow reshape warfare, and what must the Indian Army do to stay ahead? To answer this, I embarked on a journey of research and reflection, diving into the annals of military history, consulting with experts, and analysing global trends. The result is a work that blends historical context, technological insight, and strategic foresight – all tailored to the unique needs and challenges of India's armed forces.

This book is the product of countless hours spent poring over case studies, engaging in discussions with military and scientific minds, and drawing from my own observations of how technology is already beginning to alter the nature of conflict. It is my hope that the pages that follow will serve as both a resource and an inspiration for those who serve, lead, and think about the future of India's defence.

What Lies Within

The structure of *The Algorithm of War* is designed to take readers on a logical and compelling journey:

The Evolution of Technology in Warfare: We begin by looking back, tracing how technology has shaped military strategy from ancient times to the present. This historical lens reveals a timeless truth: those who master the tools of their era dominate the battlefield.

Emerging Disruptive Technologies: The heart of the book lies in its exploration of the technologies poised to redefine warfare. These are grouped into three categories: those already in use (like AI and robotics), those in prototype stages (such as advanced unmanned systems), and those still under development (including quantum computing and biotechnology). Each is examined for its potential to transform military operations.

Impact on Combat Arms: Recognizing that technology's value lies in its application, I dedicate chapters to how these innovations can enhance the capabilities of the Indian Army's combat and support arms as well as delve into how space can be leveraged to support ground operations by the Army.

A Roadmap for the Future: The book concludes with actionable recommendations, urging the Indian Army to integrate these technologies through partnerships with DRDO, ISRO, startups, and the private sector. It champions a culture of innovation, drawing lessons from global leaders and advocating for a synergy between soldiers and scientists – what I call "soldier as scientist and scientist as soldier."

Why This Matters for India

India's strategic landscape is as challenging as it is diverse. Bordered by nuclear-armed neighbours, facing asymmetric threats, and positioned in a region of geopolitical flux, the Indian Army cannot afford to lag in the global race for technological supremacy. This book is not a generic treatise on military technology; it is a focused exploration of how these advancements can address India's specific security needs. The title – *Code, Compute, Conquer* – encapsulates this vision: code will guide strategy, computing power will enable execution, and mastery of both will secure victory.

A Personal Reflection

As I wrote this book, I was struck by the human dimension of this technological revolution. Behind every algorithm and machine are the soldiers who will wield them, the leaders who will deploy them, and the innovators who will create them. This is not a story of machines replacing humans but of humans leveraging machines to achieve what was once impossible. I have sought to infuse these pages with both rigor and passion, to offer not just analysis but a call to action.

An Invitation

Whether you are a military officer, a policymaker, a technologist, or simply a reader curious about the future, I invite you to explore these ideas with me. The battlefield of tomorrow is taking shape today, and the choices we make now will determine our readiness for what lies ahead. *The Algorithm of War: Code, Compute, Conquer* is more than a book – it is a conversation, a challenge, and a vision for the Indian Army's next great leap.

Thank you for joining me on this journey.

Colonel Maneesh Parthsarthy

ACKNOWLEDGEMENTS

The creation of *The Algorithm of War: Code, Compute, Conquer – Insights for the Indian Army's Next Leap* has been a profound journey, one that would not have been possible without the guidance, expertise, and support of numerous individuals who enriched this work with their knowledge and encouragement. My deepest gratitude goes to those who have contributed to making this book a reality.

I owe an immense debt of gratitude to Lieutenant General RS Panwar, my guide and mentor, whose unparalleled expertise and vision were pivotal in shaping this book. His meticulous guidance on the content, structure, and flow of the manuscript ensured that it remained focused, coherent, and impactful. His deep insights into military strategy and emerging technologies illuminated the path forward, and his steadfast encouragement kept me motivated throughout this endeavour. This book is as much a reflection of his wisdom as it is of my efforts.

I am profoundly grateful to Major General BK Sharma, Director General, United Service Institution of India, for his steadfast mentorship and intellectual generosity. His deep understanding of warfare, strategic foresight, and nuanced perspectives on military leadership consistently enriched my thinking beyond the boundaries of my research. His ability to distil complex operational realities into strategic insights kept me grounded and focused throughout this journey. Under his visionary guidance, I gained not only academic clarity but also a richer appreciation of the evolving character of conflict. It has been an honour to learn from a leader of his stature and experience.

I am equally thankful to Major General RS Yadav, the Director of our institute, whose leadership and unwavering support fostered an environment of intellectual rigor and innovation. His belief in the significance of this project and his encouragement to explore the frontiers of military thought inspired me to aim for excellence. His distinguished career in the infantry served as a constant reminder of the real-world implications of this work, and I am deeply honoured to have had his guidance and backing.

My sincere appreciation extends to the officers and subject matter experts from the Service Headquarters and various ministries who generously shared their time and expertise. Their participation in my presentations and the valuable remarks and inputs they provided greatly enriched the content of this book, ensuring its relevance and alignment with the operational realities of the Indian Army.

I am also deeply grateful to the distinguished fellows and research scholars from various think tanks who attended my presentations and offered critical feedback.

Their diverse perspectives and incisive inputs helped refine the arguments and broaden the scope of this work, making it a more comprehensive resource for addressing the challenges of future warfare.

To my family, colleagues, and friends, thank you for your patience, encouragement, and unwavering support during the countless hours of research and writing. Your belief in this project sustained me through its challenges.

Finally, I express my profound gratitude to the Indian Army, an institution that embodies courage, sacrifice, and resilience. It is my privilege to contribute to its ongoing evolution, and I hope this book serves as a meaningful step toward preparing it for the technological leaps ahead.

Colonel Maneesh Parthsarthy

1

Technology and Warfare

"Here lies a man who was an ape
Nature, grown weary of his shape
Conceived and carried out the plan
By which the ape is now the man"
—Humbert Wolfe, *The Spectator*, 1925[1]

HISTORY OF TECHNOLOGY

The above lines by Humbert Wolfe were written as an Epitaph of the Australopithecus Africanus, the skull of the 'man-like ape' discovered by Professor Raymond Dart, who was a renowned Australian anatomist teaching at the University of Witwatersrand in South Africa at the time of the discovery. The fossilized remains came from the Taung Quarry in the process of blasting for lime. When presented with the material, Dart established that it was the face, mandible, and endocast (fossilized interior of the cranial vault) of a juvenile hominin. He based this on the anterior position of the foramen magnum and aspects of brain morphology reflected on the interior of the skull vault. Dart named his find Australopithecus Africanus, meaning "southern ape of Africa," and the specimen became known as the "Taung Child."[2] Australopithecus is said to be living on the African continent approximately 3.5 million years ago from whom the early modern humans supposedly originated.[3] Thus, the presence of technology in the life of early humans is older than the origin of the word "technology" itself and the way it has shaped human evolution. It is predated in use by the Ancient Greek word tékhne, used to mean 'knowledge of how to make things', which encompassed activities like architecture.[4] In its earliest forms, technology predates recorded history and has been in existence since early human life, right from the Stone Age to the modern era of AI and emerging disruptive technologies. The identification of the history of technology with the history of human-like species does not help in fixing a precise point for its origin, because the estimates of prehistorians and anthropologists concerning the emergence of human species vary so widely. A degree of specialization in tool making was achieved by the time of the Neanderthals (70,000 BCE); more-advanced tools, requiring assemblage of head and haft, were produced by Cro-Magnons (perhaps as early as 35,000 BCE); while the application of mechanical principles was achieved by pottery-making Neolithic (New Stone Age; 6000 BCE) and Metal Age peoples (about 3000 BCE).[5]

The earliest examples of some kind of skill and craft being applied to human endeavour can be seen from the evolution of tools which could fulfil the survival needs of the humans during the Stone Age. Perfect example of this evolution can be seen in most of museums round the world which display the variety of stone made tools used in the prehistoric times. The kind of tools developed depended on the the geography and the climate which in turn dictated the type of flora and fauna available for survival. Since the early human communities were nomadic and with sparse populations springing across large continents, the rudimentary technology and skill practiced remained largely isolated and regionalised until the advent of the Neolithic Age marked by large-scale migration of these communities. Still military technology was nowhere in the horizons of human cerebration till the time enough competing resources had been created for them to fight over. It was only once agriculture and animal husbandry took roots in human existence that organised competition for resources gave way to skirmishes, fights and battles. However, it was human physical prowess that dominated warfighting of the early ages as no considerable development in military weaponry was undertaken due to the isolated communities more focussed on survival further discouraged by no transportation and heavy human costs involved in technological innovation. The greatest discoveries that revolutionised human existence till ancient times would be stone working, fire, metals and the wheel.

Till Antiquity, warfighting continued to be dominated by human muscle and thus organised infantry units developed. However, the injection of cavalry into warfighting was a watershed moment that altered the nature of warfare from 400 CE onwards and continued to dominate the battlefields of Europe until challenged by the appearance of the crossbow in the 9th Century. It was the first of a series of technological and tactical developments that culminated in the rise of infantry elites to a position of tactical dominance. This infantry revolution began when the crossbow spread northward into areas that were peripheral to the economic, cultural and political core of feudal Europe and where the topography was unfavourable for mounted shock action and the land too poor to support an armoured elite. Within this closed military topography, the crossbow soon proved itself the missile weapon par excellence of positional and guerrilla warfare.[6]

The reasons for the crossbow's success were simple: crossbows were capable of killing the most powerful of mounted warriors, yet they were far cheaper than warhorses and armour and were much easier to master than the skills of equestrian combat. Also, it was far easier to learn to fire a crossbow than a long bow of equivalent power. Serious war bows had significant advantages over the crossbow in range, accuracy, and maximum rate of fire, but crossbowmen could be recruited and trained quickly as adults, while a lifetime of constant practice was required to master the Turkish or Mongol composite bow or the English longbow.

The crossbow directly challenged the mounted elite's dominance of the means of armed violence, a point that the religious authorities did not miss. In 1139, the second Lateran Council banned the crossbow under penalty of anathema as a weapon "hateful to God and unfit for Christians," and Emperor Conrad III of Germany (reigned 1138–52) forbade its use in his realms.[7] However, the crossbow had proved useful in the Crusades against the infidels and, once introduced, its popularity could not be stopped. This increased its acceptance among the European mounted elites, and the crossbow underwent a continuous process of technical development, further reinforced by the introduction of the Longbow, Halberd and the Pike into infantry weaponry that ended only in the 16th century, with the discovery of the harquebus and musket.[8] Another discovery which was destined to alter the fundamentals of warfighting deeply was that of Gunpowder. Chinese alchemists had laid their hands on a mixture of finely ground potassium nitrate also called saltpetre, charcoal and sulfur, which burnt rapidly once in contact with fire. Over the centuries, development of weapons that leveraged gunpowder remained slow within China, however, the Mongols were singularly responsible for the proliferation of this technology across West Asia and Europe, who worked tirelessly into the development of new weapons, muskets, rockets and canons and finally, used them against Mongol expansionist designs. Though, by the 14th Century Europa had made feverish advancement in use of Gunpowder, but the most transcendent display of its destructive power was harnessed by the Ottomans during the siege of Constantinople under the leadership of Mehemed II. This battle established the supremacy of Gunpowder as the ingredient of destructive power and catapulted Artillery to the centre stage of warfighting.

The 20th Century marked a dramatic transformation in military technologies as the power struggles between various powers aggravated over a variety of issues to include ethnicity, religion, trade, political ideology and sheer greed. It led to far greater emphasis on military means to achieve national interests, boosting innovation during the Industrial Revolution and with the militaries achieving greater precision, higher rate of fires, formidable protection, all across the three dimensions and beyond into the darker deeper space. The rifles, machine guns, canons, tanks, aircrafts and submarines saw incremental and endlessly escalating advancements. However, it involved an astronomical human cost which was evident from the colossal casualty figures of the two world wars which overshadowed various other conflicts, that continued even after the cataclysmic bombing of Hiroshima and Nagasaki. The nuclear bombs did not end the military competition, but did the exact opposite, with nations putting more and more efforts and resources into the never-ending arms race and aspiring nations creating a formidable military industrial complex. It became unclear whether conflict was fueling the hunger for technology or technology was fueling the hunger for conflict. The 21st Century is marked with military feats at an altogether unprecedented level. Whatever the human mind is capable of imagining is being pursued and being achieved to a reasonable level of success. Be it the realm of biological & chemical weapons or hypersonic weapon systems or unmanned & autonomous platforms or weaponisation of the space, technology has revolutionised warfighting across the multi domain canvas.

NATURE AND CHARACTER OF WAR

"War is more than a mere chameleon that slightly adapts its characteristics to the given case. As a total phenomenon its dominant tendencies always make war a paradoxical trinity – composed of primordial violence, hatred, and enmity…; of the play of chance and probability…; and of its element of subordination, as an instrument of policy, which makes it subject to reason alone."[9]

—Carl von Clausewitz in On War

Clausewitz, author of the most comprehensive theory of war, described war's enduring nature in the opening chapter of On War. He said and which remains true for a long period till recent history, that the nature of war largely remains constant with violence, animosity and chance integral to its existence, however, its character changes, which is a result of the relative strengths and motivation of the adversaries encompassing a mosaic of ingredients including organisation, tactics, weaponry, morale, intensity, leadership, terrain, surprise and plain luck. Thus, violence is the nature of war and the application of that violence in a particular calibrated manner manifests as the character of war. Clausewitz may have overlooked technology as an important driver for the character of war and we may not blame him for doing so, as the pace at which technology changed in his times was relatively slower than it did in the 19th and 20th centuries and shamelessly sluggish in comparison to that of the 21st century. Notwithstanding that, the various conflicts that have continued to unfold through the last two centuries have validated the Clautzwitzian Theory to a large extent given the primacy of primordial violence across the full spectrum of conflict ranging from civil unrest, civil wars, independence struggles, insurgencies, violent extremism, religious extremism, World Wars, Gulf Wars, Russia Ukraine War, Gaza War as well as the Red Sea Conflict. Irrespective of the stakeholders, whether state or non-state, violence has underscored all these conflicts. Even where the wars are being waged in multiple domains such as trade, cyber, information and space, nations are pursuing these non-contact alternatives not primarily to avoid outright violence but due to a lack of capability, opportunity and ideological support to inflict violence. Even amongst the most powerful of the nations, nonviolent means to pursue national ambitions is a matter of compulsion and not choice. Despite all the technologies available to enable non-violent contestations and achievement of national goals through means other than war, nonviolent methods only supplement, complement and reinforce the violent apparatus. As the emerging and disruptive technologies of this century alter the Character of War, the Nature of War still remains enduring with slight nudges to its true self.

TECHNOLOGY AND WAR: WHO DRIVES WHOM?

" This history may not be the most delightful to hear, since there is no mythology in it. But those who want to look into the truth of what was done in the past – which, given the human condition, will recur in the future, either in the same fashion or nearly so – those readers will find this History valuable enough, as this was composed to be a lasting possession and not to be heard for a prize at the moment of a contest."[10]

—Thucydides

For ages mankind has been striving to invent the ultimate weapon, so horrid in its effect that no one would dare to use it. A weapon that will bring the war machinery of the world to a standstill. However, the nuclear bomb couldn't achieve it, as more and more nations clamouring to join the elite nuclear club and those who formed the club constantly expanding their pool of warheads. As the nations attain nuclear capability, they go on to develop or obtain a variety of lethal weapons that allow them to wage war below the nuclear threshold with slightly less horrid results. In a manner reinforcing Thucydides' belief of the singular centrality of human nature to waging war rather than any other compelling reason. As Thucydides narrates the History of the Peloponnesian War, he repeatedly emphasizes that human beings are motivated to wage war primarily by fear, ambition, self-advantage and a desire to rule over others. He highlights the abovementioned intangible elements of human nature causing the devastating war between Athens and Lacedaemon. This holds true to this day too, as it's not technology that has been driving warfare, but largely it's the demands of the warfighting community that encourage, pursue and hasten technological advancements in the military domain.

In general, wars tend to accelerate technological development to adapt tools to solve specific military needs. Later, these military tools may evolve into non-military devices. A fairly good example of this is radar. While scientists around the world worked on using radio antennae to detect distant objects during the early part of the 20th century, Sir Robert Watson-Watt invented the first practical radar set in 1935. The British Air Ministry adopted his design and used it to detect aggressors during the early days of World War II.[11] However, in the civilian domain, radar initiated an entirely different technology leading to the invention of the most revolutionary kitchen appliances of all time. A scientist by the name of Percy L. Spencer made an interesting discovery while standing near a magnetron; a device that powers radar sets. Spencer had a chocolate bar in his pocket. The bar began to melt when Spencer stood near the magnetron. He decided to experiment with un-popped popcorn. He held the bag next to the magnetron and watched as the kernels turned into hardened, miniature white clouds.[12] And thus, the microwave oven was invented.

The peaceful repurposing of nuclear technology which we experience today is result of a purely military pursuit of inflicting irreparable and reprehensible damage to an adversary. On Oct. 4, 1957, the Soviet Union succeeded in launching the first manmade satellite, Sputnik, into the Earth's orbit. This event spurred on an intense, focused era of innovation, ultimately leading the US to create the National Aeronautics & Space Administration (NASA) a year later in October 1958.[13] Much of the competition focused on getting the United States' space technology ahead of the Soviets'. Several factors fuelled this race. One was fear, that if the Soviets could launch a rocket with a payload the size of Sputnik into orbit, it would be feasible that the country could launch a missile attack on the United States from across the globe. Even though there were plenty of scientific reasons to pursue the space race, on one level it boiled down to saber rattling between the two nations. While the motives behind the space race may not have been purely founded upon a desire to extend scientific knowledge, that in no way diminishes the accomplishments made by both countries. The space race was a symbolic conflict between both countries and put pressure on the scientists and engineers developing the systems and vehicles necessary to put men and women into space. Some of this technology later evolved into other forms and was eventually adapted to serve civilian purposes.

In the early 1960s, the US Department of Defense funded a project called the Advanced Research Projects Agency Network (ARPANET).[14] The purpose of the project was to develop the technologies and protocols necessary to allow multiple computers far separated in different locations to connect directly to one another. This would allow people to share information at unprecedented speeds. By creating a robust and flexible network, the United States could ensure that in the event of a catastrophe, access to the nation's supercomputers could remain intact. ARPANET's protocols allowed information to travel across different routes.[15] If something happened to a computer node along one route, the information could take another path to get to the right destination. The foundation for the Internet is in the protocols and designs built by the ARPANET team. And while no war directly played into its development, the threat of future conflicts did.

Conversely, over the centuries technology has been the primary source of innovation in warfighting. Though umpteen examples demonstrate how war drives technology but there have been crucial instances wherein technology has been instrumental in shaping warfare – the conduct of war. Technology drives changes in warfare more than any other factor, right from the use of bayonets and barbed wire, photo cameras, chlorine, American Tractor and telegraph to the employment of steamships, locomotives and aeroplanes. Till the World Wars, when almost the entire world was fighting, only governments and the militaries had the size and the kitty to undertake R&D. Actually, most of the technological advancements in metallurgy, communication, logistics, medicine, space and nuclear were done to fulfil the militaries requirement. As a result, the Military Industrial Complex pioneered the R&D as well as the manufacture of military hardware. However, as the Cold War set in, countries diverted their budgets towards comprehensive development and the Commercial Industrial Complex gained prominence. The advancement in information technology has democratised education, science and research. Today, the technology giants and the defense conglomerates have balance sheets larger than some countries. They can pay 7 figure salaries to young minds and carry on costly R&D. Thus, the large Mil Industrial Complex is turning into a Military Consumer Complex. Any platform required by the military can be either bought off the shelf or repurposed or ordered for production.

While political, social, and ethical factors may influence why wars are fought or place constraints on their conduct, it is technology that consistently and fundamentally reshapes how wars are actually waged. From the invention of gunpowder to the rise of drones, cyber weapons, and artificial intelligence, each technological advancement has altered the battlefield, tactics, and tempo of military operations. These shifts are not merely supportive, they are transformative, compelling changes in doctrine, organization, and even the very geography of conflict. Political leaders and military strategists respond to technological developments, not the other way around. As such, technology does not simply enhance warfare, it redefines it.

Therefore, it can be argued that technology is the primary, and arguably the only, enduring driver in the evolution of war's character. Innovations in artificial intelligence, robotics, telecommunications, biotech, and space tech are now led by the commercial sector, startups, tech giants, and universities, rather than the military. The military often adapts or modifies these civilian technologies for defense use. For example, smartphones, cloud computing, commercial satellites, and AI tools were all civilian-led innovations that now have major military relevance. So today, civilian technology often leads, and military applications follow, but the military still plays a key role in adapting, scaling, and weaponising those tools for specific use. Generally, it's a two-way street, but in recent times, the traffic flows more heavily from civil to military. Ultimately, 'Who drives whom?' is a never ending debate and it can fuel endlessly escalating duels by proponents of both the views. As of now I want to rest the case by stating that there is a balance and there have been periods when one drives the other depending upon the level of scientific exploration, industrial prowess, market dynamics and strategic competition.

Technology Presides, But Does Not Rule

Technology may be the driving force that modifies, shapes or changes warfare but it does not singularly determine the outcome of War. Had that been so, a more technologically advanced force would have come out as a natural victor in every conflict. War is never won by just one factor, neither man alone nor machine alone can guarantee victory. Machines like tanks, aircraft, or missiles give armies great power, but without skilled people to operate them, make decisions, and adapt to changing situations, those machines are useless. In the same way, brave and determined soldiers can only do so much if they don't have the right tools, weapons, or information. For example, during World War II, the German Blitzkrieg worked because it combined fast tanks, radios, planes, and highly trained troops, none of these alone could have succeeded, but together they overwhelmed their enemies. In modern wars too, like the Gulf War in 1991, it was not just advanced US technology, but also clear planning, coordination, and leadership that led to success. The success of ISIS, Al Qaeda, Taliban, Hamas and Houthi Rebels prove, how sometimes inferior force exploiting its unconventional organisation and tactics can defeat superior military force. Even in conventional conflicts sometimes leadership, generalship, purpose, popular support and pride overwhelm a technologically superior foe. Specially, the back breaking pace of military innovation in the 21st century has intensified the arms race as well as reduced the life of any technology. The current pace in unprecedented, unfathomable and probably incomprehensible in its totality. These technologies can make warfare highly interconnected, swift, transparent, lethal and precise in execution and highly selective in outcomes. This selectivity can either lower the human cost or make it incredibly exponential causing the adversary to collapse immediately. Thus, technological pursuits often come with ethical battles, whether to make the most lethal weapon system, cause max casualties, to kill or to injure, avoid collateral or not. The track record of the most developed nations in this aspect is quite full of double standards. The NPT and the Conventions on Cluster Bombs are apt examples.

However, even with all the technology in the world, it is still the human mind that decides how and when to use it. Technology provides the means, but people provide the purpose. The most powerful example of this is the use of atomic bombs on Hiroshima and Nagasaki in 1945. The technology was developed by scientists, but it was ultimately a human decision, by leaders and military commanders to use it in war. That decision changed the course of history. So, while machines and technology play a huge role, it is always a person who takes the final call.

While warfighting has historically driven technological development, through the urgent need for superior weapons, communication systems, and logistics, technology in the modern era increasingly drives changes in warfighting itself. Innovations like drones, cyber capabilities, and artificial intelligence have not only reshaped military doctrine but created entirely new domains of conflict. However, this transformation does not occur in a vacuum. Technology, in its raw form, is neutral, often developed for civilian use, scientific progress, or societal benefit. It is only when the human mind, driven by ambition, fear, or malice, chooses to repurpose these tools for conflict that they become instruments of war. The internet was not invented for cyberattacks, nor was satellite navigation created to guide missiles, yet both now serve those ends. Thus, while technology can enable new forms of warfare, it is ultimately the intent and choices of people that weaponise it. In this sense, warfare is not just shaped by what technology makes possible, but by what human judgment allows.

THE ACME OF MILITARY TECHNOLOGY

"Armies that could reach further, hit harder, and get there faster usually won, while the range-restricted, less well-armed, and slower armies lost. For this reason, a vast amount of human creative effort has been poured into extending the range, increasing the firepower, and accelerating the speed of weapons and of armies."[16]

—**Toffler**

The moot question that lies at the core of all the technological advancement taking place the world over is, what finally it achieves for the parties to conflict and mankind at large? Maybe there is no easy answer and the question cannot be dealt with objectively as every stakeholder has a unique perspective on the pursuit of a particular technology with a very subjective goal in mind. How a community, state or non-state entity intends to develop and employ a particular technology is dictated by the security imperatives, social and cultural realities, power balance, threat perception, policy, industrial and economic capacities, technological prowess and governance it experiences. However, all of them tend to justify their intentions, compulsions and motives. Mature societies try and align their programs to international norms, the powerful ones get away with justifying themselves to weak objections and the rogue ones couldn't care less. The outcomes of any military technology accomplished by a nation can be put into absolute harmony with what Thucydides alluded to a century ago – *"You know and we know, as practical men that the question of justice arises only between parties equal in strength and that the strong do what they can, and the weak suffer what they must."*[17]

Let's assume ourselves to be a perfect society with Leaders of the world having their hearts at the right place, then we can reasonably predicate that the ultimate goal of military technology and innovation would supposedly manifest into enhancing the effectiveness, efficiency and safety of military operations. It must in all probability aim to provide a decisive advantage over adversaries in terms of firepower, mobility, intelligence and also deter potential aggressors by demonstrating their ability to defend themselves and inflict significant damage if attacked. At the very same time ensuring greater force protection and minimum collateral damage. Historically, military technology has created weapons and systems to boost killing power and cause heavy casualties to adversary forces. However, in present times, a greater focus is on minimising the human cost of conflict. This has led towards increased development of precision weapons, advanced surveillance, and non-lethal options to incapacitate foes while reducing risk to civilian life. Medical advancements too have been constantly aiming at improving the survival rates of wounded troops. The increased push towards minimising the enemy and civilian casualties has been the collective concern of nations, as they become increasingly answerable towards ethics, international humanitarian laws and winning hearts & minds of the population in the war zones as well as in internally disturbed areas. Notwithstanding the ultimate goal of ensuring national security, protecting vital national interests and maintaining peace through strength, it's essential to recognize that military innovation is a complex and multifaceted process influenced by numerous factors, including technological advancements, strategic goals, operational requirements, ethical considerations and geopolitical dynamics. While reducing casualties may be a desired outcome, military forces must also maintain the capability to defend themselves and achieve their objectives in combat situations, which can sometimes necessitate the use of lethal force.

But as I said its an assumption that we live in a perfect world. All the reason of waging a *Jus Bellum* gets shunted to the back burner once war is orchestrated by humans who are prone to emotions such as anger, hatred, passion, panic, distrust and vengeance. While it's true that attributing the extremely high casualty rates of recent conflicts to longer ranges and higher lethality might oversimplify the complex factors at play, one can reasonably conclude that technological advancement and innovation have still not brought down the human cost of war. As I conclude this chapter, the conservative estimates put the human cost of war in Syria at 650000,[18] Iraq at 654000,[19] Yemen at 160000,[20] Russia-Ukraine at 340000[21] and Gaza at 31000[22] (highest ever in the 21st Century). The above figures may not be precise but even by standards of approximation, they are huge and depict the obvious bias of technological pursuits towards achieving higher kills, if not by the very nature of the technology but surely through its application.

As the War refuses to shed its brutal image, postmodern technologies are being matured and adopted to create integrated systems and platforms that enhance situational awareness and force efficiency, reduce the human cost, minimise the need for kinetic warfare and the likelihood of conflict as well as reduce the conflict between human aspirations and the nature. We find ourselves at a critical juncture where the only thing that may save the planet is harmony between our security needs and securing the environment.

Battle Field to Battle Space

During my final presentation of the research, Maj Gen BK Sharma, Director General, United Services Institution of India, asked me the question, as to "how I perceived the modern-day battlefield to be?" More than a straight forward answer, this led to a very intense discussion amongst all the learned members present in the room — physically as well as virtually (a coherent outcome of communication and space technologies). Post discussion, he handed me the book, Ready, Relevant and Resurgent: A Blueprint for the transformation of India's Military written by General Anil Chahuan, Chief of Defence Staff (CDS), which talks about the ongoing transformation of India's armed forces, focusing on jointness, integration, and self-reliance to address the challenges of 21st-century warfare and uphold national security. Though I had already completed my book by this time and had delivered my last presentation, it gave me confidence that there was some convergence of military minds, as the CDS too dwells upon, how war is rapidly evolving beyond traditional state-on-state clashes and conflict types now include grey-zone warfare, insurgencies, cyber-attacks and strategic competition in non-traditional domains. He goes on to emphasise that security issues have "diffused and proliferated" across domains beyond land, air and sea to cyber, space and the electromagnetic spectrum complicating threat assessments. Basically, the battle field has transcended from being uni dimensional to a multidimensional battle space.[23] It was after this enlightening discussion that I decided to add the succeeding paras in this chapter so as to lay the foundation of subsequent chapters. Thus, before taking a deep dive into the impact of emerging disruptive technologies on the future battlefield, it is imperative to understand the contours of the modern-day battle space itself. The conventional understanding of the battlefield, traditionally confined to physical terrains where armies clashed, has undergone a profound paradigm shift. The modern battlefield has evolved into a multidimensional battle space, encompassing land, sea, air, space, cyberspace and the cognitive domain. This transformation is driven by the integration of emerging disruptive technologies and niche innovations, such as artificial intelligence (AI), autonomous systems, cyber warfare capabilities and electronic warfare (EW). These technologies have expanded the scope of warfare, enabling conflicts to be pursued through both kinetic and non-kinetic means across all domains. They play a critical role across the various stages of conflict, while fostering increased jointness among armed forces in executing multi domain operations (MDO). The term Battle Space reflects the expanded nature of modern warfare, where conflicts are no longer limited to physical geography but extend across multiple domains.

The US Department of Defense defines battlespace as "the environment, factors and conditions that must be understood to successfully apply combat power, protect the force, or complete the mission."[24] This includes not only traditional domains (land, sea, air) but also space, cyberspace and the cognitive domain, where perceptions, narratives and decision-making are contested. The integration of disruptive technologies has blurred the lines between military and civilian spheres, physical and digital realms, and kinetic and non-kinetic operations.

The evolution from "battle field" to "battle space" represents a fundamental change in the character of war.

Death of Distance[25]

Geography largely remains the same even after substantial manmade modifications to terrain in form of defensive works, obstacles, border infrastructure, dams, military bases, subsurface military assets but both the physical movement of armies and the increased connectivity due to satellite communication and Internet of Things have shrunk timelines of military operations and their decision cycles.

Faster Kill Chain

Rapid advances in space technologies have enabled network-centric operations wherein the identification, acquisition and destruction of targets have become extremely fast and accurate. With countries increasingly expanding their space footprint, better satellite coverage will infuse greater resilience into the ISR frameworks and allow for uninterrupted swift and accurate engagements across the globe as well as into the space. This shortening of OODA Loop will make future conflicts more and more distributed and lethal.

No Safe Sanctuaries

Advances in UAS platforms, missile technologies, Hypersonic weapons and stealth systems will enable long-range precision strike capabilities and areas once thought to be relatively safe from enemy fires will become increasingly vulnerable. Cyber warfare, Electronic warfare and Cognitive Warfare will allow state and non-state actors to overcome geographical boundaries and demoralise & destabilise adversary through persistent attacks on critical infrastructure, civil amenities and military command and control centres.

Foe Without A Face

Increase participation of the civil industrial complex in the war effort will lead to proliferation of lethal weapon systems and technologies. Since most of the technologies will be dual use, with democratisation of the manufacturing material & technical knowledge as well as seamless connectivity & anonymity offered by the world wide web, non-state actors would find it increasingly convenient to procure lethal platforms and systems and leverage them at will across international borders, with or without claiming their actions.

The Physical and the Virtual: While the traditional domains of land, air, and sea remain critical, they are now completely enmeshed with the man-made domains of cyberspace and space. A military formation's physical location is now just one element of its posture; its digital signature, reliance on satellite communications and vulnerability to cyber-attack are equally important.

Kinetic and Non-Kinetic Means: Kinetic action is now often a culminating act, preceded by waves of non-kinetic effects. These include cyberattacks designed to cripple infrastructure, information campaigns to erode public will, and economic coercion to exhaust a nation's resources. The goal of non-kinetic action is to shape the battle space and achieve strategic objectives before a single shot is fired.

Multi-Domain Operations (MDO): The interconnectedness of these domains means that no single service can operate in isolation. Victory requires MDO, the seamless orchestration of effects across all domains to overwhelm an adversary. A cyber operation might disable an enemy's air defences (a cyber effect) to allow an air strike (an air effect) on a command-and-control node, enabling a ground manoeuvre (a land effect), all while being enabled by satellite intelligence (a space effect).

Several key technologies have accelerated this transformation, with recent conflicts providing stark examples of their impact.

Drones and Autonomous Systems: The Ubiquitous Eye and Sword

Unmanned Aerial Systems (UAS), from sophisticated platforms like the Bayraktar TB2 and those used in the Spiderweb attack by Ukraine[26] to cheap, commercially available FPV (First-Person View) drones, have saturated the battle space. They have collapsed the distance between reconnaissance and strike, turning tactical actions into strategic impacts.

Information Warfare and AI: The Cognitive Domain

The most profound shift has been the weaponisation of information to target the cognitive dimension of warfare, the human mind. Social media platforms, powered by AI-driven algorithms and susceptible to manipulation through bots and deepfakes, have become the primary vector for the initial stages of conflict.

Cyber Warfare: The Digital Battlefield

Cyber operations are no longer ancillary to military campaigns; they are a core component. Attacks on critical national infrastructure, power grids, banking systems, transportation networks can induce chaos and destabilize a government's ability to function.

Integrated Multi-Domain Strikes: The New Decisive Action

The most recent conflicts have demonstrated the culmination of these trends, where discrete technological advantages are woven together into a single, decisive operation that strikes across multiple domains simultaneously.

The Continuum of Conflict in the Modern Battle Space

The new technologies find their application across the continuum of conflict, designed to weaken an adversary from within before a major military commitment.[27]

Stage 1: Demoralisation

The objective is to erode the will to resist. This is a persistent, long-term effort waged in the information space. Adversaries use social media and state-controlled news outlets to sow division, promote narratives of government corruption and incompetence and question national identity. The aim is to create apathy and fracture the societal cohesion necessary to mount a unified defence.

Stage 2: Destabilisation

This stage transitions from psychological to material effects. It involves funding extremist or separatist groups, conducting targeted cyberattacks on critical services, and executing political assassinations or sabotage. The goal is to weaken the state's control and create internal chaos, making it appear incapable of ensuring the security of its citizens. The Russian support for separatists in the Donbas region from 2014 to 2022 is a textbook example of this stage.[28]

Stage 3: Creating Crisis

Here, the aggressor manufactures a pretext for direct military action. This often involves a false-flag operation, the declaration of a humanitarian crisis," or a rapid and threatening military buildup on the border. In the weeks leading up to the

February 2022 invasion, Russia made unsubstantiated claims of "genocide" in the Donbas and staged provocations, creating the casus belli for its "special military operation." This stage is designed to seize the narrative, justify intervention to domestic and international audiences, and overwhelm the adversary's decision-making process.

Stage 4: Intervention

This is the culminating kinetic stage where the full force of a joint, multi-domain military is brought to bear. The invasion of Ukraine in 2022 began with a coordinated barrage of precision missiles (air/space domain) targeting command centres and air defences, cyberattacks (cyber domain), a naval blockade (sea domain), and a multi-pronged ground assault (land domain). This stage is intended to be a swift and decisive blow, leveraging the advantages gained during the non-kinetic preparatory stages.

Underpinning the execution of these stages is a fundamental leap in decision-making velocity and complexity. The modern battle space is characterised by the evolution from the classic OODA loop (Observe-Orient-Decide-Act) to an "OODA Plus" paradigm, which has been manifested by network centricity. This new loop is fed by a sensor grid that spans all domains, fusing intelligence from satellites, drones, cyber-reconnaissance, and electronic warfare into a single, cohesive picture. To dominate in this environment, true tri-service net-centricity is non-negotiable, particularly for the Indian armed forces. This demands not only a robust and secure network but, more critically, the seamlessly networked application of capabilities where a naval sensor can guide an air force asset to prosecute a target of army interest, collapsing decision-making timelines and achieving effects with unprecedented speed and precision.

OVERVIEW OF EMERGING DISRUPTIVE TECHNOLOGIES IN WARFARE

"While the rhythm of change may appear slow,
the tempo of disruption is extremely fast."[29]

—Jay Samit

Definition and Characteristics

Emerging disruptive technologies refer to innovations that significantly alter the existing landscape of the economy, industry, or society, often leading to fundamental changes in how things are done. These technologies typically possess certain defining characteristics such as novelty, rapid advancement through improvements & iterations, uncertainty of impact and long-term implications and their potential to transform existing social and economic structures. However, there are two sides to the coin, the emerging side which implies to the nascent immature nature of the technology and the disruptive side which indicates towards a technology which has proven itself and is under full exploitation.

Emerging technologies are those that exhibit a degree of coherence over time and have the potential to have a significant impact on a wide range of socioeconomic domains. As a result of affecting the pattern of interaction between actors and institutions as well as the composition of actors and institutions, the socioeconomic domains have a significant impact on the broader domains. While this technology will have the greatest impact in the future, its application during the emerging phase is somewhat clear. The ambiguity arises from the fact that it is likely to have the greatest impact in the future.[30] Certain emerging technologies include Internet of Things, artificial intelligence, blockchain, virtual reality, drones, and 3D printing. These technologies are typically in the process of being introduced, adopted and refined and they often represent the cutting edge of scientific and technological progress.

If we study the Unified Payments Interface, it can be categorised as an emerging technology which is likely to get into the category of disruptive once it replaces the cash transactions to a large extent. It is rapidly proliferating and even getting accepted in friendly foreign countries and shall soon be the founding system of monetary transactions, until then it continues to be in the emerging territory. Thus, it is important to understand that every disruptive technology stems from an emerging phase but every emerging technology may not become a disruptive one. In an oversimplified manner, emerging technologies are the technologies representing the forefront of innovation, whereas disruptive technologies fall under the classification of emerging technologies that have matured to the point of causing significant disruption and change in the established industries, economy and social structure.

Disruptive technologies, thus, are a subset of emerging technologies that fundamentally alter existing markets, industries and businesses by replacing established products, practices, or services. They disrupt the status quo, leading to the creation of entirely new structures or the obsolescence of traditional technologies. The perfect example of a disruptive technology in the 21st Century is the smartphone revolution which disrupted the fixed-line telephone as well as camera industries by combining various functions into a single device. The combination of AI/ML and mobile communication has further disrupted the print media and radio space, changing how we communicate and access information. Ola, Uber and Rapido have disrupted the existing travel industry by offering a far more cost-effective, convenient and faster alternative through mobile apps. Social media apps, OTT video streaming platforms like Amazon Prime, Netflix have disrupted the television industry by generating on-demand content and altering content distribution and consumption. Likewise, Electric vehicles are disrupting the automotive industry by introducing environmentally friendly and efficient transportation options.

When emerging tech is absorbed by the militaries then they may impact warfare in a transcendental manner having a disruptive effect on the nature of war fighting. Some key examples include Autonomous and Unmanned systems, Cyber technology, AI/ML, Space Technology, Biotechnology and Bioengineering, Quantum Computing, Hypersonic weapons etc. These technologies can enhance reconnaissance, communication, and decision-making on the battlefield. Given the disruptive nature of AI, it has transformed systems and platforms across other technological streams and has emerged as an enabling technology all across the wider military mosaic.

The pace of military modernisation and the emergence of new capabilities arising from technology innovation will have a significant impact on the character of war and on how military operations will be conducted, including the fundamental role of the human combatant.[31] The belief that a rival is making a technological leap forward with a new weapon may induce sufficient fear to precipitate a crisis.[32] Not all technologies carry transformative potential, of course. The technologies throughout history that can lay claim to the label "revolutionary" are necessarily few – unless, of course, one's definition of a revolutionary technology is so generous that it robs the term of all utility.[33] World over, greater consensus between the experts of science and technology and the military industrial complex have accepted AI/ML, Unmanned and Autonomous Systems, Hypersonic Weapons, Directed Energy Weapons (DEWs), Cyber Warfare, Biotechnology, Robotics, Space Technology, Quantum Computing and Blockchain Technology as the technologies which qualify to be rightfully called Emerging Disruptive Technologies given their pace of development and successful exploitation in military operations, both kinetic and non-kinetic.

Most of them have figured amongst the top technologies of the 21st century and some as momentous and disruptive as the steam engine and the gunpowder. The Centre for Strategic and International Studies, Washington in 2023, went onto identify a set of technologies which were critical for the USA to maintain its Military Super Power status. After an in-depth review of dozens of important emerging technologies, researchers at CSIS on the basis of three pronged criterion of Impact on US Future Capability, requirement of government support and effective integration identified Secure and Redundant Communications, Quantum Technology, Bioengineering, Space-Based Technology, High-Performance Batteries, AI/ML and Robotics as the seven technologies that are most likely to make a significant difference in the success of the United States and its allies across the spectrum of conflict over the next decade. Combinations of these technologies are often more powerful than the sum of their parts. For example, AI/ML together with bioengineering could create radical breakthroughs, such as discovering new biological compounds, and combining space technology and quantum sensing could revolutionize intelligence work.[34]

Impact of Disruptive Technologies on Modern Warfare

"Companies survived the pandemic only by allowing wired workers to log in from home. Consumers avoided possible infection by shopping online. Specially made drones helped deliver lifesaving medicine in rich and poor countries alike. Advances in telemedicine helped scientists and doctors understand and fight the virus. Artificial intelligence helped hospitals predict how many beds and ventilators they would need at any one time. A spike in Google searches using phrases that included specific symptoms helped health officials detect outbreaks in places where doctors and hospitals are few and far between. AI played a crucial role in vaccine development by absorbing all available medical literature to identify links between the genetic properties of the virus and the chemical composition and effects of existing drugs."[35]

—**Ian Bremmer**

Ian Bremmer, an American political scientist in his 2022 book highlights the three dangers facing mankind and how the new technologies will reshape geopolitical order destabilizing our societies and livelihoods faster than we can comprehend and respond to. The growth in disruptive technologies is attributable to a host of factors primarily including advances in computing power, sensor miniaturization, wider and faster connectivity as well as technological breakthroughs in robotics, artificial intelligence, machine learning among others. Rapid advances in science and technology have led to the emergence of smarter weapons, intelligent systems and autonomous platforms that possess unheard of capabilities ranging from improved situational awareness as well as precision targeting to unmanned operations and cyber dominance.

Military applications continue to embrace these disruptions thereby forcing defense establishments and decision-makers to constantly improve and innovate strategy and policy. These specific applications of novel technologies do not only influence doctrine development but also shape force structures, strategic choices and operational doctrines besides overall national security outlook. As disruptive technologies continue to evolve and find their way into military applications, Understanding the significance of these technologies is of paramount importance to ensure the preparedness, resilience and effectiveness of military forces in the face of ever-evolving security threats and challenges. Additionally, world over, border disputes and territorial conflicts have long been significant sources of regional and global tensions. The emergence of disruptive technologies adds a new layer of complexity to these disputes, potentially altering the dynamics of conflict and necessitating a reassessment of response mechanisms. I have attempted to array few of the notable of these technologies in succeeding chapters based on the extent of their inclusion into the warfighting ecosystem.

Notes

1 News of the Second Competition. The Spectator, February 28, 1925. http://archive.spectator.co.uk/article/28th-february-1925/20/_news-of-the-second-competition.

2 Milne, Susan G. Australopithecus africanus. In The History of Our Tribe: Hominini. Milne Publishing. https://milnepublishing.geneseo.edu/the-history-of-our-tribe-hominini/chapter/australopithecus-africanus.

3 Wood, Bernard, and Brian G. Richmond. "Human Evolution: Taxonomy and Paleobiology." Journal of Anatomy 196, no. 1 (2000): 19–60.

4 Aristotle. The Nicomachean Ethics. Translated by David Ross. Oxford: Oxford University Press, Oxford World's Classics.

5 Technology in the Ancient World. Encyclopaedia Britannica. https://www.britannica.com/technology/history-of-technology/Technology-in-the-ancient-world.

6 The Horse Archer. Encyclopaedia Britannica. https://www.britannica.com/technology/military-technology/The-horse-archer.

7 Payne-Gallwey, Ralph. The Crossbow, Mediaeval and Modern, Military and Sporting: Its Construction, History & Management. London: Longmans, Green, 1903.

8 The Infantry Revolution c. 1200–1500. Encyclopaedia Britannica. https://www.britannica.com/technology/military-technology/The-infantry-revolution-c-1200-1500.

9 Clausewitz, Carl von. On War. Edited and translated by Michael Howard and Peter Paret. Princeton: Princeton University Press, 1984.

10 Thucydides. History of the Peloponnesian War. Translated by C.F. Smith.

11 Sir Robert Watson-Watt, Engineer, Inventor, and Pioneer of RADAR. Scottish Engineering Hall of Fame. https://engineeringhalloffame.org/profile/robert-alexander-watson-watt.

12 Percy Spencer, Microwave Oven, Consumer Devices. The Lemelson-MIT Program. https://lemelson.mit.edu/resources/percy-spencer.

13 Sputnik and the Dawn of the Space Age. NASA. https://www.nasa.gov/history/sputnik/index.html.

14 J.C.R. Licklider and The Universal Network. Living Internet. https://www.livinginternet.com/i/ii_licklider.htm

15 Leiner, Barry M., Vinton G. Cerf, David D. Clark, Robert E. Kahn, Leonard Kleinrock, Daniel C. Lynch, Jon Postel, Larry G. Roberts, and Stephen Wolff. "Brief History of the Internet." Internet Society, 1997. https://www.internetsociety.org/internet/history-internet/brief-history-internet/#f5.

16 Alvin and Heidi Toffler, *War and Anti-War: Survival at the Dawn of 21st Century* (New York: Warner Books, 1995), p. 30.

17 Thucydides, C.F. Smith (Translator), History of the Peloponnesian War

18 Casualties of the Syrian Civil War. Wikipedia. https://en.wikipedia.org/wiki/Casualties_of_the_Syrian_civil_war.

19 Mortality Before and After the 2003 Invasion of Iraq: Cluster Sample Survey. Brussels Tribunal (archived), September 2005. https://web.archive.org/web/20150907130701/http://brusselstribunal.org/pdf/lancet111006.pdf.

20 Yemen. ACLED Conflict Watchlist 2024. https://acleddata.com/conflict-watchlist-2024/yemen/

21 Russia-Ukraine War Report Card: Jan 2, 2024. Russia Matters. https://www.russiamatters.org/blog/russia-ukraine-war-report-card-jan-2-2024.

22 Gaza Death Toll 30,000 as Israel-Hamas War Rages. *CNN*, February 29, 2024. https://edition.cnn.com/2024/02/29/middleeast/gaza-death-toll-30000-israel-war-hnk-intl/index.html.

23 General Anil Chahuan, Chief of Defence Staff, Ready, Relevant and Resurgent: A Blueprint for the transformation of India's Military, The Nature of War and Type of Conflicts, Page 53

24 United States Department of Defense. Joint Publication 1-02: Department of Defense Dictionary of Military and Associated Terms, April 12, 2001 (as amended through October 17, 2007). https://marineparents.com/downloads/dod-terms.pdf.

25 General Anil Chahuan, Chief of Defence Staff, Ready, Relevant and Resurgent: A Blueprint for the transformation of India's Military, Military Geography, p. 11.

26 Laura Gozzi & BBC Verify, How Ukraine carried out daring 'Spider Web' attack on Russian bombers, *BBC News*, 03 June 2025, https://www.bbc.com/news/articles/cq69qnvj6nlo

27 Sharma, Maj Gen B.K. Interview by author. June 4, 2025, and June 30, 2025.

28 Conflict in Ukraine's Donbas: A Visual Explainer. The International Crisis Group. https://www.crisisgroup.org/content/conflict-ukraines-donbas-visual-explainer.

29 Jay Samit, Disrupt You!: Master Personal Transformation, Seize Opportunity, and Thrive in the Era of Endless Innovation.

30 Siddhartha Paul Tiwari, The Impact of New Technologies on Society: A Blueprint for The Future RS Global Warsaw, Poland, 2022.

31 Kott A, Alberts D, Zalman A, et al. Visualizing the Tactical Ground Battlefield in the Year 2050: Workshop Report (ARL-SR-0327), Adelphi, Maryland, U.S. Army Research Laboratory, 2015.

32 Evan Braden Montgomery, "Breaking Out of the Security Dilemma: Realism, Reassurance, and the Problem of Uncertainty," *International Security* 31, no. 2 (2006): 151–185.

33 Ash Rossiter, "Military Technology and Revolutions in Warfare: Priming the Drone Debate," *Defense and Security Analysis* 39, no. 2 (2023): 253–255.

34 Emily Harding, Harshana Ghoorhoo, Seven Critical Technologies for Winning the Next War, Report of the CSIS International Security Program, April 2023: 10-11.

35 Ian Bremmer, The Power of Crisis: How Three Threats and Our Response Will Change the World, 2022.

2

Disruption Experienced

"Battlefield invisible, consequences undeniable, any keystroke can shift the balance of power."

This category includes technologies that have already caused significant shifts in warfare. These technologies are fully developed, widely proliferated, and are being used at scale in various military operations. Their impact is not theoretical or emerging but already tangible in shaping the strategies, tactics, and outcomes of conflicts. Examples might include cyberwarfare, electronic warfare and other such technologies which have been around for a long time but new innovations have been driving their further development and widespread use across global militaries.

Key Parameters:
- *Level of Development:* Fully developed.
- *Extent of Proliferation:* Widely adopted by major and minor military forces.
- *User Exploitation:* Extensively used in real-world operations.
- *Stage of Development:* Mass-scale use.
- *Technological Evolution:* May include technologies that have been around for a while but have seen significant advancements (e.g., cybersecurity innovations).

Cyber Warfare

Never in his wildest dreams Sergey Ulasen, a software developer from Belarus had thought of himself uncovering the secret of the digital age. It all started when the technical support guys informed him about a rather unusual case: a customer in Iran reported arbitrary BSODs and computer reboots. His very first impression was that the anomalies found were due to some Windows misconfiguration or were the result of a conflict between installed applications. He finally parsed the malware and worked out that it was using zero-day vulnerabilities that allowed Stuxnet to penetrate even well-patched Windows computers and its digital certificate had been stolen. Additionally, the complexity of Stuxnet's code and extremely sophisticated rootkit technologies led his team to conclude that this malware was a fearsome beast with nothing else like it in the world.[1] Stuxnet in the rue sense announced beginning of the era of cyber warfare. Almost 58 percent of the computers in Iran had been affected and during the days Virus BlokAda, Sergey's employer worked to inform the world of such an anomaly, thousands of computers across the world had been affected by the worm. The exact numbers are debatable but it gave confidence to the hackers of the world that a sovereign nation can be attacked and defeated without firing a bullet. The world had entered the era of highly-advanced, highly-targeted, and highly-capable cyberwarfare.

Not that there were no cyber-attacks earlier than that. Prior to Stuxnet, the most sophisticated and damaging cyber-attack against an ICS was commonly regarded as the months-long attack launched by Vitek Boden against the Maroochy Shire Council sewerage control system in Queensland, Australia in 2000. A former employee of the contracting company that developed the system, the 48-year-old Australian used equipment and software he had taken from the company in order access the Industrial Control System network and alter data. His attack caused pumps to malfunction and alarms to turn off, resulting in raw sewage overflows that killed marine life and harmed the environment.[2] While not to diminish the seriousness of Boden's attack, it was nonetheless considerably less sophisticated than Stuxnet in at least three areas: access, command and control, and stealth. With respect to access, because the network controlling the centrifuges at Natanz was not remotely accessible or even connected to the Internet, Stuxnet was deployed in the form of a complex computer worm that spread across Microsoft Windows machines via USB memory sticks and local network links, exploiting several unknown (zero-day) vulnerabilities in the systems it hit and using fraudulent digital certificates to trick the systems into running its code.[3]

After the US unleashed Stuxnet, a cyber-race has ensued with countries raising armies of hackers and warriors to attack adversaries at fraction of the cost with full or limited deniability. North Korea, largely through its state-backed hacker organization, the Lazarus Group, was able to infiltrate Sony Pictures in 2014, then, in 2017, the country's Wannacry ransomware forced the UK's National Health Service into disaaray. On 23 December 2015, attackers behind the BlackEnergy malware successfully caused power outages for several hours in different regions of Ukraine. This cyber sabotage against three energy companies has been confirmed by the Ukrainian government. The power grid compromise has become known as the first-of-its-kind cyber warfare attack affecting civilians. In 2017, Russian cyberattacks brought Ukrainian banks, utilities and government agencies to their knees.

Every year, China's cyberwarfare division grows stronger from the talent scouted and zero-days identified at the Tianfu Cup – where competitive hackers tear into Google, Microsoft and Apple software. In 2021, ransomware software locked up the US's Colonial Pipeline. In August 2023, Ecuador's national election agency claimed that cyberattacks from India, Bangladesh, Pakistan, Russia, Ukraine, Indonesia and China caused difficulties for absentee voters attempting to vote online in the latest election. The agency didn't elaborate on the nature of the attacks. Pro-Hamas and pro-Israeli hacktivists have launched multiple cyberattacks against Israeli government sites and Hamas web pages in the aftermath of Hamas' attacks on Israel on October 2023. Russian and Iranian hacktivists also targeted Israeli government sites and Indian hacktivists have attacked Hamas websites in support of Israel.

In November 2023, Hacktivists stole 3,000 documents from NATO, the second time in three months that hacktivists have breached NATO's cybersecurity defenses. Hackers described themselves as "gay furry hackers" and announced their attack was retaliation against NATO countries' human rights abuses. NATO alleges the attack did not impact NATO missions, operations, or military deployments. In January 2024, Russian hackers launched a ransomware attack against Sweden's only digital service provider for government services. The attack affected operations for 120 government offices and came as Sweden prepared to join NATO. Sweden expects disruptions to continue for several weeks.[4] The list of these attacks is endlessly escalating as cyber warfare has the potential of destruction without consequences. Secondly, the cost of hardening a computer or a network than from attacking one, is almost 100 times.

Electronic Warfare

Modern militaries rely on communications equipment that uses broad portions of the spectrum to conduct military operations. This allows forces to talk, transmit data, provide navigation and timing information, and to command and control forces all over the world. They also rely on this to know where adversaries are, what adversaries are doing, where friendly forces are, and what effects weapons achieve. As a result, modern militaries attempt to dominate the spectrum through electronic warfare.[5] Electronic warfare is not new; it has been practiced in one form or another in every major conflict since the early days of the 19th century; in fact, ever since radio communications were first used in war. Early techniques were often primitive and it was only from World War II onwards that EW gained an element of sophistication and maturity. Electronic warfare has its roots in early radio and radar technologies. They can be traced back to the US Civil War in 1861. With the outbreak of the Civil War in 1861, telegraph wires became one of the most important targets. Military telegraph traffic was diverted to the wrong destinations, false orders were passed to Union commanders, and wires were cut to intercept information to Union forces. The first intentional use of radio jamming by the military occurred in 1902 during British naval exercises in the Mediterranean, then in 1903 during

US naval manoeuvres.[6] During the Russo-Japanese War of 1904–1905, the Japanese located the Russian fleet in the Tsushima Strait and transmitted the information "wirelessly" to the Japanese fleet headquarters. On July 13, 1904, Russian wireless telegraph stations successfully disrupted communication between a group of Japanese battleships. This was also the first example of electronic countermeasures. Later, the captain of the Russian warship Ural requested permission to disrupt Japanese communications with a stronger radio signal, but Russian Admiral Zinovy Rozhestvensky refused, thus allowing the Japanese to win a decisive battle at Tsushima.[7]

Electronic Warfare is a military action involving the use of electromagnetic energy to determine, exploit, reduce, or prevent the hostile use of electromagnetic spectrum as well as action which retains friendly use of electromagnetic spectrum.[8] From the perspective of military operations, there are three broad divisions of electronic warfare

- Electronic protection involves involving actions taken to protect personnel, facilities, and equipment from any effects of friendly, neutral, or enemy use of the EMS, as well as naturally occurring phenomena that degrade, neutralize, or destroy friendly combat capability.
- Electronic attack uses electromagnetic energy to degrade or deny an enemy's use of the spectrum. It involves the use of EM energy, DE, or anti-radiation weapons to attack personnel, facilities, or equipment with the intent of degrading, neutralizing, or destroying enemy combat capability and is considered a form of fires.
- EW support involves actions tasked by, or under direct control of, an operational commander to search for, intercept, identify, and locate or localize sources of intentional and unintentional radiated EM energy for the purpose of immediate threat recognition, targeting, planning, and conduct of future operations.[9]

Cognitive Warfare

Cognitive Warfare is the deliberate and systematic use of information, disinformation, psychological operations, and technological tools to manipulate the perceptions, beliefs, emotions, and decision-making processes of target populations or adversaries. It encompasses strategies that aim to influence thoughts and behaviours, exploit cognitive vulnerabilities, and dominate the information environment in both peacetime and conflict.

Cognitive warfare, in other words, concerns manipulating the perceptions and decision-making of the adversary while furthering one's strategic advantage. This manipulation has forever been part of war, starting from wars fuelled by anxiety, fear, and misinformation to current-day cyber warfare that involves hacking into a country's decision-making process. From ancient civilisations to modern superpowers, it was common to deceive, coerce through psychological warfare, spread disinformation, and manipulate the minds of the enemy as well as own leadership. Modern technologies have catapulted cognitive warfare onto a different level not only in speed but also in effectiveness. AI, machine learning, and data-driven targeting can manipulate human cognition on an unprecedented scale. Cognitive warfare blurs the lines, as peacetime and wartime distinctions are fading; influence operations occur continuously in media, social and cyberspace. What characterizes modern cognitive warfare is the sophistication of technology-based operations, permitting far more precise, targeted, and extended influence operations. While tools and technologies being employed have changed, the essence remains the same: manipulating the adversary's perception and morale.

Cognitive warfare is principally distinct from traditional warfare as it doesn't seek to physical destruction of the adversary but aims to undermine their cognitive abilities, i.e. their capacity to understand, process information, and make decisions. These days most militaries in the world, refer to a five-dimensional battlespace construct, which comprises of the land, sea, air, space, and cyberspace domains, with operations in this battlespace being referred to as Multi Domain Operations. In this five-dimensional construct, cyberspace should be understood in its wider sense as Info space, encompassing within its ambit the Electro-Magnetic and cognitive domains as well.[10] Cognitive Operations are thus a very potent means of warfare within the Informational element of Comprehensive National Power, the other two elements being Cyber Operations and Electronic Warfare.[11]

Warfare without Borders: Cognitive warfare transcends geographical boundaries. Unlike conventional forces restricted by borders or logistics, cognitive warfare can be waged globally with little cost, through digital platforms, social networks, and online content manipulation. Every individual with access to digital technology becomes a potential target, expanding the scope of warfare into the civilian realm.

Hybrid Warfare: Cognitive war integrates itself into all domains of cyber, economic, and conventional warfighting. Hybrid warfare will involve adversaries conducting coordinated campaigns that combine physical sabotage, cyber-attacks, and disinformation, blurring the line between war and peace. Thus world over the governments will have to develop strategies to counter the cognitive facet of hybrid warfare and be effectively prepared for kinetic and non-kinetic threats.

AI and Automation: Artificial intelligence will enable real-time analysis of social media, news, and other data sources, allowing for highly adaptive and responsive cognitive warfare operations. AI can detect trends, predict societal responses, and craft tailored influence operations at unprecedented speeds. Future cognitive warfare will be fought with AI-enhanced capabilities that target cognitive vulnerabilities with precision, making defense against these attacks increasingly complex.

Erosion of Trust: As cognitive warfare escalates, the credibility of media, social networks as well as government information systems will increasingly erode. The deluge of false narratives will make it very difficult for the public to differentiate between true and false information leading to increased societal polarisation and instability. Nations will need to invest heavily in instituting mechanisms to maintain social cohesion in the face of resolute cognitive attacks.

Influence on Military and Civilian Leadership: Direct targeting of the decision-making command (political masters, military leaders, and policy makers), cognitive warfare could lead to paralysis government machinery in times of crisis. Leadership training and decision-making processes will have to evolve accordingly and create redundancies to take on the onslaught of cognitive warfare, by developing strategies to mitigate the effects of disinformation, perception manipulation, and psychological coercion.

Nanotechnology

"Military applications of molecular manufacturing have even greater potential than nuclear weapons to radically change the balance of power."[12]

—Admiral (Ret.) David E. Jeremiah

By understanding advancements in nanotechnology as an example of an emerging disruptive technology, the extent of future military applications is virtually limitless. However, if advancements in nanotechnology result in the evolution of existing military capabilities, the benefits as well as consequence of new nano-weapons and warfighting capabilities may be as unprecedented as the introduction of novel warfighting capabilities that would emerge through disruptive innovation.[13]

According to the US National Nanotechnology Initiative, Nanotechnology is the understanding and control of matter at the nanoscale, at dimensions between approximately 1 and 100 nanometres, where unique phenomena enable novel applications. Matter can exhibit unusual physical, chemical, and biological properties at the nanoscale, differing in important ways from the properties of bulk materials, single atoms, and molecules. Some nanostructured materials are stronger or have different magnetic properties compared to other forms or sizes of the same material. Others are better at conducting heat or electricity. They may become more chemically reactive, reflect light better, or change colour as their size or structure is altered.[14]

The rapid emergence and confluence of nanotechnologies, biotechnologies and information technologies in the past two decades is having a huge impact on the development of military technologies.[15]

- Nanotechnology-enabled new structural materials: High strength, light weight, corrosion and heat resistant structural materials (metals, alloys, ceramics, polymers, composites).
- Thin-films made of Nano carbons that can be deposited onto surfaces for electrically active coatings.
- Nanocomposites and engineered nanoparticles for high-energy munitions.
- Nanometals and energy-absorbing nanomaterials for armaments.
- Self-Decontaminating Surfaces exploiting surface structures of nanomaterials.
- 'Smart' equipment including protective clothing, armours and ammunition; improved low observability (camouflage); sensing and mitigation of threats from landmines and chemical, biological, radiological and nuclear (CBRN) warfare agents.
- Biotechnology-enabled sensor technology, bio computing, bioengineered materials, biofuels, etc. Also military performance enhancement or wound stabilization and treatment. The darker side of application of biotechnology include the genetic modification of pathogenic materials including virus to cause bioweapons of mass extinction.
- Information technology (involving hardware, algorithms, and software) already has brought revolutionary changes in several military domains such as surveillance, communication, data transmission and cryptography. Advanced C4ISR capabilities – standing for Command, Control, Communications, Computers (C4) Intelligence, Surveillance and Reconnaissance (ISR) – provide an advantage through situational awareness, knowledge of the adversary and environment, and shortening the time between sensing and response.

Nanotechnology is poised to revolutionise future warfare by offering unprecedented advancements across various military domains, from materials science and electronics to medicine and weaponry. Its ability to manipulate matter at the atomic and molecular levels – within the nanometre scale – can enhance both offensive and defensive capabilities, fundamentally transforming how wars are fought and won. Nanostructured materials will provide the defense industry with the required means to develop lighter, flexible, more agile and more resistant military platforms, including light armorer vehicles, tanks, fighter jets and man-transportable micro unmanned air vehicles (MUAV).[16]

Nanotechnology is at the forefront of transforming future warfare as it has revolutionised the research and development of advanced metals and materials. Carbon nanotubes and graphene are nanomaterials that have extraordinary strength-to-weight ratios, and are found to be most effective in creating lightweight high strength armour. Such materials will allow of development of extremely light weight gear as well as vehicles and armament thereby reducing troop fatigue as well as enhancing overall logistics stamina of the forces by reducing wear and tear. Additionally, these materials can be induced with self-healing properties, enabling vehicles, body armour, and equipment to autonomously repair damage sustained during combat. This reduces the need for manual maintenance and increases the longevity and survivability of critical military assets. For instance, nanofibers could be used to reinforce the outer coatings of aircraft, tanks, and ships, improving their resistance to heat, pressure, and even weapons like bullets or directed energy attacks.

Nanotechnology also provides the foundation for the development of nano-drones and micro-robotics, which would play a pivotal role in future military operations. These miniaturised drones, capable of reconnaissance, surveillance, and even precise strikes, can operate in swarms to overwhelm adversaries.[17] The use of such nano-drones could provide significant advantages, especially in urban warfare, where their small size allows them to infiltrate confined spaces such as buildings, underground bunkers, and enemy installations undetected. In addition to gathering intelligence, nano-drones could perform highly targeted attacks on key personnel or critical infrastructure, minimising collateral damage. The swarm tactics enabled by nano-drones, where thousands of them can be deployed simultaneously, present a new level of sophistication in disrupting enemy defenses and complicating traditional countermeasures.

The advancements in stealth and cloaking technologies is another critical area where nanotechnology is impacting future warfare. It may be possible to manipulate and use nanomaterials to design advanced stealth coatings that minimise the detection of military equipment by systems such as radar, infrared and sonar. The use of nanomaterials that would be able to bend and absorb light and electromagnetic signals, could provide instant camouflage to soldiers, vehicles, aircraft and naval vessels, making them "invisible" to enemy detection systems. This would dramatically improve the military's operational capability to conduct covert operations in enemy territory without the risk of being identified or targeted. Nanomaterials that absorb or deflect heat could significantly reduce the thermal signatures of military platforms, such as fighter jets, tanks or submarines, making them much harder to be tracked by infrared-guided missiles and thermal sensors.[18]

Nowadays, nanosatellites are being specifically designed and launched for defense applications, leveraging their small size, low cost and versatility to conduct military operations. Generally, weighing between 1 and 10 kilograms, these satellites are deployed in the low Earth orbit (LEO) to support various military operations, including intelligence, surveillance, reconnaissance, navigation and communication.[19] Due to small size and modularity, these can be rapidly deployed for specific missions, while the modular design allows easy upgrades and customisation. This flexibility enables the integration of various sensor and communication payloads with these satellites to meet diverse military needs.

The ever-increasing capability to create smaller and smaller modular parts and electronic systems, is supporting the development of nano-enhanced weapons. Such weapons systems will benefit from increased range, accuracy and lethality. Nanomaterials could also be used to enhance the strength and heat resistance of missile casings, allowing for the development of faster, more durable and more precise missiles. Similarly, nanotechnology could improve the penetration power of projectiles and explosives, making them more effective against heavily armoured targets.[20] The development of nanotechnology-based weapons could also extend into the realm of non-lethal options, such as electromagnetic pulse weapons or nano-sized devices capable of disabling enemy electronics and communications systems without causing physical destruction.

The medical applications of nanotechnology are poised to transform battlefield medical support. Nano-medicine promises to significantly improve the treatment of injuries sustained in combat. Nanoparticles could be designed to deliver drugs or clotting agents directly to wounded areas, assisting swift recovery and reducing the risk of infections. Nanobots, tiny machines that operate at the cellular level, could perform internal repairs, such as clearing blockages in blood vessels or repairing tissue damage.[21] Furthermore, nanotechnology enabled by AI, might allow soldiers to carry nanosensors embedded in their uniforms or even within their bodies, constantly monitoring vital signs and health status. Such real time feedback could alert soldiers themselves as well as medical support teams to identify illnesses early, allowing for faster and targeted medical intervention on the battlefield.[22]

Nano-engineered materials could lead to the development of more efficient batteries, improving the endurance of electronic systems, vehicles and weapon systems during operations. For example, nano-enabled photovoltaic, thermoelectric and piezoelectric devices are on the cusp of being able to harvest electricity from solar energy, body heat and kinetic motion of a soldier's body. In many cases, these power sources could be embedded directly inside the uniform fabric, with enough power for a radio or to heat the suit or operate cooling as protection against hyperthermia/heat stroke.[23] By improving energy efficiency and reducing the logistical burden on fuel and supply chains, nanotechnology could provide military forces with greater operational flexibility, especially in extended or remote deployments.

In February 2012, scientists built the world's smallest transistor in Sydney. This was done through the accurate positioning of a phosphorus atom in a silicon crystal. This nano-device represents a breakthrough in the development of quantum computers.[24] Nanotechnology is becoming increasingly instrumental in quantum computing by enabling the creation of qubits, enhancing superconducting materials and nanoscale photonic devices. The precise manipulation of materials at the nanoscale opens new avenues for building more efficient, scalable and robust quantum systems. As research progresses, the interplay between nanotechnology and quantum computing will likely lead to significant breakthroughs, paving the way for the next generation of computing technologies, that will revolutionise various fields, from cryptography to artificial intelligence and beyond.

Through the development of light weight yet strong materials, advanced munitions and enhanced weapon systems, nanotechnology not only enhances operational performance but also improves the survivability of the personnel and equipment. Nations with advanced R&D are investing in harnessing the potential of nanotechnology, driving innovation and strategic advantages that could redefine conflicts in the 21st century. They understand that the ability to leverage nanotechnology effectively will determine their military supremacy, making it an essential focus for defense research and development.

Notes

1   Eugene Kaspersky, The Man Who Found Stuxnet – Sergey Ulasen in the Spotlight, November 2, 2011, https://eugene.kaspersky.com/2011/11/02/the-man-who-found-stuxnet-sergey-ulasen-in-the-spotlight/

2   R V Boden [2002] QCA 164; Supreme Court of Queensland: Brisbane, Australia, 2002, http://archive.sclqld.org.au/qjudgment/2002/QCA02-164.pdf (accessed on 11 July 2012).

3   Falliere, N.; Murchu, L.O.; Chien, E. W32.Stuxnet Dossier. Available online: http://www.symantec.com/connect/blogs/updated-w32stuxnet-dossier-available (accessed on 13 April 2012).

4   Significant Cyber Incidents, https://www.csis.org/programs/strategic-technologies-program/significant-cyber-incidents

5   https://crsreports.congress.gov/product/pdf/IF/IF11118

6   Price, Alfred. 1984. The History of US Electronic Warfare. Association of Old Crows.

7   "Information Warfare: History of Electronic Warfare." INFORMATION WARFARE (blog). December 7, 2009. https://ew30.blogspot.com/2009/12/such-is-reliance-onelectromagnetic-em.html.

8   Mohinder Singh, scientist, IAT, Pune, Electronic Warfare, June 1998 Popular Science and Technology Series, Defence Scientific Information and Documentation Centre, DRDO

9   https://irp.fas.org/doddir/dod/jp3-13-1.pdf joint publication 3-13.1 dated 25 January 2007.

10   https://www.usiofindia.org/pdf/20240912144421.pdf, Impact of Technology Enabled Cognitive Operations in Hybrid Warfare, USI Occasional Paper Lieutenant General (Dr) RS Panwar, AVSM, SM, VSM (Retd)

11   https://www.usiofindia.org/pdf/20240912144421.pdf, Impact of Technology Enabled Cognitive Operations in Hybrid Warfare, USI Occasional Paper Lieutenant General (Dr) RS Panwar, AVSM, SM, VSM (Retd)

12   David E. Jeremiah, "Nanotechnology and Global Security," Palo Alto, CA; Fourth Foresight Conference on Molecular Nanotechnology, 1995.

13   Military Applications of Nanotechnology: Implications for Strategic Security I Margaret E. Kosal, PhD Assistant Professor Sam Nunn School of International Affairs Georgia Institute of Technology Originally submitted January 2014

14   https://www.nano.gov/about-nanotechnology

15 N. Kumar (Ex Director, Defence laboratory, Jodhpur and CMD QNano advanced materials, Jodhpur) and A. Dixit (Indian Institute of Technology, Jodhpur) Aug 23,2022 https://www.nano werk.com/spotlight/spotid=61331.php#:~:text=Future %20warfare%20technologies%20that %20will%20be%20greatly%20influenced

16 Alain de Neve, Military Uses of Nanotechnology and Converging Technologies: Trends and Future Impacts, Royal High Institute for Defence Center for Security and Defence Studies Focus Paper 8.

17 S. Nagarajan Assistant Professor Dept of AERO- PITS, Jagannathan Panneerselvam Dept of AERO- PITS, Raj Mohmed Sadiq Batcha Dept of AERO- PITS, Nano Unmanned Aerial Vehicles - Drone or Helicopter in Military International Journal of Engineering Research & Technology Special Issue 2019 https://www.ijert.org/research/nano-unmanned-aerial-vehicles-drone-or-helicopter-in-military-IJERTCONV7IS11005.pdf

18 N. Kumar, A. Dixit, A. Mathur, A. Kumar, Nanocomposites of mixed oxides of copper (CuxO: x=1, 2) in co-polymer matrix of aniline-formaldehyde, Process of its preparation and its end use application for Anti-corrosion/fouling/microbial coatings; https://www.nanowerk.com/ spotlight/spotid=61331.php#:~:text=Future%20warfare%20technologies% 20that%20 will%20 be%20greatly%20influenced

19 Ministry of Defence, Netherlands First Dutch military nanosatellite launched successfully, 02-07-2021 https://english.defensie.nl/latest/news/2021/07/02/first-dutch-military-nanosatellite-launched-successfully

20 Maya Brehm Nanoweapons Discussion paper for the Convention on Certain Conventional Weapons (CCW) Geneva, November 2017 https://article36.org/wp-content/uploads/2020/12/ nanoweapons.pdf

21 Daitian Tang, Xiqi Peng, Song Wu, Songsong Tang, Luohu Clinical Institute, School of Medicine, Shantou University, Shantou, 28 March 2024 Autonomous Nanorobots as Miniaturized Surgeons for Intracellular Applications https://www.mdpi.com/20794991/14/7/595#:~:text=The%20 internalized %20nanorobots% 20can%20leverage,cellular%20metabolism%20with%20mini mal%20invasion.

22 Bipin Raj C, B Jnyanadeep, RV College of Engineering Soldiers Health Monitoring and GPS Tracking System. https://www.vlsisystemdesign.com/soldiers-health-monitoring-and-gps-tracking system/#:~:text=This%20project%20proposes%20a%20soldier,triggering%20alerts%20for%20 abnormal%20readings.

23 Simon Hilton Nanotechnology and Tomorrow's Infantryman, Polymer Nano Centrum https:/ /blog.polymernanocentrum.cz/nanotechnology-and-tomorrows-infantryman/

24 Alexander Chilton, Nanotechnology in Quantum Computing, 17 Sep 2014 https:// www.azonano.com/article.aspx?ArticleID=3251

3

Disruption Palpable

This category includes technologies that are rapidly advancing, with visible early-stage implementation and real potential to disrupt warfare in the near future. These technologies may still be in the prototype or limited-use phase but are nearing broader adoption. The full extent of their impact is yet to be realized, but their disruptive capacity is evident, and many military forces are actively preparing for their wider deployment. Examples could include AI-driven autonomous systems, advanced robotics, or hypersonic weapons.

Key Parameters

- *Level of Development:* Advanced but not yet fully mature.
- *Extent of Proliferation:* Gaining traction in specific sectors, but not widespread.
- *User Exploitation:* Limited, experimental, or situational use in conflicts.
- *Stage of Development:* Moving from prototype to operational trials.
- *Technological Evolution:* Early disruption visible, with strong potential for near-future breakthroughs.

Artificial Intelligence

Few years back, no one would have taken Elon Musk's statement so seriously, however, witnessing the rapid ascent of AI, not even the general public, even governments and the creators of AI seem to be overwhelmed by the promise it holds and the challenges it poses. AI, it is described loosely as "the ability of machines to perform tasks that normally require human intelligence, for example, recognizing patterns, learning from experience, drawing conclusions, making predictions, or taking action, whether digitally or as the smart software behind autonomous physical systems."[1] Artificial Intelligence (AI) refers to the simulation of human intelligence in machines that are designed to perform tasks that typically require human cognitive functions. These tasks include learning, reasoning, problem-solving, perception, language understanding, and decision-making. AI systems can be trained to analyze data, recognize patterns, and make predictions or decisions based on the information they process. AI can range from narrow or weak AI, which is designed for specific tasks like speech recognition or playing chess, to general or strong AI, which would have the capability to perform any intellectual task that a human can do.

AI is powered by algorithms and is often categorized into machine learning, where systems improve their performance over time by learning from data, and deep learning, which mimics the neural networks of the human brain to process large amounts of data. It has a wide array of applications, from autonomous vehicles and healthcare diagnostics to virtual assistants and military systems. Artificial intelligence has been a crucial tool for many nation's militaries for years. Now the war in Ukraine is driving innovation. AI is being used to sift through satellite images and drone video feeds, and that helps militaries to better understand what's happening on the battlefield, make decisions faster and then target the enemy faster and more accurately. The drones being used in Ukraine have all of the components required to build fully autonomous weapons, that can be launched over the battlefield, find their own targets and then independently attack those targets without any further human intervention. War is an accelerant of innovation. So, the longer that this war goes on, the more that we're going to see more innovation on the battlefield. We're already seeing innovative uses of drones and counter drone technologies, things like electronic warfare systems that can target drone operators, and then call it artillery strikes on the drone operator. And that kind of technology pushes militaries towards more autonomy, but it's not just confined to nation states. ISIS actually had a pretty sophisticated drone army a few years ago, and they were carrying out drone attacks against Iraqi troops very effectively.[2] AI includes both logic-based and statistical approaches.

A prominent area of AI is machine learning (ML). ML approaches use algorithms, predominantly statistical, that learn how to perform classification or problem solving without being explicitly programmed for the task domain. ML's learning ability comes from applying statistical learning algorithms, such as neural networks, support vector machines, or reinforcement learning, to expansive training data sets covering hundreds or even thousands of relevant features. For this reason, painstaking selection, extraction, and curation of feature sets for learning is often required. This limitation has been more recently addressed by deep learning, which utilizes deep neural networks with dozens of layers to not only learn classifications but also learn relevant features. This capability allows deep learning systems to be trained using relatively unprocessed data (e.g. image, video, or audio data) rather than feature-based training sets. To do this, deep learning requires massive training sets that may be an order of magnitude larger than those needed for other machine learning algorithms. When such data is available, deep learning systems typically perform significantly better than all other methods. Altogether, these advances in AI have enabled a new generation of AI applications.[3]

AI will revolutionise other emerging technologies and existing defense systems and mechanisms rapidly changing the character of war. Basically, capability of this technology to consume huge amount of data and learn through it, combined with superior computation abilities of modern computers allow military platforms to be smarter, faster and precise. Larger the data base, deeper the ML and greater the accuracy. AI due to its inherent invasive nature is going to be transform every aspect of warfighting and has already made serious inroads into intelligence, surveillance, communications, logistics, medicine, administration and is rapidly altering the fields of autonomous weapons, drones and command systems. Future generations of fighter jets, submarines, and other military vehicles, are being designed to operate with a swarm of unmanned vehicles. AI will be used to control the swarm of unmanned aerial vehicles while the pilot focuses on executing his mission. Military forces around the world are experimenting with this emerging technology as engineers work to mature artificial intelligence technology and its integration with various platforms. Though, there has been a serious debate underway considering the threat of systems which may overwhelm human decision making as more advanced AI solutions are created for the defense industry.

Robotics

This is what PW Singer, Brookings Institution researcher and author of the famous book Wired For War: The Robotics Revolution and Conflict in the 21st Century, had to say at the beginning of his address at the Carnegie Council on 05 February 2009 and I quote, "One Air Force three-star general that I met with said that it's not unreasonable to postulate the next conflict involving tens of thousands of robots. And let me be clear here, we're not saying tens of thousands of today's robots, like those that you're seeing here, but tens of thousands of tomorrow's robots, because one of the things you have in technology is Moore's Law, the fact that you can pack more and more computing power into each microchip, basically that they double in their computing power every two years. So, what that means is that within 25 years these systems will be a billion times more powerful in their computing. And I don't mean a billion like the sort of amorphous Austin Powers' "one billion."[4] I mean literally take the power of your iPhone and multiply it times a billion."

And indeed, the field of robotics have made transcendental progress since his address as today's robots laced with AI are much better in computing power, mechanical responses, fairly skilled, largely autonomous and of course realistic looking. And the robotic space has just exploded exponentially because of the

Goliath - light charge carrier, self-destructing minitank

Source: https://cs.wikipedia.org/wiki/Goliath_(lehk%C3%BD_nosi%C 4%8D_n%C3%A1lo%CS %BEe)

democratisation knowledge brought in by internet, social media and chatbots. Just like software went open source, so has warfare due to ready access to the wherewithal on the phone itself. Rudimentary robotics doesn't require a hi-tech lab to create an autonomous platform. It doesn't require the support of either the government or corporate giant. It only requires a small capital which even a school kid, provided he has the motivation and the inclination, can raise up from his pocket money. Which quite often has been the case with innovation by young minds. With such easy proliferation, the ecosystem for such solutions has matured and military has been a big beneficiary of this industry, as it has the funds, requisite manpower, governments support, perspective planning and a constant desire to bring down the human cost of war. In fact, most of the autonomous and unmanned platforms whether they look like human or otherwise, such as drones, uncrewed ground vehicles, and guided missiles; all of these things are technically robots with sophisticated components and programming. One robot getting employed in the battle field saves at least one human soldier and as the tech gets better this number shall increase dramatically. But the question is, whether this one robot is going to get inducted into the battle front as an autonomous kill machine. This requires some high level of machine learning and computing, which increases the capital requirements exponentially and also pushes the legal envelope on the subject of humans being killed by machines.

As of now the society has almost settled with the idea of humans getting killed by a robot from the sky, but facing a killer robot eye ball to eye ball will definitely raise the moral and ethical question of disparity in capability and invincibility of the human soldier. Though a stage where robots armies are fielded as Infantry units still elude the military canvas, however, robots have been employed in some of the hazardous and simple roles of mine clearing, breaching & demolition, room intervention, bomb disposal and battle field logistics since World War II. Most of the robots on the battlefield today require human input in some form, either as a controller, navigator, or to initiate a task. However, companies and organisations worldwide have been doing considerable R&D for developing and fielding robots that will completely replace soldiers and make their own decisions on the battlefield. Though most of these projects are well aware of the legal and ethical debate on full robotisation of the battlefield and at least overtly maintain that they don't support the weaponisation of the robots.

But the videos on the internet suggest differently, even if that is being done for demonstration purposes only. Robots are expanding in magnitude around the developed world. In 2013, for example, there were an estimated 1.2 million robots in use. This total rose to around 1.5 million in 2014 and is projected to increase to about 1.9 million in 2017.5 Japan has the largest number with 306,700, followed by North America (237,400), China (182,300), South Korea (175,600), and Germany (175,200). Overall, robotics is expected to rise from a $15 billion sector now to $67 billion by 2025.6.[5] With such rapid advancements, it's natural that the civil military cooperation in this field has gathered unprecedented pace, with companies fielding novel ideas for trials and improving the prototypes considerably. Also, with every iteration the size, reflexes and intelligence of the robot is being enhanced. Already the industry has graduated from tracked to four legged and now to human like biped robots. The biped Atlas made by Boston Dynamics, the world leader in robotics, manipulates the world around it: Atlas interacts with objects and modifies the course to reach its goal – delivering a bag of tools to a person waiting at the top of a multi-story scaffold. Atlas grasps, carries, and tosses the tool bag, climbs stairs, jumps between levels, and pushes a large wooden block out of its way before dismounting with an inverted 540-degree flip that project engineers have dubbed the "Sick Trick."[6] It can easily carry out most of the administrative tasks involved in warfighting and soon enough would ready to join the frontline, if the government feels so and is able to mage the public opinion, which in any case the governments have been manipulating over the years. Such a robot endowed with thermal imaging and IR make a good recce soldier on the battle field which can move undeterred through hail of fire, minefields and operate in contaminated zones which may be a no go for the human soldiers. The biggest advantage of robots is that as the power packs become more and more efficient, the robot can keep doing its task untiringly without complaints and emotional trauma, working at 100 percent efficiency over and over again. AI will allow a robotic soldier to make instantaneous adjustments to improve its chances of success. There is also the fact that high-tech sensors and cameras allow robotic soldiers to be incredibly efficient and hit targets with precision. The robot can instantly get back up and start shooting again if it is knocked down by an explosion. The human body is weak in comparison to a robot. This implies that the military can now undertake missions that previously appeared impossible or take bigger chances.

Consider urban warfare as an example. Robotic soldiers will be a game changer in counter insurgency and counter terrorism operations. Room clearance will become less unforgiving as risk to human lives will get reduced drastically and the robots will have better reflexes and situational awareness. This will also induce compelling changes at tactical and strategic levels, as the militaries will readjust their safety distances and FLOT for artillery and air strikes. Operations shall continue round the clock without interruptions of casualty evacuation and the emotional trauma that a battle field rains on the soldiers. Robots can easily take on role of first responders in most of the situations where round the clock readiness is required to give a timely response without any concern for fatigue and bio rhythm or body cycles. The governments bill on medical, post injury care, rehabilitation and welfare will also reduce and can be rerouted into robotics itself. Countries who have huge spending on border management and where the borders get activated due to intermittent duels can reduce the human cost of keeping their borders safely employing robots for border patrolling and protection of VAs and VPs.

The possibilities for employment of robots on the battle field are endless, but the bigger dilemma is the paradigm shift it brings to the character of war. The ethical principles that applied in the framework of a conflict opposing human combatants would be lost if machines were to take their place. It would therefore be necessary to reconstruct a new set of ethics designed to govern the rules of action covering the behaviour of military robots.[7] The enemy who has to fight a robotized army most certainly does not have the same perception of the situation. There is therefore a high chance that the enemy gets an even stronger feeling that he is fighting against adversaries who once again possess technological supremacy, faced with which he is sure to be defeated. If this theory proved to be true, the robotisation of the battlefield would therefore act as an accelerator of asymmetric combat in its most extreme and unexpected forms; it could then serve to reinforce the "normal," justified and efficient character of certain modes of action that we today judge to be irregular/guerrilla tactics, terrorism, suicide attacks, IT attacks, etc.[8] This flags another important issue, the falling of this technology into hands of the non-state actors and terrorist organisations. These organisations were on the heels of bonafide militaries in exploiting drones, machine guns and helicopters, which gives a clear indication that as the robotics becomes more and more acceptable and affordable, these organisations would also leverage the technology accordingly. With reduced threat to their fanatic cadres, vulnerable citizens, selective communities and government established may experience a sharp increase in deadly attacks against them. The greatest advantage these organisations will accrue is the minimal requirement of indoctrinating their robotic cadres by ideological preachers as one simple code can bring the entire cadre in sync with the organisational goals. Moreover, no mechanical system can distinguish the difference between friendly and hostile behaviour or between a civilian and a soldier.

The Geneva Convention appeals to "common sense." Try to download THAT into a computer! In South Africa in 2007, due to a "software problem", an automatic air-defense gun killed nine soldiers during a training session. Conducting the investigation and pronouncing a judgement on such accidents raises new questions.[9] It is not just with robotics but all other emerging technologies have to pass the ethical and legal scrutiny. If not pass, at least it has to stand through to such deliberations from within the scientific, political and military communities with a certain amount of overarching vigil of the public at large. An added challenge regarding fielding the robotic army is the chances of a robot becoming a POW. There is a strong chance of this robot getting reprogrammed to act as a mole or a suicide bomber and made to target its unit, fellow robots or its human controller. This might be easy to imagine than to undertake, but then what if there is another 'Stuxnet' introduced into the robot either physically or remotely by a hacker. The most hardened systems get compromised and sometimes it only takes a dissatisfied geek to get the house of cards crumbling down. AI may claim to achieve a very high degree of proficient military behaviour from the robots but they would always fall short of human reasoning in delicate situations. Delicate situations where distinction, empathy, honour, conscience and camaraderie underscore the military operation. If ethical decision-making is not completely translatable into purely algorithmic or formal terms, then an irreducible place must be kept for the human decider; he or she cannot be left out of the loop. The officer's role thus remains irreducible, for it is he or she who must in good conscience end indecision and accept or refuse the use of the weapons borne by the robot. And it is also the role of one's conscience that cannot be entirely formalized. Even if we took a purely naturalistic or reductionist standpoint that the conscience could be explained by the activation of neurons, our problem would not be solved, because even this purely naturalized consciousness could not necessarily be imitated by a digital machine, or its workings by an algorithm! So, with such issues trapping the cerebration of leaders of the world, the jobs of human soldiers appear to be safe in the foreseeable future, however, turning our face away from the imminent robotisation of the battle field will be a self-inflicted tragedy and humanity should not plan on ceding the greatness of reasoning to machines without putting in place the correct legal, ethical and control mechanisms.

Unmanned Systems

In November 2020, Azerbaijan celebrated victory, in a war which showcased the power of unmanned weapons technology. Designed to hit a wide range of targets, the manufacturer of these drones uses the obvious motto, "fire and forget". Many nations learnt their lessons fast and some joined the race to buy these autonomous weapons and there were few like Turkey who embarked on maturing the technology and giving world an alternative to US, Israel and China. In fact, Turkey in spite of being part of the EU, has successfully sold its Drones to Ukraine as well as Russia. Russia and Ukraine both have employed these autonomous systems extensively and intensified the innovation in this rapidly evolving stream of military technology. The Ukrainians have tried to develop their own long range loitering munition with an uneven success rate. In August, they attacked Russian Navy tactical jets at Saki Airbase and the Black Sea Fleet headquarters in Sevastopol with a Chinese-built, fixed-wing drone from AliExpress (costing $8,000 per unit) repurposed into a suicide drone.[10] Couple with AI, drones have been instrumental in creating an intelligence picture of the battle field through enhanced data analytics. Unmanned systems are increasingly transforming the logistic supply chains and making them more efficient, economical, simple and resilient as well as reducing the exposure of troops to hostile environment. Almost all countries with reasonable amount of defence capability and even non-state actors have been found to be clamouring for creating an inventory of autonomous systems and platforms. Houthis have been exploiting these weapons and holding the world shipping industry hostage. Cheap small drones loaded with light explosives have increased the number of actors that can contest the skies and thus, the battles now, not only include the space around ground troops, but also the space immediately above them, creating minefields in the air. The widespread use of inexpensive armed and single-use attack drones could thus make close air support and ground attack aircraft even more obsolete in conflicts with denser and capable air defenses.[11] Gone are the days when military technologies were being pursued by governments and big conglomerates, the present-day post-modern technologies are being fuelled by massive civil participation, most of the technologies being dual use.

Autonomous Systems

It's a 'start's up world' and AI enabled autonomous systems have entered the human life, culture and existence. Right from Toaster to Trucks, everything is becoming autonomous and the world is liking it, without thinking about the perils of shedding human control. Without a networked and layered AD system even the US systems failed to detect the Houthi drone attack on Saudi Oil Company facilities. The Air Defence systems today must be supported by electronic warfare and specialised counter unmanned systems (C-UAS) to defend against Swarm Drone attacks.[12] Because these technologies have become so intrinsic to human routine, it's obvious that it will have wider acceptance in the military realm and is getting the desired traction, as nations, especially the developed ones have increasingly lesser numbers of willing citizens who can be exposed to the perils of battle field. Swarming is one of the latest areas of autonomous weapons development and countries like US, China, Israel, Russia, Turkey and even India has demonstrated its competence in the field. And with the given pace of developments in autonomous technology, in the next decade or so we might find large inventories of such systems and weapons on land, water and even under water. There is extensive work being undertaken to collective harness AI, Robotics and Weapon technologies to reduce the human cost of war. Though, the cost gets reduced only for the beholder and the less developed still has to beat technology with superior skill and military genius or get killed at the hands of a faceless and heartless machine. And a machine that outpaces human soldiers can be defeated by a faster machine only. Once these systems have proliferated into the armies of the world, race to make faster versions and those which are in sync with all the sister platforms operating across land, sea and water will give rise to an autonomous battle field where machines are fighting with each other to establish human writ. Here the numbers, speed and superior autonomy will be of essence and thus the arms race. Another challenge with this level of autonomy would be that of easy escalation of conflicts given the low human cost involved, which would take care of the public opinion too, especially if the country waging a war is a developed one. Neighbours who have contested borders need to be careful with the degree of autonomy of such systems deployed along the borders in order to prevent avoidable escalation of minor and unintentional infringements. Such systems will have to be bestowed with greater power of reasoning. This is even more the case in cyber space, where there are no physical limitations to slow things down. Automated cyber wars could soon involve autonomously attacking and self-replicating cyber weapons. This speeding-up of operations could also lead to undesired chain reactions, known in the world of automated finance as "flash crashes". These happen when one automated system reacts to another, which causes a third to react, and so on. With LAWS, "flash crashes" could turn into "flash wars".[13] Notwithstanding, "flash wars" in the foreseeable future shall remain a bailiwick of the rich and the mighty.

Hypersonic Weapons

Hypersonic weapons, in their truest form, are capable of speeds above Mac five, have in-flight manoeuvrability controls, and can perform these tasks while

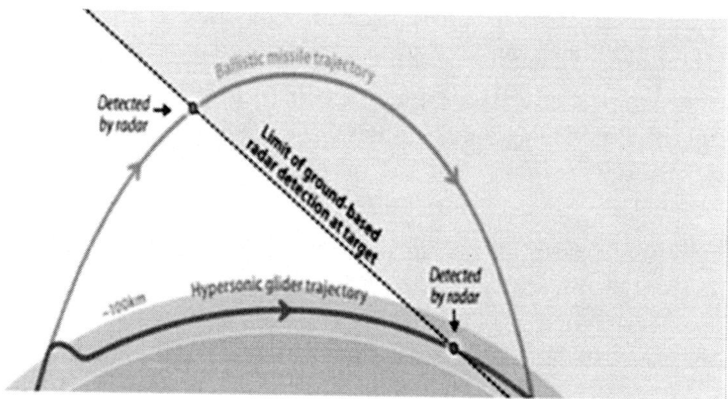

Source: CRS image based on an image in "Gliding missiles that fly faster than Mach 5 are coming," *The Economist*, April 6, 2019, https://www.econornist.com/science-and-technology/2019/04/06/gliding-missiles-that-flyfaster-than-mach-5-are-coming.

maintaining a low altitude through boost, midcourse, and terminal phases of their trajectory.[14] Advertised as a generational leap in missile technology, governments are investing billions of dollars in R&D to develop hypersonic capabilities, threaten their adversaries, and gain an unquestionable advantage in future conflicts.[15] Problems such as the heat experienced when traveling at high Mach speeds can cause plasma sheathing around the body of hypersonic, causing sensor malfunctions and computer and communication blackouts. While this is not an insurmountable problem, it is just one extra hurdle that adds time, and cost, to development. Furthermore, there are physical and material science limitations to creating an object that is durable enough to endure high G-forces, and still be light enough to perform efficiently in either gliding HGV or scramjet HCM form.[16] If we consider countries that are first to field an operational hypersonic weapon, then, Russia and China have both put operational weapons in the field. But the main question is, whether it is worth the resources that Russia and China expended on these programs in order to get ahead in the first place? Limited funds have to be prioritised when it comes to procurement, and different countries make different prioritisation choices. World governments are quickly investing in anti-hypersonic technology, such as satellites that can detect their heat signature and hardening of targets. Lastly, with the intense cost of creating even single hypersonic weapon at the moment, their mass-use capability without bankrupting a state comes into question if current manufacturing economies do not improve.[17] With the war in Ukraine placing such a premium on possessing quantity of munitions, and being able to sustain a force in combat, it always has to be remembered that a dollar fed into hypersonic development or fielding, is probably a dollar taken from elsewhere. Given their exorbitant price tag, some costing as much as an American F-35 jetfighter, HGVs are seen as a luxury weapon. Indeed, by comparison, a large number of cruise missiles, costing $2 million each, could potentially reach the same targets as a hypersonic missile at a fraction of the cost. Within the US military itself, many now suggest that HGVs could ultimately be repelled by next-generation lasers and directed energy weapons. For the moment, however, solid-state lasers remain experimental, energy deficient, limited in range and vulnerable to weather.[18]

Hypersonic weapons are an inherent risk to doctrine of Mutually Assured Destruction, as these weapons in their full form would be able to evade current missile defense systems, and their precise targeting systems may be able to take out some of a country's nuclear weaponry, removing much of the consequence for striking first and creating an inherently unstable system that rewards the aggressor. The current missile defense system is primarily reliant on spotting current ICBMs when they are in the upper atmosphere, tracking their predictable flight pattern, and striking them while in the air. Hypersonic weapons, however, can manoeuvre much closer to the ground. This means ground-based radar stations are late to alert when one is approaching, and their speed and manoeuvrability make them nearly impossible to strike out of the sky when in full boost traveling several dozen times the speed of sound.[19] Furthermore, ground-targeted hypersonics threaten to change some of the various tactics that are already being shifted by emerging technologies such as drone warfare. Spacing of bases, hardening of targets, radar warning stations and their locations, and the actual missile defenses themselves are all at risk of becoming antiquated by a successful hypersonic program in the future. Additionally, while tank warfare and slow-moving armoured columns are already proving to be less than decisive in the Ukraine conflict, their ease of "targetability" could make them even more vulnerable with the creation of hypersonic weapons systems en-masse.[20] Presently, most of the carrier groups of the Super Powers with large navies have been built and designed for traditional anti-ship offensive weapons, not at HCM/HGV speeds above Mach 5 to Mach 20. While they do have some capabilities to defend against hypersonics, SM-6 missile defense systems, these are newly deployed or retrofitted fixes. These countries rely heavily on force projection through ship presence in international and territorial waters and use aircraft carriers as floating airports for their airpower launches in any type of conflict.[21] Threat of hypersonic weapons will largely affect their established capacities for force projection across the oceans.

Although the United States, Russia, and China possess the most advanced hypersonic weapons programs, a number of other countries including Australia, India, France, Germany, South Korea, North Korea, and Japan are also developing hypersonic weapons technology. India has similarly collaborated with Russia on the development of BrahMos II, a Mach 7 hypersonic cruise missile. Although, BrahMos II was initially intended to be fielded in 2017, news reports indicate that the program faces significant delays and is now scheduled to achieve initial operational capability between 2025 and 2028. Reportedly, India is also developing an indigenous, dual-capable hypersonic cruise missile as part of its Hypersonic Technology Demonstrator Vehicle program and successfully tested a Mach 6 scramjet in June 2019 and September 2020. India operates approximately 12 hypersonic wind tunnels and is capable of testing speeds of up to Mach 13. France also has collaborated and contracted with Russia on the development of hypersonic technology. Under the V-max (Experimental Manoeuvring Vehicle) program, France is modifying its air-to-surface ASN4G supersonic missile for hypersonic flight. It successfully tested the modified version of the said missile in June 2023. France operates five hypersonic wind tunnels and is capable of testing speeds of up to Mach 21.[22] Germany successfully tested an experimental hypersonic glide vehicle (SHEFEX II) in 2012; however, reports indicate that Germany may have pulled funding for the program. German defense contractor DLR continues to research and test hypersonic vehicles as part of the European Union's ATLLAS II project, which seeks to design a Mach 5-6 vehicle. Germany operates three hypersonic wind tunnels and is capable of testing speeds of up to Mach 11.[23] In addition, South Korea reportedly has been developing a ground-launched Mach 6+ hypersonic cruise missile, Hycore, since 2018. According to Janes, South Korea is developing the missile "in response to growing concern about North Korea military modernization" and plans to eventually develop sea- and air-launched variants. Although, North Korea tested the Hwasong-8, which it identifies as a hypersonic glide vehicle, in September 2021, reports indicate that the vehicle may have reached speeds of only Mach 3. Similarly, North Korea claims to have tested a second hypersonic weapon in January 2022; however, experts believe that that weapon may instead be a manoeuvring re-entry vehicle. Finally, Japan is developing the Hypersonic Cruise Missile (HCM) and the Hyper Velocity Gliding Projectile (HVGP). According to Jane's, Japan invested $122 million in HVGP in FY2019. It reportedly plans to field HVGPs for area suppression and neutralizing aircraft carriers. HVGP is expected to enter service in 2026, with a more advanced version available by 2030, while HCM is expected to enter service in 2030. The Japan Aerospace Exploration Agency operates three hypersonic wind tunnels, with two additional facilities at Mitsubishi Heavy Industries and the University of Tokyo. According to DOD, Japan and the United States have agreed to conduct "a joint analysis focused on future cooperation in counter-hypersonic technology." Other countries – including Iran, Israel, and Brazil – have conducted foundational research on hypersonic airflows and propulsion systems, but may not be pursuing a hypersonic weapons capability at this time.[24]

Directed Energy Weapons (DEWs)

The US Joint doctrine defines directed energy (DE) as an umbrella term covering technologies that produce concentrated electromagnetic energy and atomic or subatomic particles. A DE weapon is a system using DE primarily as a means to incapacitate, damage, disable or destroy enemy equipment, facilities and/or personnel. Directed energy warfare is military action involving the use of DE weapons, devices and countermeasures to incapacitate, cause direct damage to or destroy adversary equipment, facilities and/or personnel or to determine, exploit, reduce or prevent hostile use of the electromagnetic spectrum through damage, destruction and disruption. With the maturation of DE technology, weaponised DE systems are becoming more prolific and powerful and a significant subset of the electronic warfare domain.[25] Directed-energy systems are a non-kinetic form of joint fires. Modern high-energy laser (HEL) and high-power microwave (HPM) systems are best suited for defensive mission applications. Their ultimate potential can be exploited for rapid, accurate and sustained targeting against fixed and mobile targets at long range. Directed energy weapons afford the prospect of tailored effects, from lower to higher lethality and from temporary disablement to permanent destruction. Because they may operate at range and may not feature a visible signature, they are useful for both covert and overt employment.[26] They mostly require electricity, and depending upon the range and type of target, the electricity

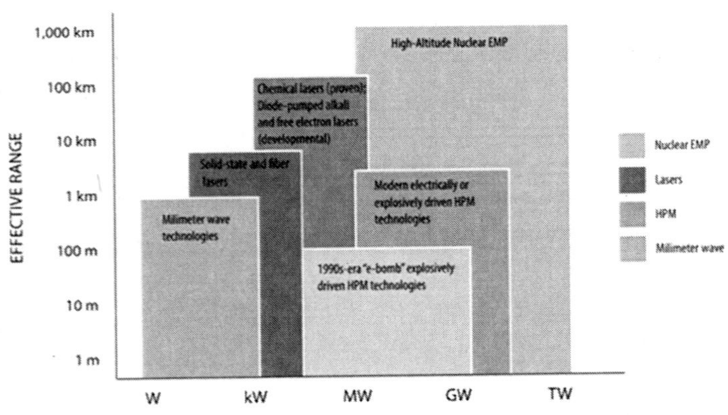

Notes: While this is a generalized representation of the effective range and power output of the various laser and radiofrequency weapon types, the actual performance of any system would in practice be system-specific and context-dependent. The nature and form of laser and radiofrequency weapon effects vary considerably. Specific meteorological conditions, atmospheric effects, weapon employment altitude and other variables also affect a system's actual performance.

Source: Jason D. Ellis, Directed-Energy Weapons: Promise and Prospects, Center for a New American Security, April 2015

requirements can be calibrated leading to a cost-effective alternative to conventional missiles and unmanned autonomous weapons. Also, the DEWs are versatile weapons and can be installed on land, air and sea-based assets.

In January 2024, during a trial at the MOD's Hebrides Range, the DragonFire laser directed energy weapon (LDEW) system conducted UK's first high-power firing of a laser weapon against aerial targets. Thought, the range of DragonFire is classified, but it is a line of sight weapon and can engage any visible target. The precision of the LDEW is equivalent to hitting a £1 coin from a kilometre away. Firing it for 10 seconds is equivalent to the cost of using a regular heater for just an hour. Therefore, it has the potential to be a long term low cost alternative to certain tasks being carried out by missiles at present. The cost of operating the laser is typically less than £10 per shot.[27] As long as you can manage electricity and keep feeding energy to the weapon system, usually you're not in any danger of running out of ammunition. However, if missiles have to be used to bring down the advanced autonomous drones of the future, then nations have to have increasingly huge stock of missiles and engage the incoming swarms of drones at exorbitant costs. It is more likely, that one of the adversaries might run out of missiles pretty sooner than the other is going to run out of drones. Furthermore, DEWs will be best suited to reinforce AD grids which would be vulnerable to autonomous swarms and low flying hypersonic missiles. The efficacy of these weapons will also depend upon their own capacity, in terms of the quantum of energy they direct on to the target, at what range and also the type of target being attacked. A target without any electronics may not be an ideal target for such systems. Most of the nations who have developed and tested these weapons successfully have been silent on their actual capabilities so as prevent the development of effective countermeasures and sometimes to hide their own challenges in the development and deployment of the technology. However, given the present state of proliferation of these systems in the most developed militaries, Rifle probably isn't going to be replaced by a Lazer Gun any time soon.

Space Technology

"The occupation of the high ground can thus mean genuine domination. Its reality is undeniable."

—Clausewitz, *On War*

In his address to the nation, Prime Minister Narendra Modi on 27 March 2019,[28] announced that India had demonstrated anti-satellite missile capability by shooting down a live satellite, describing it as a rare achievement that puts the country in an exclusive club of space superpowers. This also announced the success and continued pursuance of India's ASAT (anti-satellite) program, Mission Shakti. The ASAT test

**The ASAT missile, designated Prithvi Defence Vehicle Mark-II,
lifting off to intercept the satellite**

Source: https://twitter.com/PIB_India/status/1110910323169460224/photo/2

utilized a modified anti-ballistic missile interceptor code-named Prithvi Defence Vehicle Mark-II which was developed under Project XSV-1. The test made India the fourth country after the United States, Russia and China to have tested an ASAT weapon.[29] No doubt that India embarked on Mission Shakti as its main adversary had achieved a similar feat almost 12 years ahead, aggravating the vulnerability of the rapidly expanding Indian Space Ecosystem which consists of a total of 124 Indian satellites, including those from private operators/academic institutions. As of 2024, the government owns 23 and 29 operational satellites in LEO and GEO, respectively and Chandrayaan-2 and 3 and Aditya 1 in active mode.[30] However, the Purely defensive use of space technology by India pushes, this democratic peaceful nation into the strategic arena of "great power competition" in space. Space has emerged as the fifth recognised domain of military operations, alongside land, air, maritime, and cyber and electromagnetic (EM). Since the Cold War and the Soviet launch of Sputnik I in 1957, space was expected by many defence planners to become an active warfighting domain, though to date it has played only a supporting role to military forces in other (terrestrial) domains, including through the provision of SATCOM, Positioning, Navigation & Timing PNT and Electro-optical (space-based intelligence, surveillance and reconnaissance). Space has consequently become a critical enabler of military power projection, offering global reach, connectivity, and a 'high ground' from which to observe the terrestrial battlespace.[31]

As I write, the satellite tracking website "Orbiting Now" lists 9,596 active satellites in various Earth orbits.[32] At the turn of the 21st Century, only 14 nations operated satellites. Within the past two decades, satellites from 91 new space-faring countries reached orbit.[33] Space power is pursued by states for war, development and prestige, meeting specific needs and goals because the expense and difficulty of space activities require huge resources marshalled by a central authority. The Global Space Age has seen the creation of multiple technological infrastructures that many states have spent decades building to make their military forces more effective and high-technology economies more competitive today.[34] But space is no more simply a forum for great power grandstanding. Instead, it has become a very real battleground for economic and military competition akin to any area under strategic contestation on land or sea. It's no different from the Indo-Pacific or the South China Sea or the heavily contested lands of Eastern Europe. And the main reason for that is the democratisation of scientific research and privatisation of space exploration which has led to an unprecedented increase of stakeholders in this industry. This increased overlap between the government, military and private initiatives in space has created shared systems and technologies and a porous ecosystem that can be dual use and vulnerable to financial luring, as private companies may be inclined to provide services to the highest bidder. But still satellites serve a variety of vital functions from providing communications, internet, navigation, reconnaissance and plethora of other routine daily life functions that may appear simple and downright mundane from hiring a cab to ordering on Zomato. And militaries are just as dependent, perhaps even more reliant, on space-based systems especially for the more powerful nations who heavily undertake military operations other than war (MOOTW). Their units, squadrons and frigates operate in isolated areas far from mother bases and are heavily reliant on satellite-based communications, imagery, intelligence, positioning and navigation. And any country that wants to maintain a cohesive and integrated force capable of operations across, multiple domains cannot imagine ignoring indigenous space capability. "Indigenous" here is the key to true security, true defense and true strategic autonomy. The modern kill chain that enables military units to identify targets and then strike those with precision relies heavily on space satellites. Right from initial identification which may be for example by a reconnaissance satellite looking down from above, to a final engagement that might be by a weapon guided by GPS signals, without a well-meshed synchronised space-based support infrastructure the fighting formations are going to be deafer, dumber and blinder than they are used to being. That's the kind of dependencies that have been generated on space systems for pursuing conflicts within the earth's atmosphere. But the geo-strategic competition is pushing the envelope and there is a war breaking out in outer space itself. However, while almost every country on Earth benefits from space-based technologies, only a select few are really in a position to compete when it comes to space-based warfare. Officially, these select few countries advocate for the non-weaponisation of space.

But in the absence of some sweeping international agreement, that doesn't preclude preparing for serious space-based competition, all these countries have advanced space weapons programs and demonstrated capabilities that don't make easily to the public domain. As for the counter-space options, rather than shooting down the satellite of the adversary rendering it inoperable permanently or temporarily through electronic attack mechanisms is a preferred tactic without crossing the threshold of an outright war. Spoofing and jamming are other common techniques to confuse adversary systems especially the one with omnidirectional antennas, such as GPS receivers and satellite phones being more vulnerable to downlink jamming or spoofing than other systems. Countries like China and Russia lay huge stress on the exploitation of space infrastructure for electronic warfare against their common adversary in the NATO bloc. Russia has demonstrated its capability and affinity to indulge in space-based warfare right through its conflicts in Crimea, Georgia, Syria and now Ukraine. Thus, EW can enable you to deny an adversary some of the value that comes from being supported by space-based assets without sparking a wider confrontation.

Cyber-attacks are another potential non-destructive option. Here, instead of targeting the transmission between the satellite and the ground-based receiver, the data or the system itself is being targeted. It involves access to a control system for a satellite, or a transmission system, or any other system that might be vulnerable and diminishing, subverting and altering the operation of that system or the data that it's transmitting. Like electronic warfare, this is a relatively cheap, deniable, accessible method for undermining the advantage an opponent gets from their satellites. The cost advantage of such operations is the raison d'etre for countries to invest in hacker armies as it offsets large investments required to subvert the systems residing in orbits around the Earth. If it's impossible to compromise enemy systems from Earth, it might be possible to compromise them in space instead.

In the future, satellites might be able to directly interfere physically with the operation of the target satellite. With the number of satellites increasing by the day, the orbital paths are becoming like a congested highway and the only thing which a determined nation has to is to alter the path and the speed of its own satellite so as to turn it into a deadly projectile. Then is the more extreme option of a country with nuclear weapons launching a warhead on a long-range booster, such as an intercontinental ballistic missile (ICBM) or SLV, and probably conduct a high-altitude nuclear detonation, which would create widespread electromagnetic disruptions in space and on Earth leading to potential damage or destruction of satellites.[35] However, this is a worst-case scenario and the alternate option of not at

Firefly's Alpha rocket launches on Thursday evening from
Vandenberg Space Force Base.

Source: https://arstechnica.com/space/2023/09/firefly-and-space-force-demonstrate-ability-to-
rapidly-launch-a-satellite/

all venturing into deep space still degrading the adversary's capability through EW, cyber warfare or DEW gives a cost advantage as well as a window for creating confusion and limited deniability. Thus, some countries will be more amenable to adopting the low key options with different combinations of cost, public visibility, collateral damage, reliability and reversibility. And there's a second component here that might also be leveraged. At least such a capability gives them some deterrence if at all they don't have the sheer financial clout to own assets in the space themselves. However, this deterrence only works against an opponent for whom replacing a satellite is an uphill prospect and not to the heavy weights like US who has been developing the capability of launching Satellite constellations that are proliferated, disaggregated and distributed.

The US's Tactically Responsive Launch program aims to deny the adversary the advantage from low-cost satellite disruption actions. The US launch company Firefly put a spacecraft called "Victus Nox" into orbit within 24 hours of receiving the go command from the military, giving United States Space Force the ability to rapidly integrate capabilities and respond to aggression in tactically relevant timelines.[36] Lastly every catastrophe in the space may not be intentional, there is an ever-increasing pile of debris in space being created by the human exploration of space as well as the natural galactic phenomenon. The space race has further intensified the chances of debris collision making it a clear and present danger. Collisions between and explosions of massive derelict objects almost certainly will continue to add to the amount of space debris in orbit. As of January 2022, more than 25,000 objects of at least 10 centimetres in size were tracked and catalogued in Earth's orbit to include active satellites.[37] The primary risk to spacecraft in orbit is from uncatalogued lethal non-trackable debris (LNT), which are objects between 5 millimetres and 10 centimetres in size. An estimated 600,000 to 900,000 pieces of uncatalogued LNT are in LEO.[38] Thus with all the civilian and military ingress into outer space, it is becoming a highly contested space and increasingly vital for human advancement. This duality of space-based systems and technologies, thus, instils caution into the competitors in this arena from weaponising the space given the uncertainty shrouding such adventurism.

Notes

1   Summary of The 2018 Department of Defense Artificial Intelligence Strategy, https://media.defense.gov/2019/Feb/12/2002088963/-1/-1/1/SUMMARY-OF-DOD-AI-STRATEGY.PDF

2   Paul Scharre, How militaries are using artificial intelligence on and off the battlefield. 9 July 2023 https://www.pbs.org/newshour/show/how-militaries-are-using-artificial-intelligence-on-and-off-the-battlefield

3   Emerging Technologies Subcommittee, under Co-Chair Thad Allen, Co-Chair Cathy Lanier and Co-Chair Robert Rose Homeland Security Advisory Council Final Report on Artificial Intelligence / Machine Learning, 14 November 2019.(11-12)

4   P.W. Singer, Joanne J. Myers, Wired for War: The Robotics Revolution and Conflict in the 21st Century, February 5, 2009, https://media-1.carnegiecouncil.org/cceia/import/studio/PWSinger_WiredWar.pdf

5   Alison Sander and Meldon Wolfgang. "The Rise of Robotics." The Boston Consulting Group, August 27, 2014. https://www.bcgperspectives.com/content/articles/business_unit_strategy_innovation_rise_of_robotics/.

6   Sick Tricks and Tricky Grips, Blog. https://bostondynamics.com/blog/sick-tricks-and-tricky-grips/

7   Ronald C. Arkin, Governing Lethal Behavior in Autonomous Robots (Boca Raton, Fla: Chapman & Hall/CRC Press, 2009).

8   Didier Danet and Jean-Paul Hanon, Digitization and Robotization of the Battlefield: Evolution or Robolution? US Army Combined Arms Center, Combat Studies Institute, Fort Leavenworth, KS,66027. Jan 2014

9   Benoit Royal, Military Robots and Ethics: Ethical Problems for European Industrial, Political, and Military Leaders, US Army Combined Arms Center, Combat Studies Institute, Fort Leavenworth, KS,66027. Jan 2014

10   Rogoway, T. 2022. "Ukraine Situation Report: Shadowy Long-Range Kamikaze Drone Strikes Again." The Drive, August 20.

11   Dominika Kunertova (2023) The war in Ukraine shows the game-changing effect of drones depends on the game, Bulletin of the Atomic Scientists, 79:2, 95-102, DOI:10.1080/00963402.2023.2178180

12   Lt General Dushyant Singh (Retd), Swarm Drones - New Frontier of Warfare, Aero India 2021 Special https://www.spslandforces.com/story/?id=747&h=Swarm-Drones—New-Frontier-of-Warfare

13   Ulrike Franke, Flash Wars: Where could an autonomous weapons revolution lead us? 22 November 2018 https://ecfr.eu/article/flash_wars_where_could_an_autonomous_ weapons _ revolution_lead_us/

14 Hypersonic Weapons. Carnegie Endowment for International Peace, carnegieendowment.org/programs/npp/hypersonic.

15 The Rise of Hypersonics. Deloitte United States, www2.deloitte.com/us/en/pages/energy-and-resources/articles/rise-of-hypersonics.html.

16 Orf, Darren. "Turns Out Russia Is Overselling Its 'Unstoppable' Hypersonic Missile." Popular Mechanics, 3 Feb. 2023, www.popularmechanics.com/military/weapons/a42759933/russia-hypersonic-missile-problems.

17 Stone, Mike. "Pentagon Says Hypersonic Weapons Are Too Expensive." Reuters, 12 Oct. 2021, www.reuters.com/business/aerospace-defense/pentagon-says-hypersonic-weapons-are-too-expensive-2021-10-12.

18 Daniel Araya, Enter the Era of Hypersonic Warfare, December 13, 2023 https://www.cigionline.org/articles/enter-the-era-of-hypersonic-warfare/

19 Center for Arms Control and Non-Proliferation. "Missile Defense – Center for Arms Control and Non-Proliferation." Center for Arms Control and Non-Proliferation, 26 Oct. 2021, armscontrolcenter.org/issues/missile-defense.

20 U.S. Department of Defense. "Defense Official Says Hypersonics Are Vital to Modernization Strategy,." U.S. Department of Defense, www.defense.gov/News/News-Stories/Article/Article/2593029/defense-official-says-hypersonics-are-vital-to-modernization-strategy-battlefie.

21 Demarest, Colin. "Overmatch Secrecy Needed as China, Russia Surveil US Navy, Experts Say." *Defense News*, 26 Jan. 2023, www.defensenews.com/naval/2023/01/26/overmatch-secrecy-needed-as-china-russia-surveil-us-navy-experts-say.

22 Kelley M. Sayler, Global Hypersonic Weapons Programs, Hypersonic Weapons: Background and Issues for Congress, February 9, 2024 Congressional Research Service

23 Ibid.

24 Ibid.

25 Source: Joint Chiefs of Staff, Electronic Warfare, Joint Publication 3-13.1 (February 8, 2012), I-16.

26 Jason D. Ellis, Directed-Energy Weapons: Promise and Prospects, Center for a New American Security, April 2015

27 Advanced future military laser achieves UK first, Defence Science and Technology Laboratory and Ministry of Defence, 19 January 2024 https://www.gov.uk/government/news/advanced-future-military-laser-achieves-uk-first

28 https://www.thehindu.com/news/national/narendra-modi-announces-success-of-mission-shakti-indias-anti-satellite-missile-capability/article61568505.ece

29 https://en.wikipedia.org/wiki/Mission_Shakti#cite_note-16

30 Indian Space Situational Assessment, https://www.isro.gov.in/Indian_Space_Situational_Assessment_2022.html#:~:text=Till%202022%2C%20a%20total%20of,is%20active%20in%20lunar%20orbit.

31 Theodora Ogden, Anna Knack, Mélusine Lebret, James Black and Vasilios Mavroudis, "The Role of the Space Domain in the Russia-Ukraine War and the Impact of Converging Space and AI Technologies," CETaS Expert Analysis (February 2024).

32 https://orbit.ing-now.com/

33 https://nanoavionics.com/blog/how-many-satellites-are-in-space/#:~:text=As %20of%20March%207th%202024,to%20dominate%20low%20Earth%20orbit.

34 Dr Bleddyn Bowen, International Relations, University of Leicester, UK, Original Sin - Power, Technology and War in Outer Space, https://room.eu/article/original-sin-power-technology-and-war-in-outer-space

35 Comprehensive Nuclear Test Ban Treaty Organization; 5 November 2020; "9 July 1962: 'Starfish Prime', Outer Space"; https://www.ctbto.org/specials/testing-times/9-july-1962starfish-primeouter-space. Accessed 5 August 2020.

36 https://www.airandspaceforces.com/space-force-victus-nox-launch/

37 Union of Concerned Scientists (UCS); 1 January 2022; "Satellite Database"; https://www.ucsusa.org/nuclear-weapons/space-weapons/satellite-database. Accessed 19 March 2022.

38 McKnight, D., Macdonald, J., Arora, R., Pelton, J., Jenniges, J., and Martinez, P.; S2468-8967(20)30006-9; May 2019; "The Global Risk Continuum (GRC)"; The Journal of Space Safety Engineering, JSSE 104, International Association for the Advancement of Space Safety (IAASS) Space Safety Conference, Los Angeles, CA, May 2019.

4

Disruption Looming

"I know not with what weapons World War III will be fought, but World War IV will be fought with sticks and stones."

—**Albert Einstein**

This group comprises technologies still in the conceptual or early development stages, with the potential to radically reshape warfare once fully realized. These innovations are often on the drawing board, undergoing initial testing, or are at the experimental prototype phase. Their eventual deployment could lead to significant changes in military strategy and capabilities, but these technologies are not yet affecting the battlefield in a meaningful way. Examples might include quantum computing, advanced nanotechnology applications, or directed-energy weapons (DEWs).

Key Parameters

- *Level of Development:* Conceptual or early-stage development.
- *Extent of Proliferation:* Very limited, mostly in research or specialized labs.
- *User Exploitation:* Minimal or nonexistent in current military operations.
- *Stage of Development:* Conceptual, drawing board, or early prototype.
- *Technological Evolution:* Disruptive potential is recognized, but practical application remains years away.

Quantum Computing

The second quantum revolution is characterised by manipulating and controlling individual quantum systems (such as atoms, ions, electrons, photons, molecules or various quasiparticles), allowing to reach the standard quantum limit; that is, the limit to measurement accuracy at quantum scales. Quantum technology does not bring fundamentally new weapons or standalone military systems, but rather significantly enhances measurement capability, sensing, precision and computation power and efficiency of the current and future military technology. Most of the quantum technologies typically are technologies of dual use. Consequently, there is tremendous potential for military applications of quantum technology.[1] The mapping of quantum technologies' conceivable military applications is also important for the further assessment of threats to global peace and in the discussion of ethics policies or quantum-based preventive arms control. The future users of military quantum technologies will have to think carefully about whether, where and when to invest time and resources. The goal of the defence forces is not to develop military technology but usually only to specify requirements and their acquisition. However, they can participate significantly in development, especially if they are the end user.

The second quantum revolution will improve sensitivity and efficiency, and introduce new capabilities and sharpen modern warfare techniques rather than lead to new types of weapons. In the military sector, examples of quantum optimisations could be logistics for overseas operations and deployment, mission planning, war games, systems validation and verification, new vehicles' design and their attributes such as stealth or agility.

- **Quantum Cryptography.** The post-quantum cryptography implementation is the 'must-have' technology that should be carried out as soon as possible. The risk that hostile intelligence is gathering encrypted data with the expectation of future decryption using the power of quantum computers is real, high and present.[2] Quantum cryptography (designing of encryption schemes) uses the "spooky" phenomenon of quantum entanglement to facilitate the secure sharing of one-time pads (OTPs), which are essentially a set of cryptographic keys considered as having the highest degree of crack-resistivity. The phenomenon of quantum entanglement renders any man-in-the-middle attack infeasible, thus rendering the sharing of keys unhackable.[3]

- **Quantum Computing and Cryptanalysis.** In the related area of cryptanalysis, quantum computing, by using the superposition characteristic of quantum bits or "qubits" in conjunction with "Shor's Algorithm", enables the cracking of a large number of popular encryption algorithms which are based on prime numbers. One of the milestones often talked about in the context of progress in quantum computing is the race for achieving "quantum supremacy," which is said to be achieved when a quantum computer performs any calculation that, for all practical purposes, a classical computer cannot. It is estimated that, in order to attain this milestone, a 100 qubit quantum computer would be needed.[4] As on date, experimental quantum computers of 70 qubits have been demonstrated, and performance breakthroughs are being reported at frequent intervals.

- **Quantum Sensing.** Quantum sensing is likely to be a critical enabler for next-generation warfare strategies, allowing for more precise detection, navigation, and information gathering, while also challenging existing stealth and electronic warfare capabilities. As it becomes more mature, quantum sensing could disrupt the balance of military power by providing unparalleled situational awareness and detection abilities.

At the top will be an application for enhanced decision making, supporting military operations and functions through quantum information science, including predictive analytics and ML/AI.[5] Quantum computers are expected to play a significant role in Command and Control (C2) systems. The role of C2 systems is to analyse and present situational awareness or assist with planning and monitoring, including simulation of various possible scenarios to provide the best conditions for the best decision. Quantum computers can improve and speed up the scenario simulations or process and analyse the Big Data from ISR (Intelligence, Surveillance and Reconnaissance) for enhanced situational awareness. This also includes the involvement of quantum-enhanced machine learning and quantum sensors and imaging. Quantum information processing will probably be essential for Intelligence, Surveillance, and Reconnaissance (ISR) or situational awareness. ISR will benefit from quantum computing, which offers a considerable boost to the ability to filter, decode, correlate and identify features in signals and images captured by ISR. Quantum image processing in particular is an area of extensive interest and development.

It is expected that in the near-term situational awareness and understanding can benefit from quantum image analysis and pattern detection utilising neural networks.[6] Quantum technologies can significantly interfere in underwater warfare, with enhanced magnetic detection of a submarine or underwater mines, novel inertial submarine navigation and quantum-enhanced precise sonars. In general, in the maritime environment, sensing based on quantum photo-detectors, radar, LiDAR, magnetometers, or gravimeters can be applied.[7] Classical EW can also benefit from quantum computing, offering improved RF spectrum analysers for electronic warfare where quantum optimisations and quantum ML/AI techniques can be applied. Higher effectiveness can be reached by the processing and analysing directly of quantum data.[8]

Quantum technologies are expected to significantly improve positioning, navigation and timing (PNT) systems, especially inertial navigation. Time standards and frequency transfer (TFT) is a fundamental service that provides precise timing for communication, metrology, but also global navigation satellite system (GNSS). Although present TFT systems are well established, the performance of optical atomic or quantum clocks in combination with TFT utilizing quantum networks will keep pace with the increasing demands of the present applications (communication, GNSS, financial sector, radars, electronic warfare systems) and enables new applications (quantum sensing and imaging).[9] The development, acquisition and deployment of quantum technologies for military application will raise new, related challenges. The concept of quantum warfare will impose new demands on military strategy, tactics and doctrines, on ethics and disarmament activities and on technical realisation and deployment.

Blockchain Technology

During the 18th collective study of the Political Bureau of the Central Committee held on October 24, 2019, President Xi Jinping noted that, "We must take the blockchain as an important breakthrough for independent innovation of core technologies, clarify the main direction, increase investment, focus on a number of key core technologies, and accelerate the development of blockchain technology and industrial innovation."[10] Blockchain, also called distributed ledger technology, is a kind of user community-managed ledger technology that ensures transmission and access security with cryptography, enables consistent data storage, and prevents from any attempt to alter data or commit repudiation. A typical blockchain uses the chained block structure to verify and store data, avails the distributed node consensus mechanism to generate and update data, adopts smart contracts consisting of automated scripts to program and process data, and as mentioned above, ensures data transmission and access security with cryptography.[11]

The many benefits of blockchain make it a powerful prevention tool against cyber-attacks and its distributed nature and data integrity can pave the way for secure battlefield communications. Most defence platforms and systems are staggeringly complex. These assets are very mobile (globally) and require collaboration amongst a number of entities to keep them operational and mission-ready. Technology can help operators, manufacturers and suppliers 'harden' the supply chain and improve operational performance through the entire life cycle, from raw material to retired asset. Despite the sophistication of many weapons systems and platforms, their supply chains – both for raw material to finished product and post-production maintenance – are often still managed using traditional, and largely manual, processes.[12]

The command information system based on block chain technology can be divided into data layer, network layer, consensus layer, incentive layer, contract layer and application layer. Among them, the data layer guarantees the data or intelligence reliability, credibility and security of military information system; the network layer guarantees self-organizing and decentralized features of military information systems; the consensus layer encapsulates various types of consensus algorithms to achieve autonomous and credible decision-making; the incentive layer uses a programmable incentive mechanism to avoid all kinds of misbehaviours and achieve positive behaviour incentives; the contract layer helps automate and intelligentise military information systems, reducing the uncertainty, diversity and complexity that human and other factors bring to battle command and military management.[13]

Through the implementation of block chain technology, a modern military force can be benefited from things like Combating online attacks through cybercriminals dispatched by enemy nations or terrorist organisations. Defending vital weapons systems such as nuclear deterrents from enemies. Validating commands and intelligence in global conflicts. Management of military logistics and supply chains.[14] The development of block chain technology offers increased data confidence and data availability that can help shape future military logistics and planning. As we saw, the US intends to use it for secured databases, logistics and 3D printing, similarly to China and Russia. The EU is also eager to invest in block chain and will have the possibility to directly fund block chain technology related to military fields thanks to the upcoming Horizon Europe framework and the European Defence Fund.[15]

Using the consensus mechanism of block chain technology, we can transform from personal trust and institutional trust to machine trust. In the future battlefield, the UAV cluster command and control system will share combat command data in a decentralized manner and thus unify operations. Through the "public and private key encryption and decryption" and "digital signature" function of the block chain technology, each UAV in the cluster can serve as a network node. Distributed decision-making algorithm will ensure unified action by UAV cluster systems, and the consensus mechanism will ensure that all nodes in a distributed system agree on the objective. In real combat, each entity in the cluster will agree on operational tasks and objectives such as grouping and formatting, path planning, and barrier avoidance. The sidechain technology of block chain will allow multiple block chains to be connected to each other in a hierarchical manner. On the one hand, UAVs on different chains can act according to the protocol pre-set on the chain where they are located; on the other hand, inter-chain collaboration makes it easier to switch among diverse cluster formations. Thus, each entity in the UAV cluster can be regarded as an autonomous agent with the function of perception, reasoning and decision-making. These agents will form various decentralized autonomous clusters through smart contracts to execute optimal decisions in an autonomous mode.[16]

Block chain technology can help to determine the origin of the weapons from production to delivery and to prevent counterfeits. The date-stamped nature of the block chain enables users to track past transactions easily. For instance, in 2018, Accenture and Thales developed a system based on this technology, securing and rationalising the supply chains in the aeronautic and defence sectors. Blockchain technologies not only permit greater visibility on the material supply chain but can also improve food safety and healthcare on the battlefield. Tracking and tracing the food supply chain can prevent food-related outbreaks and help address the significant issue of the traceability of critical commodities such as pharmaceuticals.[17]

Blockchain technology offers significant advantages for military applications, particularly in enhancing security, transparency, and efficiency. Its decentralized nature ensures secure communication, data integrity, and protection from cyberattacks, making it ideal for sensitive military operations. In supply chain management, blockchain can verify the authenticity and traceability of defense equipment and reduce the risk of counterfeiting. Smart contracts streamline procurement, while secure data sharing enhances intelligence exchange between allies. Additionally, blockchain's potential extends to managing the maintenance of weapon systems, ensuring logistical efficiency, and safeguarding critical military networks, positioning it as a transformative tool for defense operations.

Biotechnology

COVID 19 was a gentle reminder to humans of the devastating potential of a biological disaster and that too a non-military one. This forced the governments and the scientific community alike, to consider the impact of planned biological military operations. Biological weapons though banned are not in any manner unimaginable and non-producible. They have been used in ancient history and can be used in the future, especially when such actions are underscored by deniability or lack of direct responsibility.

However, biotechnology in warfare is not just about biological weapons. It presents a vast canvas from nutrition, vital body functions, endurance, protection, medicine to bioengineering. Biotechnology includes a spectrum of military applications ranging from defense against biological weapons through the development of vaccines and detection systems to the controversial exploration of bio-engineered threats. It plays a vital role in enhancing soldier's performance and resilience, with research focusing on pharmaceuticals, nutritional supplements and biomedical devices. Biologically inspired materials offer possibilities for lightweight, high-strength materials, while biometric technologies aid in identification and verification processes. Additionally, biotechnology contributes to environmental monitoring, remediation efforts, and food/water security for military personnel. Medical countermeasures, such as advanced wound care products and regenerative therapies, are crucial for treating injuries in combat scenarios. Moreover, bio-inspired robotics and unmanned systems, modelled after biological organisms, are advancing reconnaissance and surveillance capabilities.

The Office of the Under Secretary of Defense for Research and Engineering (Alexandria, VA) established the DOD Biotechnologies for Health and Human Performance Council (BHPC) study group to continually assess research and development in biotechnology. The team selected four vignettes as being technically feasible by 2050 or earlier. These were ocular enhancements to imaging, sight, and situational awareness; restoration and programmed muscular control through an optogenetic bodysuit sensor web; auditory enhancement for communication and protection; and direct neural enhancement of the human brain for two-way data transfer. Although each of these technologies will offer the potential to incrementally enhance performance beyond the normal human baseline, the BHPC study group analysis suggested that the development of direct neural enhancements of the human brain for two-way data transfer would create a revolutionary advancement in future military capabilities. This technology is predicted to facilitate read/write capability between humans and machines and between humans through brain-to-brain interactions. These interactions would allow war fighters direct communication with unmanned and autonomous systems, as well as with other humans, to optimize command and control systems and operations. The potential for direct data exchange between human neural networks and microelectronic systems could revolutionize tactical communications, speed the transfer of knowledge throughout the chain of command, and ultimately dispel the "fog" of war. Direct neural enhancement of the human brain through neuro-silica interfaces could improve target acquisition and engagement and accelerate defensive and offensive systems. Although the control of military hardware, enhanced situational awareness, and faster data assimilation afforded by direct neural control would fundamentally alter the battlefield by the year 2050, the other three cyborg technologies are also likely to be adopted in some form by warfighters and civil society. The BHPC study group anticipated that the gradual introduction of beneficial restorative cyborg technologies will, to an extent, acclimatize the population to their use. The BHPC study group projected that the introduction of augmented human beings into the general population, active-duty personnel, and near-peer competitors will accelerate in the years following 2050 and will lead to imbalances, inequalities, and inequities in established legal, security, and ethical frameworks. Each of these technologies will afford some level of performance improvement to end users, which will widen the performance gap between enhanced and unenhanced individuals and teams.[18]

Biotechnology and bioengineering offer significant benefits to warfighting, including advancements in medical treatment, enhanced soldier performance, and improved battlefield capabilities. Research in biotechnology has led to the development of innovative medical treatments such as advanced wound care products and regenerative therapies, crucial for treating injuries sustained in combat. Bioengineering contributes to enhancing soldier performance through the development of pharmaceuticals and biomedical devices aimed at improving physical endurance and cognitive function. Additionally, biologically inspired materials inspired by natural structures offer lightweight, high-strength solutions for equipment durability and protection. Biometric technologies aid in battlefield security by enabling rapid identification of enemy combatants and confirmation of friendly forces, while biotechnology supports environmental monitoring and remediation efforts in military operations.

There has been a deep desire in the human race to create super humans and the field of Genome enhancement creates the possibility of creating super soldiers in the future. Though advances in this regard haven't been transcendental but the actual reason for that might not be the lack of scientific and technical prowess but ethical restraint and legal uncertainty. Also, biotechnology thrives on an ecosystem based on interactions between electronic, computer, cyber and information technologies. Quantum computing and AI/ML shall further revolutionise this space. Such a complex system remains vulnerable to loopholes getting exploited by rogue elements and thus, every step has to be carefully taken. There is a lot of vital human data that can be compromised due to the existing security threats to such exploration and research. The security hurdle is further reinforced by the ethical and moral considerations on genome enhancement and bioprinting. However, the technology offers outright relief as far as the availability of spare parts to the human body are concerned. The confidence of a soldier in battle can be well imagined if he is confident of a ready replacement for a limb that can be 3D printed for him. Not to mention the sense of relief to millions of patients who keep waiting for a transplant and fall prey to the grey market for survival.

But a soldier retires also, and if he has remained under the influence of biological enhancements during service, it would be difficult for him to integrate into natural civilian life once he retires from active duty. Even if he is psychologically strong enough to merge into civilian life, the reaction of the people around him to his artificially enhanced capabilities may not be as desired. However, the dual-use nature of biotechnology, with applications both in civilian sectors such as medicine and agriculture and in military contexts, poses challenges in balancing scientific progress with the risks of unintended harm. Legal debates centres on the adequacy of existing international agreements and treaties, such as the Biological Weapons Convention, in effectively regulating the development and use of biotechnological weapons. Furthermore, the attribution of responsibility in cases of biotechnological warfare presents difficulties in identifying perpetrators and holding them accountable under existing international law. The potential for biotechnological advancements to outpace regulatory frameworks underscores the need for ongoing dialogue among policymakers, scientists, ethicists, and civil society to ensure that biotechnology is used responsibly and by ethical and legal standards, thereby mitigating risks to global security and human well-being.

Generative AI

Generative AI is a transformative technology that goes beyond traditional artificial intelligence by enabling machines to create new data, designs, or strategies based on patterns learned from existing datasets. Typical examples of generative AI systems include image generators (such as Midjourney or Stable Diffusion), large language models (such as GPT-4, PaLM, or Claude), code generation tools (such as Copilot), or audio generation tools (such as VALL-E or resemble.ai). Using the term "generative AI" emphasizes the content-creating function of these systems. It is a relatively intuitive term that covers a range of types of AI that have progressed rapidly in recent years.[19]

Unlike normal AI, which typically analyses and reacts to given inputs, generative AI can produce entirely new outputs, such as designing complex systems, simulating scenarios, or generating creative solutions. Recent advances in generative models have been driven by three major developments:

- The availability of huge amounts of training data in the form of human language available on the internet (and in curated datasets of internet or user-generated content).
- Improvements in the underlying neural network models and the algorithms used to train them.
- Rapid growth in the amount of computational power that leading actors have used to train these models, which allows for the creation of larger, more sophisticated models.

In many cutting-edge applications, acquiring sufficient computational power to train a model is the most expensive of these components, and the relative capability of different models tends to roughly correspond to how much computational power was used to train them.[20]

So far, generative AI enables us to automate and augment cognitive and creative tasks, including some of the tasks done by doctors, designers, musicians, marketers, and more. Physical jobs such as building, assembly line work, cleaning, and so on generally remain unaffected by the wave of generative AI transformation. But combine generative AI with robots, and that may change. We could see AIs working in all sorts of sectors.[21]

This ability of Generative AI to innovate and adapt has vast implications for warfare, where generative AI is being applied to weapon systems, strategy development, and autonomous platforms. Globally, several countries are leveraging generative AI to enhance their military capabilities. In the United States, the Pentagon's DARPA (Defense Advanced Research Projects Agency) is exploring generative AI through projects like ACE (Air Combat Evolution), where AI algorithms generate new air combat strategies and help autonomous drones learn aerial manoeuvres in ways that mimic human creativity.[22] Similarly, the Navies are attempting to employ GENAI to generate optimal ship designs, considering complex variables like hull design, armament placements, and manoeuvrability in contested waters.[23] In China, generative AI is being integrated into autonomous and intelligentised weapon systems, enabling them to adapt to different environments, whether urban or wilderness, without extensive pre-programming.[24] The Chinese military is also investing in AI-driven cyber warfare capabilities, where generative AI can produce new malware strains or adaptive attack vectors to breach security systems more efficiently.[25]

Russia is another key player, utilizing generative AI for creating novel electronic warfare techniques, navigation and drone based targeting. AI-generated algorithms can dynamically jam enemy communications and radar systems in more effective and unpredictable ways, making electronic warfare systems harder to counter.[26] Additionally, the Israeli military's bombing campaign in Gaza used a previously undisclosed AI-powered system named Lavender, that at one stage identified 37,000 potential targets based on their apparent links to Hamas, according to intelligence sources involved in the war.[27]

This technology has the potential to be used for a variety of purposes in warfare, including:

- *Cyber warfare*: Generative AI can initiate cyber-attacks at predetermined intervals against enemy computer networks, thereby disrupting enemy communications, disabling their weapons systems, and compromising sensitive data at the most opportune and critical moments during the conflict.
- *War-gaming*: Generative AI generated themes and narratives can be used simulate different war scenarios, allowing military planners to test and refine their strategies. This potentially will assist in risk mitigation and faster response in real-world conflicts.
- *Disinformation campaigns*: Generative AI could be used to create fake news articles, videos, and social media posts that could be used to sow discord and chaos among enemy populations without them realising the origins of the content.
- *Autonomous weapons*: Generative AI can enable autonomous weapons to make their own decisions about who to target and when to fire. This would minimise human interference and speed up target identification, acquisition and engagement.
- *Logistics*: Generative AI can be harnessed to improve the efficiency of military logistics, such as warehousing, tagging and transport of supplies. This aims to reduce the strain on military resources and improve the efficiency of administrative functions.

The potential benefits of generative AI in warfare are clear as this technology has the potential to:

- Make warfare more efficient and less reliant on human judgment.
- Reduce collateral damage.
- Accelerate conflict resolution by improving the chances of success.
- Save human cost by enabling unmanned operations.
- Reduce the overall cost of warfare.
- Improve the accuracy of ISR.
- Enhance the effectiveness of training.

However, there are also several risks associated with generative AI in warfare. These risks include:[28]

- Destabilize societies and lead to conflict.
- New arms race and make it more difficult to control the use of force.
- Devastating impact on critical infrastructure.
- Blurred distinction between war and peace.
- Unintended civilian casualties.

Generative AI if not ethically employed can be pursued to create weapons that are far destructive than their present day avatars, which poses a significant challenge to society and humanity at large. This technology can accelerate the development of autonomous lethal weapon systems which function without human intervention. This raises ethical concerns about the accountability in warfare as the machines get to make life and death decisions without human oversight based on prewired algorithms. The capability to generate hyper realistic deep fakes and launch active disinformation campaigns will have a destabilising effect on the public already suffering the conventional vagaries of war. Further, its ability to constantly learn and evolve cyber warfare tools shall lead to highly destructive cyber-attacks crippling critical civil infra structure such as energy grids, financial systems and essential services.

While generative AI offers immense potential for innovation and operational efficiency in warfare, its disruptive nature also presents profound ethical and societal challenges. The development of autonomous weapons, advanced cyber threats and disinformation techniques fuelled by AI, demand international regulation and careful consideration to avoid destabilizing impacts on global security and peace.

Notes

1   Krelina EPJ Quantum technology for military applications, 2021 https://doi.org/10.1140/epjqt/s40507-021-00113-y

2   Wolf SA et al. The changing face of data security: 2020 Thales data threat report. Thales. 2020

3   Lieutenant General (Dr) RS Panwar, AVSM, SM, VSM (Retd), *Apr 28, 2020*, Disruptive Military Technologies Part I: Classification, AI & Robotics and Quantum Technologies https:/

/futurewars.rspanwar.net/disruptive-military-technologies-an-overview-part-i/

4   Graham Carlow, *Quantum Supremacy is Coming: Here's What You Should Know*, Quanta Magazine, 18 Jul 2019, Accessed 28 Feb 2021.

5   Wilson JR. The future of artificial intelligence and quantum computing. Military & Aerospace Electronics. Aug. 2020. URL: https://www.militaryaerospace.com/computers/article/14182330/future-of-artificial-intelligence-and-quantum-computing (visited on 01/27/2021).

6   Middleton A, Till S, Steele M. Quantum Information Processing Landscape 2020: Prospects for UK Defence and Security. DSTL/TR121783. June 2020.

7   Lanzagorta M, Uhlmann J. Space-based quantum sensing for low-power detection of small targets. In: Ranney KI et al., editors. Radar sensor technology XIX and active and passive signatures VI. 2015. Bellingham: SPIE, https://doi.org/10.1117/12.2183326.

8   Dunjko V, Taylor JM, Briegel HJ. Quantum-Enhanced Machine Learning. Physical Review Letters. 2016; 117(13). https://doi.org/10.1103/physrevlett.117.130501.

9   Giovannetti V, Lloyd S, Maccone L. Quantum-enhanced positioning and clock synchronization. Nature.2001; 412(6845):417–9. https://doi.org/10.1038/35086525

10   Y Zhu, X Zhang, Zh Y Ju and Ch Ch Wang, A study of blockchain technology development and military application prospects, The 2020 Spring International Conference on Defence Technology

11   Y Zhu, X Zhang, Zh Y Ju and Ch Ch Wang, A study of blockchain technology development and military application prospects, The 2020 Spring International Conference on Defence Technology

12   https://www.pwc.com/gx/en/industries/aerospace-defence/publications/blockchain-can-transform-defence-assets.html

13   Y Zhu,A study of blockchain technology development and military application prospects, Journal of Physics: Conference Series, Volume 1507, The 2020 Spring International Conference on Defence Technology, ICDT Spring, 20-24 April 2020, Nanjing, China https://iopscience.iop.org/article/10.1088/1742-6596/1507/5/052018

14   https://www.analyticsinsight.net/how-blockchain-is-being-used-by-global-militaries/

15   Alessia Cornella, Linda Zamengo, Alexandre Delepierre, Georges Clementz. Blockchain in defence: a breakthrough? https://finabel.org/wp-content/uploads/2020/09/FFT-Blockchain.pdf

16   Y Zhu,A study of blockchain technology development and military application prospects, Journal of Physics: Conference Series, Volume 1507, The 2020 Spring International Conference on Defence Technology, ICDT Spring, 20-24 April 2020, Nanjing, China https://iopscience.iop.org/article/10.1088/1742-6596/1507/5/052018

17   Alessia Cornella, Linda Zamengo, Alexandre Delepierre and Georges Clementz, European Army Interoperability Center, Finabel September 2020

18   Emanuel, Peter (CCDC CBC); Walper, Scott (NRL); DiEuliis, Diane (NDU); Klein, Natalie (AMRDC); Petro, James B. (USD(R&E)); Giordano, James (GU), Cyborg Soldier 2050: Human/Machine Fusion and the Implications for the Future of the DoD, 10-2019, v

19   https://cset.georgetown.edu/article/what-are-generative-ai-large-language-models-and-foundation-models/

20   Deep Ganguli et al., "Predictability and Surprise in Large Generative Models," 2022ACM Conference on Fairness, Accountability, and Transparency, June 2022, 1747–1764, https://doi.org/10.1145/3531146.3533229.

21   https://www.forbes.com/sites/bernardmarr/2024/03/05/the-future-of-generative-ai-6-predictions-everyone-should-know-about/

22   OUTREACH@DARPA.MIL 4/17/2024 ACE Program Achieves World First for AI in Aerospace Defense Advanced Research Projects Agency https://www.darpa.mil/news-events/2024-04-17

23   Sahil Thakur, Navneet V Saxena, and Prof Sitikantha Roy, Naval Construction Wing,

Department of Applied Mechanics, IIT Delhi, India Generative AI in Ship Design https://arxiv.org/pdf/2408.16798

24 Elsa B. Kania "AI weapons" in China's military innovation April 2020 https://www.brookings.edu/articles/ai-weapons-in-chinas-military-innovation/

25 Nathan Beauchamp, Mustafaga, Exploring the Implications of Generative AI for Chinese Military Cyber-Enabled Influence Operations: Chinese Military Strategies, Capabilities, and Intent Testimony presented before the U.S.-China Economic and Security Review Commission on February 1, 2024 https://www.rand.org/content/dam/rand/pubs/testimonies/CTA3100/CTA3191-1/RAND_CTA3191-1.pdf

26 Samuel Bendett APRIL 2024 The Role of AI in Russia's Confrontation with the West Center for a New American Security https://s3.us-east-1.amazonaws.com/files.cnas.org/documents/Russia-AI_2024-final.pdf

27 Bethan McKernan, Davies, 03 Apr 2024 The machine did it coldly': Israel used AI to identify 37,000 Hamas targets https://www.theguardian.com/world/2024/apr/03/israel-gaza-ai-database-hamas-airstrikes

28 https://www.linkedin.com/pulse/generative-ai-next-war-navigating-uncharted-waters-modern-ido-katz/

5

AI and Autonomy: The Force Behind the Forces

"The question is no longer whether robots will be used in war. The question is how, where, and under what rules."

—Peter W. Singer

In recent years there has been a compelling need to treat AI and Autonomy as a distinct and dedicated area of analysis within military transformation. AI has emerged not merely as a technological enabler but as a pervasive and foundational force fundamentally reshaping the character of warfare across domains. It has infused every facet of military thought, R&D, force structuring and operational design, catalyzing a rapid evolution from automatic systems to robotic platforms and onward to fully autonomous combat capabilities. What distinguishes this transformation is the cognitive leap AI enables, allowing machines not only to execute tasks but to develop human power of reasoning, prediction and intuitive action.[1]

As a result AI and autonomy have moved beyond niche applications to become core drivers of capability development influencing doctrine, training and even ethical considerations. From intelligent sensor networks and autonomous unmanned systems to predictive logistics and decision-support tools, AI is now woven into the fabric of modern combat power. Given its sweeping impact it is prudent to treat AI and autonomy not as a subtopic within specific arms and services but as a central pillar of contemporary military capability with dedicated space to explore its principles, architecture, applications and challenges. This chapter therefore offers a comprehensive examination of AI and autonomy as cross-cutting enablers that transcend individual services and combat arms.

An Intelligent Inflection

In the context of war-fighting, AI and autonomy are distinct but complementary concepts. AI refers to systems that are designed to simulate human intelligence, allowing them to perform tasks like learning from data, pattern recognition and decision-making. In military applications, AI is used to for processing large amounts of data, analyzing battlefield conditions and providing insights or recommendations to assist human operators in decision making. For example, AI might be employed in surveillance and reconnaissance platforms to identify threats, predict enemy movements, or assist in logistics planning. Thus, AI typically enhances human decision-making and requires human oversight. While it can automate certain tasks, it does not operate independently and still relies on human judgment to make the final decision.

In contrast, autonomy refers to the capability of a system to perform tasks independently, without direct human intervention. Autonomous systems in military context are designed to execute missions and make decisions based on pre-programmed rules, algorithms or real-time feedback from their environment. These systems can operate without continuous human input, making decisions on navigation, targeting and engagement. For instance, autonomous drones can fly to a target, identify it and even launch an attack, all without a human operator. The level of autonomy can vary, some autonomous systems may require minimal human oversight, while others, like fully autonomous vehicles or robots, can operate entirely without human intervention. The key difference is that autonomy allows systems to act independently, with the ability to adapt to changing conditions without waiting for human commands.

The primary distinction between the two lies in control and independence. AI often serves as a tool for decision support, enhancing human capabilities, on the other hand, autonomy focuses on the independent operation of platforms. AI is primarily used to process and analyze data, provide actionable insights and automate certain tasks, but human operators usually remain in control of critical decisions. However, autonomous systems are capable of making their own decisions and executing missions on their own, even in a dynamic and complex environment. Autonomy does not necessarily require human involvement once the system is deployed, although human intervention may still be needed in cases of unforeseen complex scenarios.

Both AI and autonomy are critical to modern warfare, with AI enhancing situational awareness and decision-making for humans, while autonomy allows for independent, rapid action in mission execution. The combination of both technologies could potentially revolutionize future warfare, where AI facilitates better decision support, while autonomous platforms execute complex missions with minimal human control.

The important question in this regard is whether we would prefer to develop and test safe and controllable autonomous systems, especially those with the capacity to take human life. Incremental progress in such military system development is occurring in many countries in air, ground, on water and underwater vehicles with varying degrees of success. Several types of autonomous helicopter that can be directed with a smartphone by a soldier in the field are in development in the US, in Europe and in China. Autonomous ground vehicles such as tanks and transport vehicles are in development worldwide, as are autonomous underwater vehicles. In almost all cases, however, the agencies developing these technologies are struggling to make the leap from development to operational implementation.[2] What seems clear is that a new AI arms race is underway, and there is probably little that can be done to stop it. In an open letter in late March, more than 2,000 technology leaders and researchers, including Elon Musk and Steve Wozniak, called on labs worldwide to pause in training up the newest digital intelligence models because of cascading fears that they could lead to disaster for the human race.[3]

The power of human induction i.e. the ability to form general rules from specific pieces of information is critical in a situation that requires both visual and moral judgment and reasoning. For humans, induction that drives such judgments is necessary to combat uncertainty. Computer algorithms – especially those that are data-driven like typical algorithms that fall under the category of AI, are inherently brittle, which means that such algorithms cannot generalize and can only consider the quantifiable variables identified early on in the design stages when the algorithms are originally coded.[4] Replicating the intangible concept of intuition, knowledge-based reasoning and true expertise is, for now, beyond the realm of computers. There is significant research currently under way to change this, particularly in the machine learning/AI community, but progress is slow.[5]

Even if the role of AI is limited to supporting human decision-making and humans make the final decision, there is still a risk of human judgment being dominated by AI. Many studies often cite the downing of a US Navy Aircraft F/A-18 by a US surface-to-air missile unit in the 2003 Iraq War. In this incident, the automated system of the US surface-to-air missile Patriot misidentified a friendly aircraft as an enemy aircraft. The human operators, who had to make a decision in just a few seconds, fired the missile in accordance with the automated system's indication, killing the pilot of a friendly F/A18.[6] Therefore, even if AI continues to alter, transform and dominate the scientific discourse, military innovation and military operations, the outcome of wars will still be largely influenced by the agility, perseverance, intelligence and creativity of humans fighting and directing those wars.

AI Enabling Robotics

While robotics and AI are both advanced technologies with growing applications in various industries, it is AI, not robotics, that primarily enables decision support systems. A Decision Support System (DSS) is designed to assist decision-makers in analyzing data, evaluating alternatives, and making informed decisions in complex situations. It is a technology that is widely used in business, healthcare, military operations and many other fields. AI focuses on providing cognitive capabilities, such as reasoning, learning and decision-making, which are central to the functioning of DSS, whereas robotics typically refers to physical machines designed to perform tasks autonomously or with human guidance.

AI can help analyse complex, multi-dimensional datasets, make predictions and recommend the best course of action based on statistical models, machine learning algorithms or optimisation techniques. For example, in a military context, an AI enabled Decision Support System can evaluate battlefield data received from various surveillance platforms in real-time, process intelligence reports and suggest tactical manoeuvres or resource allocations. AI can also study historical data to predict future outcomes, allowing decision-makers to consider various scenarios before making decisions.

Essentially, robotics provides the physical means to execute decisions, while AI provides the intelligence and decision-making power behind those decisions. Robotics involves the design and operation of machines that perform physical tasks, while the Decision Support System powered by AI is focused on cognitive tasks. For instance, a robot may be used to execute actions on an assembly line in an ordinance factory or carry reinforcements on the battlefield, but it would be AI that will drive the decision-making process behind the robot's actions. Robotics may perform tasks such as moving objects, navigating terrain or interacting with different environments, but these actions are often pre-programmed or driven by AI models that interpret data and decide what actions the robot should take.

AI in DSS is capable of learning from past decisions, adapting to new information and continuously improving its decision-making process over time. This is where AI's ability to apply machine learning and data analytics makes it a powerful tool for decision support. In contrast, robotics, even if equipped with sensors and automation software, will not inherently possess the ability to analyze data, learn from the analysis and modify decisions. It relies on the inputs and programming provided by human operators or AI systems, but without AI, it would simply carry out tasks in a predefined manner without the cognitive ability to evaluate alternatives or optimize performance.

AI: Making Borders Intelligent

AI-driven surveillance systems have revolutionised border security by enabling real-time monitoring, threat detection, and predictive analysis. Countries have leveraged AI technologies to enhance its border surveillance capabilities, particularly in regions prone to infiltration and cross-border terrorism. There are a plethora of simultaneous lines of development being pursued for AI enabled border control solutions. These

AI technologies come in multiple forms and can include algorithms designed to evaluate travellers' nuanced and almost imperceptible emotional expressions, biometric analysis of fingerprints and facial recognition, and scanner software that can differentiate humans from wildlife in remote border sections. Many of the systems are based on surveillance tools that were already being used for the same purpose but now have become increasingly automated so that computer systems and not human beings are involved in making initial enquiries about possible threats and response of the authorities to these threats. Artificial intelligence promises to supercharge this surveillance, making tools more powerful and capable of processing and interpreting more data than in the past.[7]

A RAND Europe study commissioned by Frontex, the EU border agency, and released in 2021 underscored this interest and found that AI could potentially be used in five different areas: situational awareness and assessment; information management; communication; detection, identification, and authentication; and training and exercise. The study also identified multiple potential barriers, including technological weaknesses; perception of high costs and commercial barriers; insufficient understanding and awareness of AI; lack of skills and expertise; constrained access to relevant technologies; and potential ethical, human-rights, and regulatory issues.[8]

iBorderCtrl's technology is founded on "affect recognition science," which itself is a field still under considerable debate and differing interpretations and acceptability. Affect recognition claims to expose truths about someone's personality and emotions through the analysis of their facial features. Proponents argue that emotions are "fixed and universal, identical across individuals, and clearly visible in observable biological mechanisms regardless of cultural context". According to them, studying faces "produces an objective reading of authentic interior states".[9] As the developments in AI and supporting hardware enable greater efficiency and accuracy it will lead to increased automation and reduction of human intervention in border control as the automated immigration control system will integrate e-gate hardware, document scanning and verification, facial recognition and other biometric verification to facilitate faster processing of travellers on border crossing while enhancing security through the integration of various AI-enabled tools.

These functions will be integrated as part of automated control points that will help establish whether the person is the rightful owner of relevant documents and thus automatically determine whether someone can pass through a border according to pre-defined rules. The system will be able to alert border guards to any potential issues or non-compliance with these pre-defined rules. Research continues to advance recognition models to enable systems to function in non-perfect conditions (e.g. low lighting, travellers wearing glasses, physical position, etc.). Technologies including iris scanning and facial recognition are not able to achieve fool proof accuracy given the current biometrics systems available, despite increased efficiency enabled by AI.

In the next few years, the development of AI-based sensors and extraction algorithms is anticipated to lead to enhanced capabilities, with models producing the same results that currently only exist for biometrics data-collection in well-controlled settings. It is necessary to establish a comprehensive understanding of the technical advantages as well as risks associated with novel AI-enabled solutions, particularly with regards to complex technical systems and surveillance technologies. This is likely to have implications for how organisations ensure they are set up to integrate these systems and manage the new risks that they bring. Learning about and using these systems will make it possible to mitigate and cope with any risks identified in relation to the system.

The integration of AI into conventional border control systems have made the systems efficient, responsive and indefatigable reducing the tremendous mental and physical drain on human soldiers who had to ensure 24x7 security at the border posts and fencing. Israel's army introduced an AI enabled robotic weapon system called the Smart Shooter which comprises of a twin gun turrets installed on top of the guard tower overlooking refugee camps in the West Bank. The residents of the refugee camp confirmed that the system is faster than the Israeli soldier and the range of the weapon had also got enhanced. The Israeli Army stated that the weapon doesn't fire live round and it can only fire tear gas, stun grenades, and sponge-tipped bullets. But the fundamental point here is the integration of AI with the weapon which allows it to be controlled by a software or a system of softwares, what type of munition is being fired, that is a matter of choice.[10]

Similarly, AI is being increasingly incorporated into surveillance systems for identification, tracking and disseminating information to central control rooms where a clear intelligence picture is being created by AI enabled GIS systems and proliferated in real time back to troops on ground as well as automated surveillance and engagement platforms. In majority of border control and management functions that involve unmanned systems, robotic platforms, or fully autonomous systems, AI is increasingly emerging as an enabling technology.

Powering The Infantry

AI and ML are at the forefront of revolutionising military operations by enabling intelligent decision-making, predictive analytics, and autonomous systems. AI can process massive amounts of battlefield data in real time, providing soldiers and commanders with actionable insights. For example, AI-powered systems can analyse drone footage, satellite images, and sensor data to predict enemy movements, identify threats, and optimize tactical decisions. As is the case for other equipment oriented arms and services, AI/ ML will have a transcendental impact on infantry operations too, by providing actionable intelligence at the most opportune moment to the last unit of Infantry Force deployed in the battlefield.

The US Army's Joint All-Domain Command and Control (JADC2) AI-driven system integrates data from multiple sources to provide a comprehensive view of the battlefield, allowing for faster and more informed decision-making. Automatic machine-to-machine transactions will extract, consolidate and process massive amounts of data and information directly from the sensing infrastructure.[11] Another such initiative is the Project Maven, launched by the US Department of Defense, that uses AI to analyse video footage from drones to identify insurgent activities, reducing the time and manpower needed for surveillance and target acquisition. By 2020, the US Army was running live ammunition exercises using the computer algorithm at the 18th Airborne Corps in North Carolina, the largest operational test bed for Project Maven.[12]

AI and ML will increasingly enable autonomous decision-making on the battlefield, by supercharging the Battle Field Management Systems and enabling the entire ecosystem of the Infantry Soldier allowing him to operate with greater independence and efficiency. These technologies will also enhance human-machine teaming, where soldiers work alongside AI-driven systems to maximize combat effectiveness.

Generative AI: Revolutionizing Infantry Warfare

Generative AI, as a subset of artificial intelligence focuses on creating new content or solutions based on existing data, is poised to revolutionise future infantry warfare. This technology leverages deep learning algorithms to generate everything from realistic simulations and strategic plans to adaptive responses in real-time combat scenarios. Succeeding paras explore how generative AI will impact various aspects of infantry warfare, from training and decision-making to battlefield operations and logistical support.

Enhanced Training and Simulation

Generative AI has the capability to create highly realistic training environments and scenarios that can be made adaptable to the actions of soldiers in real-time. Unlike traditional simulations, which are often pre-programmed and static, generative AI is capable of developing dynamic environments that evolve based on the trainee's decisions, providing a more immersive, realistic and challenging training experience.

Unique and adaptive combat scenarios can be developed instantly, simulating variety of enemy tactics, battle conditions and unanticipated enemy courses of actions. AI can assess performance of individual soldiers and also build personalized training packages that target weaknesses or skills that require improvement in a certain soldier or his subunit training. This tailored approach ensures that each soldier is optimally prepared for the realities of combat.

Training future infantry soldiers on complex, multi-layered training environments that reflect the chaos of real world combat will develop better critical thinking, decision making skills and adaptability in soldiers under combat pressure.

Real-Time Decision Support

Generative AI could be used as a form of real time decision support system for infantry units. By integrating large amounts of data from multiple sensors, platforms and intelligence networks, AI based training can produce tactical recommendations and predictive models of enemy action, along with courses of action to optimise responses. This would be an unbiased tool to assist the Commanders during fluid battle conditions.

Generative AI could recommend a variety of tactical manoeuvres to the Commanders based on the current battlefield scenario, producing multiple potential options for each course of action.

Such systems can be programmed to conduct Dynamic Threat Assessment of the battlefield and generate updated threat assessments, helping infantry units to stay ahead of enemy decision cycle and adapt their strategies on the go.

Intelligent Battlefield Simulations

Generative AI powered systems will be able to create and simulate an actual battlefield context, as well as adversary forces that can mimic human behaviour. These simulations can be used for training, operational planning and rehearsals. The AI can create a variety of scenarios, including counteractions by the enemy, the movement of reinforcements and even chemical and radiological scenarios, to prepare and allow infantry units to practice in variety of contingencies.

Military planners could use generative AI to design unpredictable scenarios based on the behavioural patterns of adversary leadership and throw a challenge at own seasoned commanders. Pre mission AI generated contingency planning can help commanders to ascertain chinks in their armour and refine their plans before the actual launch. For example, if an infantry unit is planning to seize a particular enemy position or a feature, the AI can simulate the operation under various conditions, such as different enemy force compositions or night versus day with different weather conditions.

Logistics and Supply Chain Optimization

Beyond tactical applications, generative AI can transform military logistics by optimising supply chains and forecasting logistical needs by studying the historical data as well as real time consumption patterns. This facilitates infantry units to have the necessary supplies whenever they actually need them, decreasing downtime and maximising operational effectiveness.

Generative AI can forecast supply requirements and create on Predictive Supply Management solutions to generate optimized delivery schedules that factor in elements such as weather, terrain and enemy activity. This reduces the risk of supply shortages in critical situations. Taking inputs from the overall operational picture, AI can generate and adjust resupply routes on the go, ensuring that supplies reach the frontline units while avoiding potential threats like ambushes or airstrikes. Generative AI will allow for increased automation and optimisation of military logistics, reducing human error and ensuring that infantry units remain well supplied and ready for extended operations in any environment.

Psychological Warfare and Information Operations

Generative AI is being leveraged more frequently in psychological operations and information warfare for content creation with the intent to persuade through disinformation campaigns and social media campaigns including deep fakes, that undermine the enemy's psyche or impact public perception. Generative AI is capable of generating believable content and rapidly disseminating it, opening the possibility of altering the psychological domain of warfare in ways not experienced ever before.

Generative AI can use automated content creation to produce targeted propaganda or misinformation campaigns that exploit social or

cultural weaknesses in enemy groups or civilian populations. For example, AI-generated deep fakes could be used to sow distrust among enemy leaders or create confusion about ongoing operations, aptly demonstrated during Operation SINDOOR. AI can also be used defensively to detect and counter enemy information operations, producing truthful content or debunking misinformation more rapidly than human analysts can.

Human-Machine Teaming

Generative AI will play a very crucial role in human-machine teaming, allowing AI and human soldiers to work together to achieve collective mission objectives. AI can improve and refine new strategies and tactics that humans conceive, further reinforcing human intuition and experience. Generative AI can be used to control swarms of autonomous drones that support infantry units by undertaking reconnaissance, electronic warfare missions or direct attacks on enemy positions.

AI enabled system can be designed to act as a co-pilot in vehicles and aircrafts, generating navigation routes, assisting target acquisition and recommending evasive manoeuvres, while the human operator focuses on broader mission objectives.

Generative AI is set to revolutionize infantry warfare by enhancing training. decision-making, battlefield simulations, logistics, and psychological operations. Its ability to generate new, adaptive solutions and strategies in real-time will make it an indispensable tool for future military operations. By leveraging generative AI, infantry units will be better prepared, more adaptive, and more effective in both traditional and unconventional warfare scenarios. As this technology continues to advance, it will reshape the nature of combat, making future warfare more dynamic, unpredictable. and data-driven.

Intelligent and Mechanised

Embedding AI and autonomy in armoured fighting vehicles has initiated a significant new set of technological advances. AI will support the central functions of navigation, target acquisition and threat identification, which will lessen the cognitive burden on their crews and improve the timeliness and coherence of their decision making. Unmanned systems, like the US Army Robotic Combat Vehicle platform, are a scalable version of this trend, in which an increasing range of autonomous and semi-autonomous vehicles, in particular, drones, are capable of conducting reconnaissance or strike missions without the direct intervention of human operators.[13]

These technologies allow for complex, yet coordinated manoeuvres, once thought impossible, facilitating new forms of swarm tactics and autonomous engagements. The development of autonomous and unmanned armoured vehicles is a landmark technological advancement in warfare. Unmanned Ground Vehicles (UGVs) in particular are increasingly being integrated into conventional mechanised forces, enhancing operational flexibility, limiting human operators' exposure to risk and providing greater lethality on the battlefield.

Impact on AFV Design

AI-Enabled Combat Vehicles: AI enabled systems are being integrated into AFVs for direct assistance to operators with navigation, target acquisition and threat identification. These systems give their crew operational abilities and allow rapid decision making without excessive cognitive fatigue, especially when they encounter multiple targets and engagements during an intense battle.

Unmanned and Autonomous Vehicles: Increased employment of unmanned and autonomous AFVs is the most revolutionary innovation in mechanised warfare. These UGVs can operate in dangerous environments without exposing human crews. Like the Autonomous Tactical Light Armour System (ATLAS) Collaborative Combat Variant (CCV) developed by BAE Systems Australia, which is meant to work together with manned tanks and infantry fighting vehicles. The vehicle is equipped with a new, lightweight, highly automated medium-calibre turret called Vantage ATS that is armed with an M242 Bushmaster 25mm cannon, providing lethal fire to a range of 2,500 m. A ready-use ammunition capacity of 260 rounds allows for multiple engagements and high battlefield persistence.[14] These UGVs are capable of target acquisition, reconnaissance, and fire support and are come fitted with advanced sensors, communication, and weapons systems. The goal is to enhance the effectiveness of armoured units by providing them with additional capabilities that can be deployed with minimal risk to human personnel.

Robotic Combat Vehicle (USA): A program under the US Army to develop AI-enabled, unmanned ground vehicles capable of performing various combat functions.[15]

Shturm Remote-Controlled Tank (Russia): This next-generation main battle tank is equipped with advanced AI systems for target identification and autonomous operation. T-72 tank was chosen as the platform for the Remote-Controlled Tank (RTK) due to its cost-effectiveness, reliability, protection features, and mobility. The T-72 Remote-Controlled Tank (RTK) is equipped with weaponry that includes the RPO-2 "Shmel-M" flame rocket launcher, 30mm automatic cannons, and 220mm thermobaric unguided rockets from the TOS-1A Solntsepyok heavy flamethrower system.[16]

Impact on Tactics

AI allows autonomous vehicles to coordinate with each other and exploit large-scale swarming to overcome enemy forces. Autonomous and unmanned AFVs are increasingly used for reconnaissance missions, gathering intelligence without exposing personnel to danger. This allows for safer and more informed tactical decisions.

Systems enabled with artificial intelligence are being incorporated in to all elements of armoured forces for enhancing their operational capabilities. AI algorithms can crunch through troves of sensor data collected from drones and other systems, as well as local information and deliver timely analysis back to a commander. This process spans from static target identification and enemy movement prediction to logistic and maintenance schedule optimization. The development of autonomous systems, drones and especially UGVs is one of the key uses of AI in armoured forces. Such systems are capable to function either on their own or cooperatively with manned vehicles: they can a variety of mission, ranging from reconnaissance and target identification to even direct assault. Artificial Intelligence (AI) helps in improving the performance of already existing systems like the fire control systems and targeting system as it allows faster and more accurate computing.

The Rise of Intelligent Firepower: A Deep Dive into the Impact of AI on Artillery

Artificial Intelligence does not only make artillery systems incrementally better but marks a turning point in the very nature of artillery warfare. These changes derive from the ability of AI to handle increasingly greater amounts of data, learn from experience, and make complex decisions at a speed that no human operator can surpass. In this section, we discuss the many ways that AI is impacting the artillery, with the potential to transform targeting systems, improve servicing procedures, and the critical ethical challenges raised by increased autonomy in weapon systems.

Autonomous Targeting: Redefining Precision and Speed

The dynamism of AI can facilitate artillery targeting with a shift from human-driven to automated systems, which can quickly identify, prioritize, and engage targets with unprecedented accuracy. Such a shift is due to certain capabilities of AI:

- *Advanced sensor fusion*: AI systems use data fusion to correlate clues originating from disparate sources – radar, acoustic sensors, drones, and forward observers. This generates a coherent real-time battlefield picture that allows for target identification and tracking, even in complicated and cluttered environments.
- *Predictive targeting*: By utilizing historical data, terrain information, and enemy movement patterns, AI can predict future enemy locations and movements. This allows for pre-emptive strikes, greatly reducing the sensor-to-shooter timeline and leaving opponents little time to react.
- *Target Prioritization and Selection*: An AI can conduct an automated assessment of numerous targets with respect to a predefined rules of engagement and strategic goals, before prioritizing high-value targets and ensuring minimum collateral damage. It is crucial in dynamically evolving combat situations which require speed and an informed perspective.

The birth of autonomous targeting tools is set to greatly boost artillery effectiveness because of:

- *Highly Accurate Engagements Resulting in Minimal Collateral Damage*: AI-based systems for artillery lead to pinpoint accuracy and avoid civilian casualties and collateral damage, a critical factor in modern warfare.
- *Faster Response Times*: Autonomous targeting drastically reduces the decision-making cycle and allows very rapid responses to threats, thus giving an edge to such systems in fast-paced combat scenarios.

- *Improved Survivability*: Besides allowing one to minimise the need for Artillery Observers, giving artillery units a fair chance of survival on the battlefield.

AI Enabled Fire Control Systems

Fire control systems enhanced with artificial intelligence aren't quite autonomous artillery platforms, but they are an important part of the larger trend toward automation. AI processes targeting data much faster and with less potential for human error than human operators, computes trajectories, corrects for environmental variabilities, and anticipates enemy movements. AI-powered artillery responds to changing battlefield conditions quicker and more effectively, boosting mission success rates and lessening the risk for collateral damage.[17] The new Reconnaissance-Fire System (ROS) allows combined-arms units to conduct operations in real time and greatly increases the speed and accuracy of Russian fires on the future battlefield. The ROS is a network-centric capability offering vastly enhanced target acquisition and strikes across the range of Russian systems capable of targeting ground targets and especially benefits artillery systems.[18]

AI-Powered Predictive Maintenance

Beyond targeting, AI also transforms artillery maintenance, evolving from reactive repairs to predictive, proactive maintenance models. This evolution enables the maximization of operational availability and combat readiness of increasingly complex artillery systems. Achieved through:

- *Sensor Data Analysis*: AI algorithms allow monitoring of various critical parameters for artillery systems under operation. They gather, analyze, and assess sensor data regarding temperature, vibration, pressure, and overall performance indicators of the systems concerned.
- *Anomaly Detection*: AI is capable of establishing performance parameter baselines within the model. When working in their real-world situation, these algorithms hold the potential to recognize flaws before failure happens.
- *Predictive Modeling*: Drawing from historical maintenance data, patterns of operational use, and environmental considerations, AI estimates when a component may fail and can schedule maintenance before a failure occurs. Consequently, this reduces downtime and cuts costs and ensures that artillery systems are always ready for action.

The advantages for AI predictive maintenance go far beyond cost savings and uptime:

- *Enhanced Safety*: By identifying potential failures before they occur, AI-driven maintenance reduces the risk of catastrophic equipment malfunctions that could endanger personnel or compromise missions.

- *Improved Operational Efficiency*: Proactive maintenance minimizes unplanned downtime, ensuring that artillery units are always prepared for deployment and maximizing their operational effectiveness.
- *Extended Lifespan of Equipment*: By optimizing maintenance schedules and minimizing wear and tear, AI-driven approaches can extend the lifespan of expensive artillery systems, maximizing return on investment.

Development of Autonomous Artillery Systems

Although not widely employed, autonomous artillery systems epitomize an emerging and far-reaching transformation in the field. While these systems are provided with considerable autonomy for several applications, the use of AI and robotics allows them to accomplish complex fire missions quickly, smoothly, and with high precision as compared to human controls.

With an array of functions ranging from target acquisition, artillery system control, mobility, and resupply, autonomous artillery systems are basically traditional artillery platforms equipped with AI, advanced sensors, and robotics-they can operate all functions with little or no direct human involvement.

For example, the Russian 2S35 Koalitsiya-SV is a potent self-propelled howitzer that features such semi autonomous capabilities as automatic load and firing along with AI-assisted fire control systems. It is an advanced system that can engage targets located far away and with precision. In addition, the autonomous capability diminishes the crew's size, thereby enhancing efficacy and survivability.[19]

Advantages of Autonomous Artillery

Faster, more precise, and responsive, autonomous artillery systems are advantages against standard platforms. These systems can process vast amounts of data in real-time and adapt more easily to changing battlefield environments, increasing the interaction and effectiveness of fire missions.

Autonomous artillery like the Robotic one will be able to operate in environments too dangerous for human crews, such as those contaminated by chemical, biological, or radiological threats: by totally removing humans from the immediate line of fire, this at once lessens casualties and ensures greater survival of artillery units.[20]

AI: The Wind Beneath Wings

Artificial Intelligence, represents a paradigm shift in technological capability, encompassing the use of machine learning, neural networks, and sophisticated algorithms to endow systems with the ability to perform tasks that traditionally required human intelligence – tasks such as decision-making, pattern recognition, and adaptive learning. In the realm of military aviation, AI's influence is both broad and profound, touching every facet of operations across helicopters, UAVs and autonomous systems.

Helicopters

In helicopters, AI acts as a force multiplier, assisting pilots in making better decisions by analysing data in real-time. It enhances situational awareness by processing information from sensors and networks, offering a clearer battlefield overview. AI reduces pilot workload by automating routine tasks and helps with predictive maintenance by analysing equipment data to foresee failures, minimizing downtime. The Sikorsky S-97 Raider, used by the US Army, is an example of AI improving helicopters. The S-97 integrates algorithm driven avionics that optimize flight performance and pilot efficiency, enabling rapid, agile manoeuvres in contested environments – a testament to how machine learning is elevating the helicopter's role on the modern battlefield.[21]

UAVs

AI's impact is equally transformative, enabling a level of autonomy and sophistication that redefines their operational scope. AI transforms UAVs by enabling them to navigate complex routes without human input. It allows UAVs to identify and prioritize threats in real-time and adjust their missions' mid-flight as conditions change. AI also lets multiple UAVs work together as a team. The MQ-9 Reaper, used by the US military, is a prime example. It excels at intelligence, surveillance, reconnaissance and precision strikes, using AI to autonomously identify targets and navigate contested airspace with minimal human assistance.[22]

Autonomous Systems

AI is nothing short of foundational, serving as the core technology that enables these platforms to operate independently of human oversight. AI is vital for autonomous systems, allowing them to operate without human control. It enables them to execute missions by processing data and making real-time decisions, conducting threat assessments, and using adaptive learning to improve over time. The Boeing MQ-28 Ghost Bat, developed in collaboration between Australia and the United States, stands as a pioneering example.[23] Designed as a "loyal wingman" to complement manned aircraft, the Ghost Bat uses AI to execute autonomous missions, seamlessly integrating with human pilots to enhance overall mission effectiveness. This capability underscores AI's role in ushering in a new era of human-machine teaming, where autonomous systems amplify the reach and resilience of military aviation.

Artificial intelligence is essential to the development of autonomous combat aircraft and systems. AI provides machine learning algorithms that allow it to analyze massive amounts of data, recognize structures and match suitable weapon to target in real time optimising combat power and its effectiveness. AI-driven autonomy allows an aircraft or a drone to operate autonomously and establish an understanding of the threats present on the battlefield and respond without the need for continuous human oversight.

Another key application of AI in military aviation is in the development of autonomous drones. These drones can perform a wide range of missions, including ISR, target acquisition, electronic warfare, and direct combat. AI enables these drones to navigate complex environments, identify and track targets and execute precise strikes with minimal human input.

For example, the US Air Force's Skyborg program which is focused on developing AI driven drones that can operate alongside manned aircraft, providing additional capabilities and support in contested environments. These drones can autonomously carry out missions such as electronic warfare, suppression of enemy air defences (SEAD) and close air support (CAS), enhancing the overall effectiveness of military aviation forces.[24]

Advantages of AI Powered Combat Aircraft

AI powered combat aircraft offer several advantages over traditional manned platforms. First, AI driven autonomy allows for faster decision making and response times, as AI can process and analyse data more quickly than human operators. This is very crucial in highly contested environments, where split second decisions can determine the success of a mission.

Second, AI powered aircraft can fly in hazardous and extremely dangerous environments, such as an area with the highest level of enemy air defences or extreme weather conditions. By eliminating the human pilot requirement, the aircraft can accomplish missions that would typically put personnel at serious risk.

Lastly, AI-driven autonomy allows for greater operational flexibility. Autonomous aircrafts can be deployed in large numbers, creating a swarm effect that can saturate enemy defences and achieve objectives that would be difficult for manned aircraft to accomplish, as seen in the latest Iranian drone strikes on Israel aimed at overwhelming their missile defence systems.

The development and deployment of AI powered combat aircraft and autonomous systems present several challenges. One of the primary concerns is the reliability and security of AI-driven systems. Ensuring that AI can operate effectively in complex and contested environments, while also being resistant to cyberattacks and electronic warfare, is critical to the success of these systems.

AI and Air Defence

AI and machine learning will bring a revolution in air defence by greatly cutting down on reaction times and augmenting target discrimination. The future for these systems lies in using deep learning algorithms to analyse their sensor data in real-time and provide predictive analytics and the ability for autonomous decision-making in multi-threat environments. AI imbedded in sensor fusion networks, together with interceptor systems, could provide a more cohesive sensor-to-shooter chain and reduce the burden on human operators. Additionally, as AI systems mature, it would provide the means for developing adaptive strategies that could learn from battlefield data and dynamically optimise defensive postures. The rise of autonomous systems and AI will enhance response speed and accuracy of the air defence architecture; however, such systems provide lethal decisions as an output of an algorithm, which leads to valid ethical concerns. These require thorough human oversight and establishment of clear rules of engagement within which an efficient yet accountable system can operate. Promises of such a trend posit great capability but require robust governance.

The development of new ground-based air defence systems is being driven by the need to identify and neutralise novel threats and proliferation of cutting edge technologies. The Integrated Air and Missile Defence Battle Command System (IBCS),[25] developed by the United States, connects various systems such as Patriot launchers, THAAD batteries, and Sentinel A4 radars into one architecture.[26] The system uses AI to integrate sensor data, allowing for dynamic allocation of weapons providing the Missile Segment Enhancement (MSE) interceptors the capability to neutralise varied targets from drone swarms to intercontinental ballistic missiles. Meanwhile, the S-500 Prometey from Russia has incorporated the new 77N6-N missile and 76T6 radar designed for hypersonic targets, with ranges of up to 600 kilometres giving the system advanced long-range air defence capabilities.[27]

Emerging technologies are key drivers. AI enhances systems like the NASAMS, where the Fire Distribution Center integrates data from the AN/TPQ-64 radar to optimize AIM-120 AMRAAM engagements, reducing human error and reaction time.[28] Quantum radar, still experimental, promises to detect stealth aircraft using quantum entanglement, potentially complementing systems like the US's AN/SPY-6 radar on land-based Aegis Ashore platforms.[29] Hypersonic threats are spurring projects like the US Glide Breaker,[30] which pairs a kinetic interceptor with space-based Hypersonic and Ballistic Tracking Space Sensor (HBTSS) sensors,[31] aiming to neutralize high-speed glide vehicles mid-flight.

Ground-based tactical air defence is standing at the intersection of traditional missile technology and the digital revolution. As these systems increasingly incorporate AI-driven decision support, advanced radar and sensor networks, and new forms of interceptor technology, they will continue to create a potent line of defence against the modern mosaic of airborne threats. Overcoming the technological challenges inherent to this integration will require sustained investment in research and development, close collaboration between military and industry, and the continuous refinement of operational doctrines to ensure that ground-based air defence remains robust, agile and effective in the face of ever-evolving adversaries.

AI and Autonomy in Space

The integration of AI and autonomous systems into space operations represents a transformative leap forward, promising to enhance the sophistication and efficiency of military missions with minimal human oversight. In due course, the space will be operated by AI-driven platforms taking centre-stage, analyzing petabytes of satellite data; optimizing mission planning through simulations and responding to threats in real-time. AI can enhance space situational awareness by rapidly processing imagery and signals intelligence, enabling commanders to anticipate adversary actions and allocate resources more effectively.

Autonomous satellites, capable of independent navigation, self-repair, or defensive manoeuvres, could maintain operational continuity even under duress, ensuring uninterrupted support for ground forces. Yet, these benefits are tempered by inherent risks. The increased reliance on AI introduces vulnerabilities to cyberattacks that could corrupt data or hijack autonomous systems, potentially turning them against friendly forces. Furthermore, the delegation of decision-making to machines raises ethical questions, particularly in scenarios involving lethal force or strategic escalation. The Army must therefore invest in robust cybersecurity protocols and establish clear guidelines for human oversight, ensuring that AI and autonomy amplify rather than undermine its operational effectiveness. By harnessing these technologies judiciously, the Army can position itself to thrive in a future where speed, precision, and adaptability define success on the battlefield.

Ethical Implications: Navigating the Moral Minefield

While the potential benefits of AI in artillery are unquestionable, an increase in autonomy in weapons systems presents profound moral dilemmas that call for deliberation and international dialogue.[32] Important dilemmas arise:

- *Accountability and Responsibility*: In cases where there is a decision taken to kill, it becomes complex to attribute accountability in the event of unintended consequences or errors. Establishment of demarcation regarding accountability of the acts of autonomous weapons is highly crucial.
- *Risk of Unintended Escalation*: Due to their speed and autonomy, AI systems may lead to unintended escalation of conflicts, especially when human oversight is limited.

- *Potential for Bias and Discrimination*: AI algorithms are as good as the data upon which they were trained. If the data used to train the AI demonstrates existing biases, those very biases will persist into operational use-and could well even become amplified-in regard to targeting decisions.[33]

To deal with these ethical questions, a multi-pronged approach is required.

- *International Regulations and Treaties*: It will be very important to create international regulations and treaties that govern the development, deployment, and use of autonomous weapons to prevent the development of an AI arms race and to protect responsible usage.
- *Human Oversight and Control*: The critically important role of human control over decisions, especially for those involving lethal force must be maintained. While the widespread deployment of AWS is likely inevitable, there is still time to influence where and how it is deployed, and human-independent deployment is not technically trivial.[34]

Notes

1   General Anil Chauhan, Chief of Defence Staff, Artificial Intelligence (AI) Transforming Warfare, Page 179, Ready, Relevant and Resurgent A Blueprint for the Transformation of India's Military

2   M.L. Cummings, Artificial Intelligence and the Future of Warfare International Security Department and US and the Americas Programme, January 2017,10

3   Michael Hirsh, How AI Will Revolutionize Warfare, APRIL 11, 2023. https://foreignpolicy.com/2023/04/11/ai-arms-race-artificial-intelligence-chatgpt-military-technology

4   Smith, P. J., McCoy, C. E. and Layton, C. (1997), 'Brittleness in the design of cooperative problem solving systems: The effects on user performance', IEEE Transactions on Systems, Man and Cybernetics, Part A: Systems and Humans, 27(3), pp. 360–371

5   M.L. Cummings, Artificial Intelligence and the Future of Warfare International Security Department and US and the Americas Programme, January 2017

6   Michael C. Horowitz, Lauren Kahn, and Laura Resnick Samotin. "A High-Reward, Low-Risk Approach to AI Military Innovation," Foreign Affairs, May/June 2022, https://www.foreignaffairs.com/articles/united-states/2022-04-19/force-

7   Hannah Tyler, The Increasing Use of Artificial Intelligence in Border Zones Prompts Privacy Questions. February 2, 2022

8   Erik Silfversten, Luke Huxtable, Linda Slapakova, Pauline Paille, and Kevin Martin. Artificial Intelligence-Based Capabilities For The European Border And Coast Guard. 17/03/2021 https://www.frontex.europa.eu/assets/Publications/Research/Frontex_AI_Research_Study _20 20 _final_report.pdf

9   Lucien Begault, Automated technologies at EU borders and the future of Fortress Europe. 27/03/2019 https://www.euronews.com/2019/03/27/automated-technologies-at-eu-borders-and-the-future-of-fortress-europe-view?utm_medium=Social&utm_source=Facebook#Echo box=1553694747

10   Roselyne Min with AP Israel deploys AI-powered robot guns that can track targets in the West Bank 17/10/2022 https://www.euronews.com/next/2022/10/17/israel-deploys-ai-powered-robot-guns-that-can-track-targets-in-the-west-bank

11   Summary of the Joint All-Domain Command & Control (JADC2) Strategy, March 2022 Department of Defense https://media.defense.gov/2022/Mar/17/2002958406/-1/-1/1/SUMMARY-OF-THE-JOINT-ALL-DOMAIN-COMMAND-AND-CONTROL-

STRATEGY.pdf

12   Saleha Mohsin 1 March 2024 https://www.bloomberg.com/news/newsletters/2024-02-29/inside-project-maven-the-us-military-s-ai-project

13   Congressional Research Service The Army's Robotic Combat Vehicle (RCV) Program July 23, 2024 https://crsreports.congress.gov/product/pdf/IF/IF11876

14   *Peter Felstead* BAE Systems Australia unveils new highly autonomous armed UGV 11 September 2024 https://euro-sd.com/2024/09/major-news/40327/baes-australia-unveils-new-ugv/

15   Rojoef Manuel APRIL 4, 2024 US Taps Consortium for Army's Robotic Combat Vehicle Software Prototype https://thedefensepost.com/2024/04/04/us-robotic-combat-vehicle-software/#:~:text=The%20RCV%20is%20part%20of,%2C%20medium%2C%20and%20heavy%20variants

16   Russia develops AI-powered unmanned T-72 Shturm tanks for deployment in Ukraine. 25 Sep, 2024 Defense News Army 2024 https://armyrecognition.com/news/army-news/army-news-2024/exclusive-russia-develops-ai-powered-unmanned-t-72-shturm-tanks-for-deployment-in-ukraine

17   Future Trends in Artillery Technology: Innovations and Impacts," Total Military Insight, accessed February 25, 2025, https://totalmilitaryinsight.com/future-trends-in-artillery-technology/

18   McDermott, R. 2023. "The Technological Transformation of Russian Conventional Fires." Journal of Slavic Military Studies 36 (July): 241–70. doi:10.1080/13518046.2023.2283962

19   Sakshi Tiwari, Russia Set To Unleash Its 'Most Lethal' Artillery Against Ukraine; RuMoD Says 1st Batch In Final Stage Of Production, January 31, 2024 https://www.eurasiantimes.com/russia-set-to-unleash-its-most-lethal-artillery-gun/

20   Baker, D. Should We Ban Killer Robots? John Wiley & Sons, 2022.

21   Sydney J. Freedberg Jr, Sikorsky Be Nimble: S-97 Raider Shows Off For Army FARA, June 25, 2019, https://breakingdefence.com/2019/06/sikorsky-be-nimble-s-97-raider-shows-off-for-army-fara/#:~:text=But%20the%20S%2D97%20can,still%20if%20they%20stay%20level.

22   Joseph Trevithick, MQ-9 Reaper Flies With AI Pod That Sifts Through Huge Sums Of Data To Pick Out Targets, The War Zone, Sep 5, 2020, https://www.twz.com/36205/reaper-drone-flies-with-podded-ai-that-sifts-through-huge-sums-of-data-to-pick-out targets#:~:text=MQ%2D9% 20Reaper%20Flies%20With,when%20they%20had%20taken%20place.

23   Boeing MQ-28 Ghost Bat, https://www.boeing.com/defence/mq28#overview

24   Woodrow Bellamy III, Air Force Research Lab Conducts First Flight Test of Skyborg Autonomy Core System, Air Force Research Lab, Kratos, Skyborg, U.S. Air Force, May 10, 2021 https://www.aviationtoday.com/2021/05/10/air-force-research-lab-conducts-first-flight-test-skyborg-autonomy-core-system/

25   Missile Defense Project, "Integrated Air and Missile Defense Battle Command System (IBCS)," Missile Threat, Center for Strategic and International Studies, November 3, 2016, last modified June 7, 2021, https://missilethreat.csis.org/defsys/ibcs/.

26   The Terminal High Altitude Area Defense (THAAD) System, Congressional Research Service, October 17, 2024, https://crsreports.congress.gov/product/pdf/IF/IF12645

27   Missile Defense Project, "S-500 Prometheus," Missile Threat, Center for Strategic and International Studies, May 4, 2017, last modified July 1, 2021, https://missilethreat.csis.org/defsys/s-500-prometheus/.

28   Naval Air Systems Command. "Advanced Medium-Range Air-to-Air Missile (AMRAAM)." Naval Air Systems Command. Accessed March 11, 2025. https://www.navair.navy.mil/product/AMRAAM.

29   Naval Sea Systems Command. "NSWC D.D. - AN/SPY-6(V)1 Radar and Aegis Baseline 10 Test Team Earns Top Honor." Naval Sea Systems Command. Accessed March 11, 2025. https://www.navsea.navy.mil/Media/News/Article-View/Article/3867461/nswcdds-anspy-6v1-radar-and-aegis-baseline-10-test-team-earns-top-honor/.

30   Elizabeth Howell, "DARPA's Glide Breaker Hypersonic Interceptor Enters New Phase."

Space.com. Accessed March 11, 2025. https://www.space.com/darpa-glide-breaker-hypersonic-interceptor-new-phase.

31 Northrop Grumman. "Hypersonic and Ballistic Tracking Space Sensor Satellites." Northrop Grumman. Accessed March 11, 2025. https://www.northropgrumman.com/space/hypersonic-and-ballistic-tracking-space-sensor-satellites

32 Philip Feldman, Aaron Dant, Harry Dreany War Elephants: Rethinking Combat AI and Human Oversight,May 1, 2024

33 Bergman, R. and Fassihi, F. The scientist and the A.I.- assisted, remote-control killing machine. The New York Times, September 2021.

34 Riley Simmons-Edler, Ryan P. Badman, Shayne Longpre, Kanaka Rajan AI-Powered Autonomous Weapons Risk Geopolitical Instability and Threaten AI Research 31 May 2024

6

Unmanned, Unseen, Unstoppable: The Drone Revolution

"We are witnessing the most profound change in warfare since the introduction of the rifle and railroads… drones are at the center of it."

—**General Mark Milley**

Modern day warfare is undergoing a fundamental transformation in the manner how conflicts are being initiated, pursued and settled. Drones have evolved from being a mere supporting element to becoming a standalone instrument of war, that decisively influences the outcome of battles and in some case has remained the sole means of conducting violence against adversary. While the majority of drones currently deployed by militaries and non-state actors remain piloted systems, the increasing convergence of artificial intelligence and robotics is rapidly giving rise to semi and fully autonomous drones. This technological evolution is not only transforming the very nature of military engagements across multiple domains but also redefining the character of war.

The Azerbaijan-Armenia War: A Watershed Moment in Drone Warfare

The Nagorno-Karabakh conflict of 2020, between Azerbaijan and Armenia was a watershed moment in military history, convincingly establishing drone technology as a means to alter the balance of power on the battlefield. Azerbaijan's strategic deployment of Turkish-made Bayraktar TB2 drones and Israeli-made Harop loitering munitions effectively neutralised Armenia's Soviet-era air defense systems and armoured formations.[1] The conflict showcased how relatively affordable drone technology could overcome traditional military advantages of heavy armour and well dug defensive positions.

Azerbaijan's use of drone operations exhibited sophisticated tactical innovations that would influence military thought around the globe. The Azeri forces used modified Soviet era AN-2 aircraft as decoy drones, to locate the positions of the Armenian air defenses, who were then subsequently destroyed via precision strikes from TB2s.[2] This use of tactical deception was a unique innovation in how drone technology could be creatively adapted to exploit enemy weaknesses and achieve strategic objectives at minimal cost.

The effect of drone warfare in Nagorno-Karabakh was unprecedented in terms of precision. The Azerbaijanis used their drones to strike their opponents' supply lines, logistics hubs, and overall command structure even further from their frontline. The integration of drones with traditional artillery and ground forces created a combined arms approach that proved devastatingly effective against Armenian defensive positions. Open-source reporting indicates that drones contributed to the destruction of hundreds of Armenian tanks, fighting vehicles, artillery pieces and air-defense systems.

The psychological effect of continuous surveillance and precision drone strikes had a profound effect on Armenian effectiveness and morale. The Armenians never enjoyed freedom of movement or the ability to establish secure positions, for fear of drone attacks. This demonstrated how drone technology could achieve strategic effects beyond mere physical destruction by creating an environment of constant threat and uncertainty for enemy forces.

The US-Taliban War

Following the US withdrawal from Afghanistan in August 2021, the Taliban has demonstrated remarkable innovation in developing autonomous drone capabilities using abandoned NATO infrastructure.[3] Intelligence reports indicate that the Taliban has repurposed former British SAS bases in Logar Province as primary drone testing sites, with successful test flights of suicide drones being conducted at facilities once used by UK special forces.[4] The Taliban's drone program, active for at least two years, has transformed Camp Phoenix near Kabul, previously a major US logistics and training hub, into a secretive manufacturing facility where Taliban engineers study abandoned Western military hardware to reverse-engineer drone design.[5]

The Taliban's drone development program has received significant international technical support, with intelligence sources citing assistance from countries including Iran, Turkey, China, Russia, Belarus, and Bangladesh. Current drone prototypes are reportedly modeled after advanced systems such as the US-made MQ-9 Reaper and the Iranian-designed Shahed-136, with Taliban engineers attempting to increase range and payload capacity while experimenting with new explosive loads and detonation techniques.[6]

Post-Withdrawal US Drone Operations

Despite the complete withdrawal of US ground forces, American drone operations have continued over Afghan airspace, demonstrating the effectiveness of over-the-horizon capabilities. The precision strike against al-Qaeda leader Ayman al-Zawahiri in central Kabul on July 31, 2022, served as a master class in intelligence and operational capacity, proving that US intelligence could still operate effectively in Afghanistan without permanent ground presence.[7] The operation tracked Zawahiri for months, establishing patterns of his routine and activity, and managed to carry out the strike with reportedly no civilian casualties or injuries.

However, US drone operations in Afghanistan have also resulted in significant civilian casualties, highlighting the challenges of autonomous targeting systems. The August 29, 2021 drone strike in Kabul, conducted during the final days of the US evacuation, killed ten innocent civilians including seven children, with the youngest victim being just two years old.[8] US intelligence had tracked an aid worker's car for eight hours, believing it was linked to ISIS-K militants, but the investigation found that what appeared to be explosives being loaded into the vehicle were actually containers of water.

The Israel-Hamas-Hezbollah War of 2024

Hamas demonstrated sophisticated drone warfare capabilities during the October 7, 2023 attack on Israel, employing commercial drones as a pivotal component of their assault strategy.[9] The first wave of small, explosive-laden commercial drones targeted surveillance infrastructure including observation towers, cameras, sentries, and communication systems along the Gaza border, effectively blinding the Israel Defense Forces and creating confusion that allowed Hamas fighters to breach defensive positions.[10] Hamas utilized two primary types of drones during the operation: commercial off-the-shelf quadcopters equipped with explosives, commonly DJI models, and newly developed fixed-wing Zouari drones deployed as loitering munitions.[11]

The tactical sophistication of Hamas drone operations included the deployment of 35 Zouari drones, named after Tunisian aerospace engineer Mohammed Zouari who pioneered Hamas' drone program before his assassination in 2016.[12] These drones function similarly to Iranian Shahed systems, loitering over targets before striking with explosive payloads. Hamas also employed drones as aerial munitions platforms, modifying RPG-7 warheads to be dropped on Israeli tanks, armoured vehicles, and civilian targets, with Palestinian Islamic Jihad maintaining a dedicated drone operations room during the attack to facilitate real-time coordination.[13]

Israeli Counter-Drone Technologies and AI Integration

Israel's response to the drone threat has involved significant technological adaptations and the development of advanced counter-drone systems. Israeli forces have deployed GPS anti-jamming technology through companies like InfiniDome to prevent mini-drones from being intercepted by simple jamming techniques used by Hamas fighters.[14] Israeli troops have been supplied with mini surveillance drones capable of flying at very low altitude and entering buildings and tunnels to determine safety for ground forces.[15]

The Israel Defense Forces have extensively employed artificial intelligence in their drone targeting systems, with programs such as "The Gospel" generating suggestions for buildings where militants may be operating, and "Lavender" programmed to identify suspected Hamas members for assassination.[16] The "Where's Daddy?" system reportedly tracks militant movements by monitoring their phones to target them at their homes, with air strikes potentially killing entire families or apartment buildings. These AI-powered systems have dramatically accelerated the pace of targeting, with current systems able to generate in one week what previously required 250 days of work by 20 intelligence officers.[17]

The Israel-Iran Conflicts of 2024

Iran's April 13, 2024 attack on Israel, codenamed Operation True Promise, represented the largest attempted drone strike in history, designed to overwhelm anti-aircraft defenses through massive scale deployment.[18] The operation involved approximately 170 drones, over 30 cruise missiles, and more than 120 ballistic missiles launched toward Israel and the Israeli-occupied Golan Heights.[19] Iran utilized a sophisticated tactical approach, saturating the Iron Dome and David's Sling systems with a first wave of hundreds of HESA Shahed 136 loitering munitions to clear the way for cruise and ballistic missiles in the second wave.[20]

The attack included 185 of the newer and faster-flying, jet-propelled Shahed 238 loitering munitions, demonstrating Iran's advancement in drone technology and tactical coordination.[21] Coalition forces, operating under the codename Iron Shield, successfully destroyed 99 percent of the incoming weapons, with most intercepted before reaching Israeli airspace. However, at least nine Iranian missiles struck two Israeli airbases, causing minor damage to the Nevatim Airbase while it remained operational.[22]

Israel's Operation Rising Lion and Preemptive Drone Operations

Israel's June 13, 2025 attack on Iran, Operation Rising Lion, marked the largest assault on Iranian territory since the Iran-Iraq War of the 1980s, employing sophisticated drone and precision weapon systems.[23] The Israel Defense Forces and

Mossad utilized more than 200 Israeli aircraft to strike approximately 100 targets, while Mossad agents had pre-positioned precision weapons and established covert drone bases near Tehran to disable air defenses and secure air superiority.[24] The operation targeted key nuclear facilities including the Natanz nuclear facility, where the pilot fuel enrichment plant housing advanced centrifuges was reportedly destroyed.[25]

Israeli forces employed pre-positioned explosive drones and precision weapons that had been smuggled into Iran by Mossad operatives, targeting air defenses and missile infrastructure to create corridors for manned aircraft operations.[26] The sophisticated planning included targeting the Isfahan Nuclear Technology/Research Center, damaging uranium metal production facilities and infrastructure for reconversion of enriched uranium, alongside strikes on IRGC military bases and command centers across multiple regions.[27]

Iran's Retaliatory Drone Operations

Iran's response to Operation Rising Lion, codenamed Operation True Promise III, involved launching over 150 ballistic missiles and more than 100 drones against Israeli targets on the evening of June 13, 2025.[28] The retaliatory strikes targeted military and intelligence infrastructures but also caused civilian casualties, marking a significant escalation in direct confrontation between the two nations.[29] Iranian forces launched the drones from multiple locations, with Israeli air and naval forces intercepting at least 20 drones within a single hour according to Israel Defense Forces reports.[30]

The Iranian drone attacks demonstrated continued reliance on swarm tactics designed to saturate Israeli air defense systems, though Israeli forces successfully intercepted the majority of incoming unmanned aerial vehicles.[31] The IDF released video footage showing Israeli forces downing several drones over the Mediterranean and Negev regions, while maintaining air superiority through continued strikes on Iranian radar arrays and air defense systems.[32] The confrontation highlighted how drone warfare has become central to the strategic competition between Israel and Iran, with both nations investing heavily in offensive drone capabilities and counter-drone technologies.[33]

The evolution of drone warfare from the Azerbaijan-Armenia conflict to Ukraine's Spider Web operation demonstrates the rapid transformation of military technology and tactics. The transition from human-piloted to fully autonomous systems represents more than a technological advancement, it signifies a fundamental shift in how warfare is conceived and conducted. Non-state actors have proven remarkably adept at leveraging these technologies, democratising advanced military capabilities and challenging traditional power structures.

Technological Innovation and Autonomous Capabilities

The Spider Web operation showcased advanced autonomous navigation capabilities that allowed drones to operate deep within heavily defended Russian territory. Reports suggest that artificial intelligence solutions were used to program the drones to strike Russian aircraft at their most vulnerable points, using museum pieces and technical documentation as training data for the targeting algorithms.[34] The drones utilized ArduPilot autopilot software that enabled autonomous navigation even when human operators temporarily lost communication signals.[35]

This operation demonstrated how cheap, commercially available technology could be transformed into sophisticated military platforms capable of penetrating advanced air defense systems. The ability of these drones to function without GPS signals while evading radar detection and electronic warfare systems represents a significant leap in drone capabilities.[36] The success of Operation Spider's Web has fundamentally altered strategic thinking about deep strike operations and the vulnerability of previously secure military installations.

Non-State Actors and the Democratization of Drone Technology

Non-state actors have increasingly leveraged drone technology to challenge conventional military forces and achieve strategic objectives previously beyond their capabilities. Currently, over 65 non-state actors are known to possess drone capabilities, with this number continuing to rise due to the unregulated nature of civilian drone technology.[37] The accessibility and affordability of commercial drone systems have enabled insurgent groups, terrorist organizations and criminal networks to acquire significant military capabilities at minimal cost.

The proliferation of drone technology among non-state actors poses escalating security concerns for traditional military forces. These groups have demonstrated remarkable innovation in adapting commercial drones for military purposes, including weaponisation with improvised explosive devices, precision targeting systems and surveillance capabilities. The flexibility and operational simplicity of drone systems make them particularly attractive to non-state actors who lack the resources and infrastructure required for conventional military equipment.[38]

Iranian-backed groups have become particularly sophisticated in their use of drone technology across multiple conflict zones. The Iranian Shahed series of drones, including the Shahed-136 kamikaze drone, have been extensively used by proxy forces and supplied to allied nations for military operations.[39] These systems demonstrate ranges of up to 2,400 kilometers and can carry warheads of 15-50 kilograms, providing non-state actors with significant strike capabilities.[40]

Hezbollah's Advanced Drone Capabilities

Hezbollah has demonstrated increasingly sophisticated drone capabilities throughout 2024, with the organization possessing approximately 2,000 unmanned aerial vehicles of varying types by 2021, with capabilities likely enhanced since then.[41] The October 13, 2024 drone strike on an Israeli Defense Forces training base near Binyamina marked a significant escalation, with a Hezbollah kamikaze drone killing four soldiers and wounding 67 others in the highest casualty toll from a drone attack since the war began.[42]

The Binyamina attack showcased Hezbollah's tactical innovation, with the organization launching three Mirsad-1 drones simultaneously alongside short-range rockets and precision rockets targeting Haifa to overwhelm Israeli air defenses.[43] One drone was shot down by the Israeli Navy and another by the Iron Dome, but the third evaded Israeli Air Force jets and helicopters that fired at it twice, while electronic warfare measures also failed to force the drone to lose its bearings.[44] The drone dropped off radar 48 kilometers northeast of Acre and was assumed to have crashed, but reappeared briefly on screens before striking its target without triggering warning alarms.[45]

Hezbollah has also demonstrated unprecedented reach with drone operations, becoming the first organization to target the Tel Aviv area with attack drones when it launched a swarm of drones at the Bilu military base south of Tel Aviv in November 2024.[46] The organization has employed Iranian-made drones including the Mirsad-1 with a range of 124 miles and the Ayoub, a derivative of Iran's Shahed-129 with a range of 1,000 miles, as well as Iranian systems such as the Karrar, Mohajer, and Sammad models.[47]

ISIS and Tactical Innovation

The Islamic State demonstrated remarkable innovation in weaponising commercial drone technology during its territorial control period. ISIS developed specialized drone modification workshops that converted civilian quadcopters into precision bombing platforms capable of targeting specific individuals and vehicles.[48] The organization's drone program included dedicated personnel responsible for manufacturing and modifying commercially produced drones for military applications.

Coalition airstrikes specifically targeted ISIS drone development facilities and technical personnel, recognizing the significant threat posed by the organization's drone capabilities.[49] The elimination of key ISIS drone developers and their facilities significantly disrupted the group's ability to develop and deploy weaponised drone systems.[50] However, the technical knowledge and tactical innovations developed by ISIS have continued to influence other non-state actors seeking to weaponise drone technology.

BEYOND AERIAL PLATFORMS: MULTI-DOMAIN DRONE OPERATIONS

Underwater and Subsurface Operations

The myth that drones are exclusively aerial platforms has been thoroughly dispelled by recent military developments across multiple domains. Underwater drones have emerged as critical assets for naval warfare, mine detection and submarine operations.[51] The U.S. Navy's Orca XLUUV represents a new generation of extra-large unmanned underwater vehicles designed for long-duration anti-submarine warfare missions.[52] These systems can operate autonomously for extended periods while gathering intelligence and tracking hostile submarine activity.

Ukraine's development of the Marichka underwater drone demonstrates the expanding capabilities of subsurface autonomous systems. The 6-meter-long unmanned underwater vehicle has a range of up to 1,000 kilometers and is designed for reconnaissance, transportation and direct attack missions.[53] Ukrainian forces have reportedly used Marichka drones to conduct precision strikes against the Crimean Bridge, demonstrating the strategic impact of underwater drone technology.[54] The drone's ability to loiter underwater for hours while remaining undetected by most radar and sonar systems represents a significant advancement in subsurface warfare capabilities.

Surface and Maritime Operations

Surface drone operations have revolutionized naval warfare through the deployment of unmanned surface vehicles capable of conducting autonomous missions.[55] Ukraine has extensively used naval drones equipped with Starlink satellite communication systems to conduct attacks against Russian naval assets in the Black Sea.[56] These surface drones initially appeared to be reconnaissance platforms but have proven capable of carrying significant explosive payloads for kamikaze-style attacks against enemy vessels.[57]

The US military has developed the Manta Ray underwater drone system that harnesses ocean currents and wave energy for propulsion, enabling extended autonomous operations without refueling. This enormous underwater platform can remain deployed for very long periods while conducting surveillance and reconnaissance missions.[58] The system represents a new approach to autonomous underwater operations that reduces logistical requirements while extending operational capabilities.

Ground-Based Autonomous Systems

Unmanned ground vehicles have evolved from simple remote-controlled platforms to sophisticated autonomous systems capable of complex military operations. The

Israeli Elbit Systems ROOK represents a new generation of heavy-duty unmanned ground vehicles designed for urban warfare and border protection missions.[59] This 6x6 platform can carry 1,200 kilograms of payload while operating in fully autonomous beyond-line-of-sight mode. The system's ability to transport multiple sensors and weapon systems while maintaining high mobility makes it suitable for diverse military applications.[60]

Military forces worldwide are developing ground-based autonomous systems for logistics, reconnaissance and direct combat operations. The US Army's Multi-Utility Tactical Transport system provides infantry units with autonomous cargo transport capabilities, reducing the logistical burden on ground forces.[61] These systems can navigate complex terrain while carrying supplies and equipment to forward positions without requiring human operators in dangerous areas.

Human-Centric to Machine-Centric Operations

The transition from piloted to autonomous drone systems represents a fundamental shift in military operational paradigms. Traditional warfare relied heavily on human decision-making and direct control of weapon systems, but autonomous drones can now make tactical decisions and engage targets without continuous human oversight.[62] This shift enables military forces to conduct operations at scales and speeds previously impossible with human-controlled systems.

Autonomous drone systems can process battlefield information and respond to threats faster than human operators, providing significant tactical advantages in rapidly evolving combat situations. The ability of these systems to operate continuously without fatigue or psychological stress enables sustained military operations over extended periods.[63] This represents a fundamental change from traditional military operations that were limited by human endurance and decision-making capabilities.

Strategic Implications and Future Warfare

The proliferation of autonomous drone technology is reshaping strategic military planning and force structure development. Military organisations must now consider how to integrate autonomous systems with traditional forces while maintaining effective command and control.[64] The ability of autonomous drones to operate in contested environments where human-controlled systems would be vulnerable provides new options for military commanders facing sophisticated enemy defenses.

The emergence of drone swarm tactics and coordinated autonomous operations represents a new form of warfare that could overwhelm traditional military defenses. These systems can adapt their tactics in real-time based on enemy responses, creating dynamic battlefield situations that require entirely new defensive approaches.[65]

Military forces worldwide are investing heavily in counter-drone technologies and tactics to address the growing threat posed by autonomous drone systems.

Unmanned Systems for Unmanned Spaces

UAVs, commonly known as drones, have transformed border patrol operations by offering aerial reconnaissance, surveillance, and rapid response capabilities. Even in India, deployment of UAVs along its borders has enhanced situational awareness, border surveillance, and response to security threats. US Department of Homeland Security has been employing Small Unmanned Aircraft Systems (sUAS) that conduct overt and covert reconnaissance, surveillance, and target acquisition. Interdiction and apprehension efforts include; damage assessment, disaster relief and search and rescue. The sUAS mission primarily involves remote sensing of using a variety of sensor payloads such as day, night, and thermal imaging capabilities and since 2019, the sUAS program has grown to average 28% of US Custom and Border Patrol's direct air support hours flown, and accounts for 48% of air-assisted apprehensions.[66]

The Unmanned Ground Vehicles (UGV) are also being increasingly becoming a platform of choice especially over difficult terrain and odd hours. Given the long endurance of the UGV, they are being teamed with drones with the high speed and manoeuvrability, providing a capability that is ideal for the needs of surveillance, reconnaissance, real-time situational awareness, and force protection. The unique capabilities of both platforms can be combined to give an expanded patrol and surveillance dataset, and enable more powerful AI use cases for mapping, navigation, object and person recognition, object and person tracking and scene reasoning.[67] These systems are readily available, are relatively easy to operate, and thus do not require special training or skills to operate. Simple training is likely all that will be needed to upskill border guard personnel in how to operate these drones.

The development of AI will also reduce the requirements for training as drones will be able to operate autonomously with humans in the loop. This reduces the logistics required to field, operate and maintain such systems. In fact, a swarm of these systems can be easily controlled by a single soldier thereby reducing the operational manning cost to the field army. The only prerequisite will be to provide appropriate maintenance and technical support to ensure they remain serviceable and able to operate effectively. In the next 5–10 years, there are likely to be significant advances in the integration of a range of AI technologies that will improve unmanned capabilities and provide real-time situational awareness to border guard patrols, including full-motion video, automatic target detection and geolocation.

There are also likely to be improvements in the ability of sUAS to operate fully autonomously through AI-enabled and computer-vision-based precision landing capability, which enables a sUAS to launch from and land on static as well as moving platforms, such as ground vehicles. Finally, sUAS will be equipped with real-time on-board processing of imagery and video, as well as neural networks for enhanced object detection and classification capability. The most important component for this capability is the information and data gathered by the UAS. There will be a requirement to invest time and effort into training the AI-enabled UAS to capture and analyse information appropriately for the needs of border security. Given that most of these systems come from a few countries who have almost monopoly in this field, there is the risk that these units might automatically send information back to their manufacturer. Thus, there is a need to institute appropriate security measures such as secure VPNs and firewalls to make sure that the information gathered remains with the operator.

Infantryman's Arrow

Drones and UAS are providing infantry units with unprecedented situational awareness and precision strike capabilities. Small, portable drones are becoming standard equipment for infantry units, used for reconnaissance, target acquisition and even direct attacks on enemy positions.

The Israeli Defense Forces (IDF) Skylark UAV is a lightweight, hand-launched drone used by infantry units for real-time reconnaissance and surveillance. It provides live video feeds to soldiers on the ground, significantly enhancing their situational awareness.[68] The US Army's Black Hornet Nano, is a tiny, palm-sized drone used for close reconnaissance in urban environments. The same is used by the Indian National Security Guard for recce in high density urban settings during day and night. It allows soldiers to scout buildings and alleys without exposing themselves to danger.[69]

With the advent of emerging technologies and refining of hardware and software, Drones will continue to evolve, becoming more autonomous, longer-ranged, and capable of carrying more sophisticated sensors and weapons. Swarm drones, which can operate in coordinated groups, will provide overwhelming force, exploit the weaknesses in enemy defences and act in a distributed manner but achieve collaborative effects, thereby changing the dynamics of infantry engagements.

Enhancing Manoeuvre of Mechanised Columns

Nonetheless, challenges remain in integrating UGVs into an armoured force for their intended purposes. Due consideration must be given to communication latencies, potential compromises regarding cybersecurity and

the reliability of autonomous systems in a dynamic battlefield. In addition, ongoing debates that surround autonomy also include, the ethical and legal implications of their use in combat situations, particularly in the context of the use of lethal force.

In the future, the UGVs are expected to further evolve into increasingly autonomous and sophisticated machines. UGVs of the future may see themselves much more independent in their modes, operating independently for extended periods, as part of Combat Groups and Combat Teams. These can operate as part of Recce and Support units providing ISR inputs and also in tethered mode to the tanks and AFVs there by extending the firepower of mechanised columns beyond direct firing platforms. This has significant implications on how mechanised forces are organised and employed, leading to new doctrines and operational concepts.

Making Artillery Versatile

Drones and UAVs are being integrated into artillery operations increasingly. Drones have become a major player in modern artillery operations by providing real time ISR, target acquisition and battle damage assessment (BDA). Drones give artillery units a bird's-eye view of the battlefield and make artillery units more accurate and effective and assists them in engaging targets with enhanced precision and situational awareness.

The US Army's RQ-11 Raven is a small, hand-launched UAV used by artillery units for reconnaissance and target acquisition. The Raven provides live video feeds and GPS coordinates to artillery units, allowing them to adjust their fire based on real-time information.[70] This capability is particularly valuable in complex environments, such as urban areas or dense forests, where traditional targeting methods may be less effective.

Integration of drones into artillery operations presents challenges such as airspace management, communication and vulnerability from enemy air defense. Additionally, the growing battlefield presence of drones poses the risk of IFF failures, accidental collisions and interference with other military assets.

Nonetheless, the integration of drones into artillery operations shows promising developments. Advances in AI and machine learning and sensor based technologies will further strengthen their collaborative employment alongside artillery systems, especially in conjunction with future developments in swarm technology

Russia's loitering munitions, the Lancet Kamikaze Drones targeted Ukrainian artillery systems and air defenses, as seen in the Battle of Zaporizhzhia (2023).[71] Ukraine's indigenously developed Punisher drone carried explosives to strike Russian forces in Kherson and Kyiv, crippling resupply efforts.[72]

Revolutionising Aviation

Unmanned Aerial Vehicles (UAVs), commonly known as drones, are crucial in modern military operations. They have revolutionized how forces conduct surveillance and combat missions, offering new levels of flexibility, persistence, and precision.

The role of UAVs in military operations has expanded significantly over the past two decades. Initially used for reconnaissance and surveillance, they now acquire targets, engage in electronic warfare and take part in direct combat. Because UAVs can be controlled remotely, military personnel can perform missions in dangerous areas without risking their lives. This is particularly helpful in asymmetric warfare, where enemies leverage unexpected tactics and challenging terrain to their advantage.

The Rise of Tactical and Strategic UAVs

UAVs have transformed air warfare by democratizing airpower. Low-cost platforms can be mass-produced and deployed in large numbers, allowing forces to saturate enemy defences and execute precision strikes.

Bayraktar TB2: The Bayraktar TB2, developed by Turkey, is a medium-altitude, long-endurance (MALE) UAV that has transformed modern aerial combat by blending capabilities of reconnaissance and precision strike. It has an operational endurance of more than 24 hours and an approximate range of 300km.[73] The UAV can loiter above the contested zones and provide real-time intelligence and precision-guided munitions, such as the Mini Akilli Mühimmat (MAM-L) laser-guided bombs.[74] Its low cost and versatility have turned the tables in conflicts like the Armenia-Azerbaijan, Libya and Ukraine, where it has been deployed extensively to destroy troop concentrations, mechanised forces, command and control centres, logistic echelons and even air defence systems. By operating in environments considered to be risky ground troops as well as for manned aircrafts, the TB2 has forced adversaries to rethink their air defence strategies, particularly on how to counter affordable, unmanned platforms that challenge traditional paradigms.

Shahed-136: Extensively used by Russian forces in Ukraine, these Iranian drones have been used in swarm attacks against critical infrastructure. They are cost-effective and disposable, making them ideal for overwhelming attacks, forcing adversaries to invest in countermeasures.[75]

Adapted Commercial Drones: In many instances, commercially available drones (such as DJI Mavic models) have been modified for combat roles. Armed with explosives, these FPV drones provide a low-cost means to attack enemy positions and are used by both insurgent groups and state militaries to supplement more advanced platforms.[76]

Advancements in Drone Technology

The technological advancements in drones have significantly enhanced their capabilities, making them more versatile and effective in various military roles. The new generation of drones feature improved navigation systems, AI enabled autonomous operation and sophisticated payloads that increase their effectiveness, endurance and resilience.

Another important development is the ability of drones to operate in swarms. Swarms are groups of drones that operate together to achieve a goal, with each drone communicating and coordinating with the other drones in the swarm. Swarming enables drones to overwhelm enemy defences, conduct joint missions and jam enemy communications and/or radar systems.

While UAVs and drones offer various advantages, there are challenges that come with their use. One challenge is the easy susceptibility to electronic warfare and cyber warfare, which allows the enemy to disrupt communications, or take control of the drones. Additionally, the proliferation of drones has raised concerns about airspace management and the potential for accidental collisions with other manned and unmanned platforms. As more nations develop and deploy drones, the need for standardized airspace regulations and de-confliction measures becomes increasingly important.

Moving ahead, UAVs and drones will play an increasingly larger role in military operations. Advances in AI and autonomous systems will enhance their capacity to handle complex situations and smaller, more agile drones will be especially useful in urban and jungle warfare, where manoeuvrability and stealth are crucial.

The Strategic Impact of UAVs and Autonomous Systems

Ukraine's Operation Spider's Web marked a significant strategic shift in modern warfare by using long-range drones launched from within Russian territory to strike deep inside multiple Russian airbases. The attack reportedly damaged or destroyed a substantial portion of Russia's strategic bomber fleet including Tu95s and early warning aircraft, weakening its ability to conduct long-range missile strikes.[77] This demonstrated the power of low-cost asymmetric tactics where small drones caused disproportionate damage to billion-dollar assets. The psychological impact was equally severe, causing panic, disruption, and massive internal resource diversion within Russia. The operation showcased Ukraine's technological ingenuity and operational reach while forcing global militaries to re-examine their vulnerability to drone swarm tactics and to invest in better counter-drone systems. It also strengthened Ukraine's bargaining position by proving that it could hit critical assets far beyond the battlefield, all while rewriting the cost-benefit logic of strategic warfare.

Such semi autonomous and autonomous systems have not only redefined technological capabilities, they have also forced a re-evaluation of military strategy:

Non-Linear Application: Drones whether dumb, semi autonomous or fully autonomous has destroyed the notions of linearity, by giving the capability to strike varied depth and height within the battle space or far from the actual area of operations.

Mass Deployment and Attritability: The low cost of UAVs means that large numbers can be fielded, creating a "swarm" effect that overwhelms enemy defences. This has led to a strategic emphasis on using expendable systems to create redundancy and maintain pressure on adversaries.

Enhanced Situational Awareness: Autonomous systems can quickly analyse vast amounts of data to give commanders a clear view of the battlefield. This helps them make quicker decisions and allows for more precise targeting.

Force Multiplication: When used with manned vehicles, autonomous systems boost effectiveness. They extend the reach of manned fighters, provide constant surveillance and can absorb enemy fire in hostile areas. The concept of a "loyal wingman" illustrates how these systems blend human judgment with machine precision.[78]

Galactic Synergy

The integration of space technologies has profoundly transformed the landscape of drone warfare by extending its reach, precision, survivability and autonomy.[79] At the core of this transformation lies satellite-based navigation communication and surveillance systems, which enable drones to operate seamlessly across vast distances and in denied or contested environments. Global Navigation Satellite Systems such as GPS GLONASS, Galileo and BeiDou provide real-time positioning and navigation, that allows drones to maintain course, execute complex flight paths and deliver munitions with extraordinary accuracy.[80] This level of precision has been critical in expanding the strategic utility of drones from tactical battlefield roles to long-range strategic missions such as targeted strikes deep into enemy territory.

Equally important are satellite communication systems that empower drones with persistent command and control links, even beyond line of sight. This ensures that unmanned systems can be operated from distant bases, while maintaining full situational awareness and mission flexibility. The increasing use of high bandwidth, low latency satellite constellations, particularly in low Earth orbit, will continue to enable real-time video streaming, secure data relay and dynamic re-tasking of drones mid-mission.[81] Such capabilities have made it possible for commanders to respond rapidly to evolving battlefield conditions and to integrate drone operations seamlessly with manned platforms and broader command networks.

Space-based ISR or Intelligence Surveillance and Reconnaissance platforms now play a pivotal role in cueing drone missions. High-resolution electro-optical infrared and synthetic aperture radar satellites detect and track enemy movements across all weather conditions and terrain. These inputs guide drones toward targets optimize their flight profiles and increase mission success rates while reducing exposure to threats. Furthermore space assets help overcome the challenges posed by electronic warfare environments where drones might otherwise lose contact or navigation. By integrating satellite data drones can operate in GPS-denied or jammed conditions using alternative positioning or terrain-matching algorithms sourced from space sensors.

The convergence of artificial intelligence and space technologies is also pushing drone warfare toward greater autonomy. Space-based cloud computing and edge-processing systems may soon allow drones to analyse ISR feeds in-flight, identify patterns and make decisions without waiting for ground control input.[82] This has major implications for swarm operations, autonomous targeting and time-sensitive missions. As space assets continue to miniaturise and proliferate, smaller drones may directly tap into satellite networks instead of relying on ground relays, thereby enhancing operational agility.[83]

Moreover, space launch capabilities influence drone warfare by enabling rapid deployment of small ISR satellites or drone command relays, tailored to specific theatres of operation.[84] Responsive launch and in-orbit servicing technologies are emerging as force multipliers allowing real-time reconfiguration of satellite support for drone missions.[85] Countries that can rapidly deploy and adapt space assets, will gain decisive advantages in drone warfare, especially in regions, where terrestrial infrastructure is lacking or vulnerable.[86]

In essence, space technologies have extended the operational depth, intelligence reach and tactical flexibility of drone warfare. They have fused the aerial and orbital domains into a seamless command and control continuum, allowing unmanned systems to operate with greater confidence in hostile environments. As the lines between air, space, cyber and electromagnetic domains blur, the synergy between space and drone systems is becoming the backbone of modern military power. Future conflicts will likely see this integration deepen as space enabled drone warfare evolves into a key pillar of strategic deterrence, precision strike and real-time situational dominance.

Technological Evolution and Market Growth

The global drone warfare market is experiencing unprecedented expansion, projected to grow from $25.2 billion in 2024 to $41.9 billion by 2030, representing a compound annual growth rate that reflects the increasing strategic importance of unmanned systems in modern military operations.[87] This growth is driven by the integration of artificial intelligence, advanced stealth capabilities and the development of networked multi-domain operations that position drones as integral components of future military strategies rather than standalone platforms.[88]

The convergence of AI and robotics is fundamentally transforming drone capabilities, enabling autonomous operations with minimal human intervention.[89] AI enabled drones can now process vast amounts of real time data, including battlefield imagery, weather patterns, and enemy movements, using this information to make split second tactical decisions that enhance operational effectiveness in dynamic combat environments. These systems can detect acoustic signatures ten times quieter than previous models while operating up to forty times faster than human analysts, demonstrating the revolutionary potential of machine learning integration.[90]

Autonomous Operations and Swarm Intelligence

The future of drone warfare will be characterized by fully autonomous systems capable of complex mission execution without continuous human oversight. Current developments in Ukraine demonstrate this trajectory, with Ukrainian forces purchasing 10,000 AI-enhanced drones in 2024 as a preliminary step toward broader adoption of advanced autonomous capabilities.[91] These systems utilize convolutional neural networks, recurrent neural networks, and reinforcement learning algorithms integrated directly into onboard software, enabling real-time processing and decision-making even when communication links are compromised.

Autonomous navigation systems now combine stereo cameras, LiDAR, and radar sensors to enable obstacle avoidance and terrain navigation without human input. The integration of machine learning allows drones to identify and classify potential threats while adapting to changing battlefield conditions, with AI algorithms processing sensor data to make tactical decisions in milliseconds.[92] This technological advancement represents a fundamental shift from human controlled to machine-autonomous warfare, enabling military operations at scales and speeds previously impossible.

Swarm Technology and Coordinated Operations

Drone swarm technology represents one of the most significant developments in autonomous warfare capabilities, with the global swarm intelligence market expected to reach $447.2 million by 2030 at a compound annual growth rate of 40.47%.[93] Swarm systems can increase the lethality of drone attacks by up to 50% while reducing losses to enemy fire by the same percentage, providing substantial tactical and economic advantages. These coordinated systems leverage collective intelligence to share information and adapt tactics in real-time, enabling synchronized operations across multiple domains simultaneously.

The development of underwater drone swarms demonstrates the expansion of autonomous capabilities beyond aerial platforms. Companies like Helsing have developed swarming underwater surveillance drones capable of operating for up to three months, with hundreds of units deployed simultaneously to patrol underwater environments.[94] French company Arkeocean has created micro-AUV swarms that can establish "underwater barriers" of protection, using acoustic listening devices to record vessel signatures and create persistent surveillance networks.[95]

Multi-Domain and Space-Based Operations

The future of drone warfare extends into space-based operations, with nations developing satellite-based defensive and offensive capabilities. The United States' proposed Golden Dome system, projected to cost $175 billion, aims to create the world's first active combat infrastructure deployed in orbit, fundamentally altering the nature of space warfare.[96] This system would feature satellites equipped with sensors, interceptors, and directed-energy weapons capable of neutralizing threats during their boost or mid-course phases.

Recent developments demonstrate the reality of space-based drone operations, with both China and India conducting high-speed satellite manoeuvres that replicate aerial combat training in space. These "dogfights" involve satellites manoeuvring at orbital speeds of 28,800 kilometers per hour, requiring technical sophistication similar to autonomous drone operations but at unprecedented velocities.[97] The integration of space-based platforms with terrestrial drone systems will create comprehensive multi-domain warfare capabilities.[98]

Hypersonic and Advanced Propulsion Systems

The development of hypersonic drone technology represents the next frontier in unmanned warfare systems. Companies like Kratos are developing drones capable of flying at speeds above Mach 5, with projected costs less expensive than comparable systems.[99] These hypersonic platforms combine the advantages of traditional drone operations with unprecedented speed capabilities, enabling rapid response and deep strike operations previously impossible with conventional systems.

Quantum computing integration promises to revolutionize drone navigation and AI processing capabilities. Quantum-powered systems can process multiple flight paths simultaneously and select optimal routes in real-time, significantly improving decision-making speed and accuracy.[100] This technology enables drones to analyze complex urban environments and select the safest navigation paths while reducing collision risks and improving operational reliability.

Communication and Network Evolution

The emergence of 6G communication networks will fundamentally transform drone operations by enabling edge computing capabilities that reduce onboard processing requirements. Future lighter-than-air drones will benefit from 6G edge computing resources, allowing most information processing and AI automation to be located on the network side rather than within the drone structure.[101] This approach reduces weight and energy consumption while extending flight times and operational capabilities.

Satellite-enabled communication systems provide global coverage and enhanced security for drone operations beyond visual line of sight. Advanced satellite connectivity enables piloting and navigation adjustments for UAVs conducting missions deep into hostile territory, with encrypted communication channels that are less susceptible to cyber attacks than terrestrial networks.[102] These systems allow operators to maintain constant control over drone flight paths even when thousands of kilometers away from the operational area.

STRATEGIC CHALLENGES AND VULNERABILITIES

Cybersecurity and Electronic Warfare

The proliferation of drone technology has created significant vulnerabilities that adversaries are actively exploiting. Commercial drones suffer from fundamental security weaknesses, including clear text storage of sensitive information and plain text password storage, making them susceptible to cyber attacks and electronic warfare countermeasures.[103] The conflict in Ukraine has demonstrated how Russian electronic warfare equipment can effectively disrupt drone operations, with "an approximate distribution of at least one major system covering each 10 kilometers of front".

Drone systems are particularly vulnerable to communication disruption and navigation interference. Simple WiFi de-authentication attacks can sever communication channels between drones and controllers, while GPS jamming and spoofing techniques can misdirect navigation systems.[104] These vulnerabilities have resulted in drone operators becoming targets themselves, as demonstrated by DJI drones broadcasting unencrypted location information that reveals operator positions to enemy forces.[105]

Cost Asymmetry and Defense Challenges

The economic dynamics of drone warfare present significant strategic challenges, with inexpensive drones capable of inflicting massive damage on high-value targets. A \$1,000 drone can potentially destroy a \$200 million aircraft, creating cost asymmetries that favor offensive operations over defensive measures.[106] Traditional kinetic defense systems are expensive and inefficient against drone swarms, with missile-based interceptors costing millions to neutralize drone swarms worth hundreds of thousands of dollars.

Modern militaries are developing multi-layered defense strategies that combine various technologies to address cost asymmetries. These systems integrate AESA radars, electro-optical sensors, acoustic detectors, and AI-powered sensor fusion for detection, followed by directed energy weapons, electronic warfare systems, and interceptor drones for neutralization.[107] Countries like India are investing heavily in hard-kill anti-drone systems, recognizing that traditional electronic jamming is becoming less effective against pre-programmed, jam-resistant autonomous drones.[108]

Ethical and Legal Implications

Autonomous Decision-Making and Accountability: The transition to fully autonomous drone systems raises profound ethical questions about machine decision-making in lethal operations.[109] The development of AI systems capable of identifying and engaging targets without human authorization challenges traditional principles of military ethics and international humanitarian law. Current Israeli systems like "Lavender" and "The Gospel" demonstrate the acceleration of AI-powered targeting, with programs capable of generating targeting suggestions in one week that previously required 250 days of human analysis.[110]

The psychological impact on drone operators remains a significant concern, even as systems become more autonomous. The phenomenon of "drone stare" contributes to dehumanization by abstracting individuals from their contexts and reducing them to mere targets.[111] This abstraction process facilitates remote monitoring and destruction while potentially normalizing violence against perceived adversaries.

International Law and Regulatory Frameworks: The legal framework governing autonomous drone warfare remains complex and often ambiguous. International humanitarian law principles of distinction and proportionality become increasingly difficult to apply as AI systems make targeting decisions with minimal human oversight.[112] The secrecy surrounding many autonomous drone operations makes it challenging to assess their legality and effectiveness, creating calls for greater transparency and accountability mechanisms.

Looking BLOS

Operation Sindoor in May 2025 marked a turning point for India's military capabilities, showcasing its advancements in autonomous warfare and reinforcing its strategic position in South Asia. By leveraging cost-effective, networked drone systems, India shifted regional aerial power dynamics, demonstrating the ability to dominate battlespaces without risking manned aircraft. Drones enabled calibrated, precise strikes that avoided escalation, establishing a new deterrence model for limited conflicts.[113] The operation highlighted India's indigenous innovation, with over 550 startups delivering combat-ready solutions, aligning with the 'Make in India' and 'Aatmanirbhar Bharat' initiatives.[114] Real-time intelligence from the National

Technical Research Organisation (NTRO) and Indian Space Research Organisation (ISRO) ensured accurate targeting, enhancing operational success.[115]

The operation accelerated India's push for self-reliance, driving increased investment in Defence Research and Development Organisation (DRDO) projects like the Autonomous Flying Wing Technology Demonstrator (AFTD) and Archer-NG UAV.[116] It also underscored the need for advanced technologies, such as artificial intelligence, electronic warfare, and laser-based systems, to counter evolving drone threats. India's defense strategy is now evolving to incorporate new military doctrines, international collaborations, and public-private partnerships, ensuring readiness for future drone-centric conflicts.

However, this transformation brings significant challenges including cybersecurity vulnerabilities, cost asymmetries favoring offensive operations, and complex ethical questions about autonomous lethal decision-making. Military organizations worldwide must develop comprehensive strategies that address these challenges while harnessing the transformative potential of autonomous drone systems. The nations that successfully navigate these technological and strategic challenges will likely dominate the future battlefield, while those that fail to adapt risk strategic obsolescence in an increasingly autonomous warfare environment.

Notes

1   Shantanu K Bansal, How Azeri Military Innovated its Drone Tactics to Win the Nagorno-Karabakh War. IAD News. Accessed June 18, 2025. https://iadnews.in/how-azeri-military-innovated-its-drone-tactics-to-win-the-nagorno-karabakh-war/#:~:text=Liotering%20Munitions%20(Aka.,(SAM)%20site%20in%20Shuskakend

2   Shantanu K Bansal, How Azeri Military Innovated its Drone Tactics to Win the Nagorno-Karabakh War. IAD News. Accessed June 18, 2025. https://iadnews.in/how-azeri-military-inno vated-its-drone-tactics-to-win-the-nagorno-karabakh-war/#:~:text=Liotering%20Munitions%20 (Aka.,(SAM)%20site%20in%20Shuskakend

3   Kabul Now. Taliban Building Suicide Drone Arsenal at Former NATO Bases, Report Says. June 8, 2025. https://kabulnow.com/2025/06/taliban-building-suicide-drone-arsenal-at-former-nato-bases-report-says/.

4   Kabul Now. Taliban Building Suicide Drone Arsenal at Former NATO Bases, Report Says. June 8, 2025. https://kabulnow.com/2025/06/taliban-building-suicide-drone-arsenal-at-former-nato-bases-report-says/.

5   Kabul Now. Taliban Building Suicide Drone Arsenal at Former NATO Bases, Report Says. June 8, 2025. https://kabulnow.com/2025/06/taliban-building-suicide-drone-arsenal-at-former-nato-bases-report-says/.

6   Kabul Now. Taliban Building Suicide Drone Arsenal at Former NATO Bases, Report Says. June 8, 2025. https://kabulnow.com/2025/06/taliban-building-suicide-drone-arsenal-at-former-nato-bases-report-says/.

7   Miller, Aaron David. What the Al-Qaeda Drone Strike reveals About U.S. Strategy in Afghanistan. August 2, 2022. https://carnegieendowment.org/posts/2022/08/what-the-al-qaeda-drone-strike-reveals-about-us-strategy-in-afghanistan?lang=en.

8   BBC News. US Military to End Afghanistan Operations on August 31. September 18, 2021.

https://www.bbc.com/news/world-us-canada-58604655.

9   Dohainstitute.org. Remote Warfare: Evaluating the Combat Effectiveness of Drones in the Context of the War on Gaza. November 21, 2024. https://www.dohainstitute.org/en/Lists/ACRPS-PDFDocumentLibrary/remote-warfare-evaluating-the-combat-effectiveness-of-drones-in-the-context-of-the-war-on-gaza.pdf.

10   Rassler, Don, and Yannick Veilleux-Lepage. On the Horizon: The Ukraine War and the Evolving Threat of Drone Terrorism. March 2025. https://ctc.westpoint.edu/on-the-horizon-the-ukraine-war-and-the-evolving-threat-of-drone-terrorism/.

11   Dohainstitute.org. Remote Warfare: Evaluating the Combat Effectiveness of Drones in the Context of the War on Gaza. November 21, 2024. https://www.dohainstitute.org/en/Lists/ACRPS-PDFDocumentLibrary/remote-warfare-evaluating-the-combat-effectiveness-of-drones-in-the-context-of-the-war-on-gaza.pdf.

12   Dmytro Kaniewski, "Hamas: Learning about Drone Warfare from the War in Ukraine," Deutsche Welle, October 20, 2023.

13   Rassler, Don, and Yannick Veilleux-Lepage. On the Horizon: The Ukraine War and the Evolving Threat of Drone Terrorism. March 2025. https://ctc.westpoint.edu/on-the-horizon-the-ukraine-war-and-the-evolving-threat-of-drone-terrorism/.

14   New Indian Express. GPS War: Israel's Battle to Keep Drones Flying and Enemies Baffled. February 23, 2024. https://www.newindianexpress.com/world/2024/Feb/23/gps-war-israels-battle-to-keep-drones-flying-and-enemies-baffled.

15   New Indian Express. GPS War: Israel's Battle to Keep Drones Flying and Enemies Baffled. February 23, 2024. https://www.newindianexpress.com/world/2024/Feb/23/gps-war-israels-battle-to-keep-drones-flying-and-enemies-baffled.

16   Serhan, Yasmeen. Gaza, Ukraine, and the New Era of AI Warfare. December 18, 2024. https://time.com/7202584/gaza-ukraine-ai-warfare/.

17   Serhan, Yasmeen. Gaza, Ukraine, and the New Era of AI Warfare. December 18, 2024. https://time.com/7202584/gaza-ukraine-ai-warfare/.

18   Wikipedia. April 2024 Iranian Strikes on Israel. Accessed June 19, 2025. https://en.wikipedia.org/wiki/April_2024_Iranian_strikes_on_Israel.

19   Ynet and Calcalist. Drone and ballistic missile attack from Iran: Most intercepted, minor damage to IDF base, April 14, 2024. https://www.calcalist.co.il/local_news/article/hygslvog0.

20   Rothwell, James. Iran's Shahed-136 Drones Used by Russia Have a Range of 2,000 km. April 13, 2024. https://www.telegraph.co.uk/world-news/2024/04/13/iran-shahed-136-drones-used-russia-range-2000/.

21   Ben-Yishai, Ron (14 April 2024). "Zero hits: Israel's historic achievement against Iran's drones and missiles" https://www.ynetnews.com/article/rkrrw7yxr

22   Klippenstein, Ken; Boguslaw, Daniel (15 April 2024). "U.S., Not Israel, Shot Down Most Iran Drones and Missiles" https://theintercept.com/2024/04/15/iran-attack-israel-drones-missiles/

23   Lam, Lana, Sofia Ferreira Santos, Jaroslav Lukiv, and Nathan Williams. Israel-Iran: How did latest conflict start and where could it lead? BBC News, June 12, 2025. https://www.bbc.com/news/articles/cdj9vj8glg2o.

24   Fabian, Emanuel. "IDF: 200 Aircraft Involved in Opening Strikes on Iran, with Several Top Officials Killed." The Times of Israel, June 13, 2025. https://www.timesofisrael.com/liveblog_entry/idf-200-aircraft-involved-in-opening-strikes-on-iran-with-several-top-officials-killed/.

25   Etson, Udi. "After Natanz: Why the IDF Won't Strike Iran's Live Reactor at Bushehr." The Jerusalem Post, June 13, 2025. https://www.jpost.com/middle-east/iran-news/article-857645.

26   Report: Mossad Carried Out Covert Sabotage Operations Against Iranian Air Defenses, Long-Range Missiles." The Times of Israel, June 13, 2025. https://www.timesofisrael.com/

liveblog_entry/report-mossad-carried-out-covert-sabotage-operations-against-iranian-air-defenses-long-range-missiles/.

27 Explosions Heard in Natanz, Home to Key Nuclear Site, Says Iranian State Television." The Times of Israel, June 13, 2025. https://www.timesofisrael.com/liveblog_entry/explosions-heard-in-natanz-home-to-key-nuclear-site-says-iranian-state-television/.

28 "Live: Explosions Reported in Iran Amid Israel Tensions." Al Jazeera, June 13, 2025. https://www.aljazeera.com/news/liveblog/2025/6/13/live-explosions-reported-in-iran-amid-israel-tensions?update=3771470.

29 Cornwell, Alexander, Parisa Hafezi, and Steve Holland. "Israel Says It Strikes Iran Amid Nuclear Tensions." Reuters, June 14, 2025. https://www.reuters.com/world/middle-east/israel-says-it-strikes-iran-amid-nuclear-tensions-2025-06-13/.

30 Nath, Santushti. "Video Shows How Israeli Forces Intercepted 20 Iranian Drones in an Hour." NDTV, June 15, 2025. https://www.ndtv.com/world-news/iran-israel-conflict-video-shows-how-israeli-forces-intercepted-20-iranian-drones-in-an-hour-8675467.

31 Nath, Santushti. "Video Shows How Israeli Forces Intercepted 20 Iranian Drones in an Hour." NDTV, June 15, 2025. https://www.ndtv.com/world-news/iran-israel-conflict-video-shows-how-israeli-forces-intercepted-20-iranian-drones-in-an-hour-8675467.

32 Bob, Yonah Jeremy. "IDF Intercepts Massive Drone Attack from Iran, Pummels Tehran's Air Defenses." The Jerusalem Post, June 13, 2025. https://www.jpost.com/israel-news/defense-news/article-857598.

33 Bishara A Babbah, Iran-Israel Drone Competition and the Changing Nature of Warfare in the Middle East. Arab Center Washington DC. 13 Oct 2022. https://arabcenterdc.org/resource/iran-israel-drone-competition-and-the-changing-nature-of-warfare-in-the-middle-east/.

34 Kateryna Bondar, How Ukraine's Spider Web Operation Redefines Asymmetric Warfare. Center for Strategic and International Studies. June 2, 2025. https://www.csis.org/analysis/how-ukraines-spider-web-operation-redefines-asymmetric-warfare.

35 Bhati, Divya. "Ukraine Used 20-Year-Old Open Source Software in Its Operation Spider Web Against Russia." India Today. June 4, 2025. https://www.indiatoday.in/technology/news/story/ukraine-used-20-year-old-open-source-software-in-its-operation-spider-web-against-russia-2735459-2025-06-04.

36 Operation Spiderweb Shows How Small $30K Drones Deliver Big Blows to Russia's $7 Billion Fighter Jets. The Economic Times. June 3, 2025. https://economictimes.indiatimes.com/news/defence/operation-spiderweb-shows-how-small-30k-drones-deliver-big-blows-to-russias-7-billion-fighter-jets/articleshow/121586687.cms?from=mdr.

37 Clausen, Maria Louise. "Non-State Armed Groups in the Sky." Danish Institute for International Studies. April 15, 2024. https://www.diis.dk/en/research/non-state-armed-groups-in-the-sky#:~:text=The%20proliferation%20of%20inexpensive%20and,will%20continue%20to%20rise%20unchecked.

38 Drone Warfare 2025." Solace Global. Accessed June 18, 2025. https://www.solaceglobal.com/report/drone-warfare-2025/#:~:text=What%20Risks%20Does%20Drone%20Warfare%2 0Pose%20in,drone%20attacks%20using%20improvised%20explosive%20devices%20(IEDs).

39 Rubin, Uzi. Iran's Drones Tip the Balance of Power in Mideast. Jerusalem Institute for Strategy and Security. January 25, 2022. https://jiss.org.il/en/rubin-irans-drones-tip-the-balance-of-power-in-mideast/.

40 Shahed 136 Loitering Munition Kamikaze Suicide Drone Technical Data. Army Recognition. March 12, 2025. https://armyrecognition.com/military-products/army/unmanned-systems/unmanned-aerial-vehicles/shahed-136-loitering-munition-kamikaze-suicide-drone-technical-data.

41 Frantzman, Seth. "Hezbollah Drone Strikes IDF Base as Israel Uncovers Weapons in Lebanon." Long War Journal, October 14, 2024. https://www.longwarjournal.org/archives/2024/10/hezbollah-drone-strikes-idf-base-as-israel-uncovers-weapons-in-lebanon.php.

42 "2024 Hezbollah Drone Strike on Binyamina." Wikipedia, June 19, 2025. https://

en.wikipedia.org/wiki/2024_Hezbollah_drone_strike_on_Binyamina.

43   "2024 Hezbollah Drone Strike on Binyamina." Wikipedia, June 19, 2025. https://en.wikipedia.org/wiki/2024_Hezbollah_drone_strike_on_Binyamina.

44   Berman, Lazar. *Hezbollah UAV That Killed 4 Dropped Off Radar, Then Briefly Reappeared: Probe.* Times of Israel, October 14, 2024. https://www.timesofisrael.com/hezbollah-uav-that-killed-4-dropped-off-radar-then-briefly-reappeared-probe/.

45   Berman, Lazar. *Hezbollah UAV That Killed 4 Dropped Off Radar, Then Briefly Reappeared: Probe.* Times of Israel, October 14, 2024. https://www.timesofisrael.com/hezbollah-uav-that-killed-4-dropped-off-radar-then-briefly-reappeared-probe/.

46   *Hezbollah Attack Drones Target Tel Aviv Army Base as Israel Pounds Lebanon.* Al Jazeera, November 7, 2024. https://www.aljazeera.com/news/2024/11/7/hezbollah-attack-drones-target-tel-aviv-army-base-as-israel-pounds-lebanon.

47   Dohainstitute.org. Remote Warfare: Evaluating the Combat Effectiveness of Drones in the Context of the War on Gaza. November 21, 2024. https://www.dohainstitute.org/en/Lists/ACRPS-PDFDocumentLibrary/remote-warfare-evaluating-the-combat-effectiveness-of-drones-in-the-context-of-the-war-on-gaza.pdf.

48   U.S. Central Command. Strikes Kill ISIS Drone Experts. U.S. Central Command, September 29, 2017. https://www.centcom.mil/MEDIA/NEWS-ARTICLES/News-Article-View/Article/1329189/strikes-kill-isis-drone-experts/./ centcom.mil+11

49   U.S. Central Command. Strikes Kill ISIS Drone Experts. U.S. Central Command, September 29, 2017. https://www.centcom.mil/MEDIA/NEWS-ARTICLES/News-Article-View/Article/1329189/strikes-kill-isis-drone-experts/./ centcom.mil+11

50   U.S. Central Command. Strikes Kill ISIS Drone Experts. U.S. Central Command, September 29, 2017. https://www.centcom.mil/MEDIA/NEWS-ARTICLES/News-Article-View/Article/1329189/strikes-kill-isis-drone-experts/./ centcom.mil+11

51   InsideFPV Blog. *The Rise of Underwater Drones for Naval Defence.* InsideFPV, March 7, 2025. https://insidefpv.com/blogs/blogs/the-rise-of-underwater-drones-for-naval-defence.

52   InsideFPV Blog. *The Rise of Underwater Drones for Naval Defence.* InsideFPV, March 7, 2025. https://insidefpv.com/blogs/blogs/the-rise-of-underwater-drones-for-naval-defence.

53   Global Espreso. *Ukraine's Marichka Underwater Drones Threat to Crimean Bridge and Russian Fleet.* Global Espreso, June 16, 2025. https://global.espreso.tv/military-news-ukraines-marichka-underwater-drones-threat-to-crimean-bridge-and-russian-fleet.

54   The Military Show. *Ukrainian Underwater Drone 'Marichka' Hits Crimea Bridge in HUGE BLAST: The Military Show.* YouTube video, June 4, 2025. https://youtu.be/ZURfKBMEXQM.

55   Mukherjee, Tuneer. *The Proliferation of Drones in Naval Warfare.* Observer Research Foundation, 2025. https://www.orfonline.org/expert-speak/the-proliferation-of-drones-in-naval-warfare/.

56   Sutton, H.I. "First Image of Ukraine's Sidewinder Armed Magura V7 Surface Drone." *Naval News,* May 4, 2025. https://www.navalnews.com/naval-news/2025/05/first-image-of-ukraines-sidewinder-armed-magura-v7-surface-drone/

57   Sutton, H.I. "First Image of Ukraine's Sidewinder Armed Magura V7 Surface Drone." *Naval News,* May 4, 2025. https://www.navalnews.com/naval-news/2025/05/first-image-of-ukraines-sidewinder-armed-magura-v7-surface-drone/

58   Tucker, Patrick. *Giant military manta ray drone passes first ocean test.* Defense One, May 1, 2024. https://www.defenseone.com/technology/2024/05/giant-military-manta-ray-drone-passes-first-ocean-test/396244/

59   Elbit Systems. *ROOK: Elbit Systems Networked Warfare Autonomy Unmanned Ground Vehicles.* May 01, 2025. https://www.elbitsystems.com/networked-warfare/autonomy/unmanned-ground-vehicles/rook/.

60   Ibid

61   Multi Utility Tactical Transport (MUTT) UGV. Army Technology. July 8, 2020. https://

www.army-technology.com/projects/multi-utility-tactical-transport-mutt-ugv/.

62   Moti, Altaf. How AI-Powered Drones Are Shaping the Future of Warfare. Eurasia Review. January 2, 2025. https://www.eurasiareview.com/02012025-how-ai-powered-drones-are-shaping-the-future-of-warfare-oped/.

63   Peter, Hemanth. Autonomous Weapons Systems: The Future of Military Operations August 6, 2024. https://thehyperstack.com/blog/autonomous-weapons-systems-future-military-operations/

64   Strupczewski, Wojciech. Machine Wars: Integrating Autonomous Systems into Military Frameworks. Karve International. August 6, 2024. https://www.karveinternational.com/insights/machine-wars-integrating-autonomous-systems-into-military-frameworks.

65   Military Drone Swarm Intelligence Explained. SDI. Accessed June 18, 2025. https://sdi.ai/blog/military-drone-swarm-intelligence-explained/.

66   Neesa Sweet. Use of Small UAS on the U.S. Border on the Rise. November 2, 2023 https://insideunmannedsystems.com/use-of-small-uas-on-the-u-s-border-on-the-rise/

67   Mike Ball. Autonomous Drone-Launching Security UGV Under Development. 28 Nov 2020 https://www.unmannedsystemstechnology.com/2020/11/autonomous-drone-launching-security-ugv-under-development/

68   Unmanned Aerial Vehicle (UAV) Administration Ministry of Defense, Israel https://english.mod.gov.il/About/Innovative_Strength/Pages/Unmanned_Aerial_Vehicle_ Administration.aspx

69   Indian Defence News October 18, 2018 Eye in the Sky: NSG Acquires Black Hornet Nano – World's Smallest Spy Cam UAV https://www.indiandefensenews.in/2018/10/eye-in-sky-nsg-acquires-black-hornet.html

70   RQ-11 Raven Unmanned Aerial Vehicle, July 22 2021 https://www.army-technology.com/projects/rq-11-raven/

71   Zaluzhnyi, V. Modern positional warfare and how to win in it. https://athenalab.org/wp-content/uploads/2023/11/ZALUZHNYI_ FULL_VERSION-2.pdf, 2023.

72   Julian Borger, "The Drone Operators Who Halted Russian Convoy Headed for Kyiv," The Guardian, March 28, 2022, https://www.theguardian. com/world/2022/mar/28/the-drone-operators-whohalted-the-russian-armoured-vehicles-heading-for-kyiv

73   Bayraktar TB2, BAYKAR, https://baykartech.com/en/uav/bayraktar-tb2/

74   MAM-L lightweight Smart Micro Munition, Roketsan, https://www.roketsan.com.tr/en/products/mam-l-smart-micro-munition

75   How are 'kamikaze' drones being used by Russia and Ukraine?, BBC, 29 December 2023, https://www.bbc.com/news/world-62225830

76   How are 'kamikaze' drones being used by Russia and Ukraine?, BBC, 29 December 2023, https://www.bbc.com/news/world-62225830

77   Samya Kullab, A surprise drone attack on airfields across Russia encapsulates Ukraine's wartime strategy, 02 June 2025, https://apnews.com/article/what-to-know-ukraine-drone-attack-russia-bombers-2d01b23341e2289882760b9f121431d4

78   Tim Fish, Uncrewed ambitions of the Loyal Wingman, November 1, 2022 https://www.airforce-technology.com/features/uncrewed-ambitions-of-the-loyal-wingman/?cf-view

79   Abhik. Drones Add a Whole New Dimension to Warfare: Lesson from the War in Gaza. Fins India. October 30, 2024. https://finsindia.org/drones-add-a-whole-new-dimension-to-warfare-lesson-from-the-war-in-gaza.html.

80   Reim, Garrett. Space-Based Cellular Services Will Soon Extend Drone Warfare's Reach. Aviation Week. March 05, 2025. https://aviationweek.com/space/commercial-space/space-based-cellular-services-will-soon-extend-drone-warfares-reach.

81   NASA. Ground Data Systems and Mission Operations. NASA. February 05, 2025. https://www.nasa.gov/smallsat-institute/sst-soa/ground-data-systems-and-mission-operations/#:~:text=It%20is%20common%20for%20a,more%20with%20DTE%20ground%2 0stations.

82   Rishika, Sai. Drones, Data, and Cloud: How Real-Time Analytics Shape the Battlefield.

NASSCOM Community. October 25, 2024. https://community.nasscom.in/communities/cloud-computing/drones-data-and-cloud-how-real-time-analytics-shape-battlefield#:~:text=They%20can%20track%20the%20movements,enemy's%20locations%20and%20possible%20tactics.

83 Ground Control. Satellite-Enabled Drones in Modern Warfare. Ground Control. Last modified October 18, 2024. https://www.groundcontrol.com/blog/satellite-enabled-drones-in-modern-warfare/#:~:text=Satellite%20connectivity%20is%20a%20reliable,time%20video%20footage%20from%20UAVs.

84 Gur, Eli. How Satcom Is Used for BVLOS Drone Flights. Elsight. Last modified February 19, 2024. https://www.elsight.com/blog/how-satcom-is-used-for-bvlos-drone-flights/#:~:text=This%20makes%20the%20technology%20ideal%20for%20drones,disaster%20zones%20where%20infras tructure%20 has%20been%20destroyed.

85 NASA. NASA Satellite Servicing Project Report. NASA, October 2010. https://www.nasa.gov/wp-content/uploads/2023/10/nasa-satellite-servicing-project-report-0511.pdf#:~:text=Viable %20plans%20can%20be%20put%20into%20place,supporting%20future%20ventures%20 to%20m ore%20distant%20destinations.

86 Easley, Miakyla. Space Force GPS 3, SpaceX Falcon 9 Rapid Response Launch. DefenseScoop. Last modified May 30, 2025. https://defensescoop.com/2025/05/30/space-force-gps-3-spacex-falcon-9-rapid-response-launch/.

87 Business Wire. Drone Warfare Global Strategic Research Report 2025: Global Market to Reach $41.9 Billion by 2030 – Demand for Energy-Efficient Drones Strengthens Business Case for Long-Endurance Combat Missions. March 27, 2025. https://www.businesswire.com/news/home/20250327802947/en/Drone-Warfare-Global-Strategic-Research-Report-2025-Global-Market-to-Reach-$41.9-Billion-by-2030

88 Business Wire. Drone Warfare Global Strategic Research Report 2025: Global Market to Reach $41.9 Billion by 2030 – Demand for Energy-Efficient Drones Strengthens Business Case for Long-Endurance Combat Missions. March 27, 2025. https://www.businesswire.com/news/home/20250327802947/en/Drone-Warfare-Global-Strategic-Research-Report-2025-Global-Market-to-Reach-$41.9-Billion-by-2030

89 Inside FPV. "Future Trends in Defence Drone Technology." December 13, 2024. https://insidefpv.com/blogs/blogs/future-trends-in-defence-drone-technology.

90 Gosselin Malo, Elissabeth. "AI Company Helsing Unveils Swarming Underwater Surveillance Drones." Defense News, March 13, 2025. https://www.defensenews.com/global/europe/2025/05/13/ai-company-helsing-unveils-swarming-underwater-surveillance-drones/.

91 Naik, Arjun. "Drone Trends Shaping the Future of Defence and Beyond in 2025." Business World, January 24, 2025. https://www.businessworld.in/article/drone-trends-shaping-the-future-of-defence-and-beyond-in-2025-545884.

92 Bondar, Kateryna. 2025. Ukraine's Future Vision and Current Capabilities for Waging AI Enabled Autonomous Warfare. Policy report, Center for Strategic and International Studies March 6, 2025. https://www.csis.org/analysis/ukraines-future-vision-and-current-capabilities-waging-ai-enabled-autonomous-warfare.

93 Maris Tech. 2024. *Drone (UAV) Swarm Technology: Transforming Military Potential.* July 29, 2024. https://www.maris-tech.com/blog/drone-swarm-its-impact-at-the-military-scene-maris-tech.

94 Gosselin Malo, Elissabeth. "AI Company Helsing Unveils Swarming Underwater Surveillance Drones." Defense News, March 13, 2025. https://www.defensenews.com/global/europe/2025/05/13/ai-company-helsing-unveils-swarming-underwater-surveillance-drones/.

95 Martin, Tim. "French Underwater Drone Swarm Maker Bids to Reshape Maritime Surveillance Missions." Breaking Defense, February 7, 2024. https://breakingdefense.com/2024/02/french-underwater-drone-swarm-maker-bids-to-reshape-maritime-surveillance-missions/

96   The Economic Times. 2025. "After Drones, a New Warfare Straight out of Science Fiction: Space War, Drone Battles to Space Dogfights – How Modern Warfare is Shifting." *The Economic Times*, June 3, 2025. https://economictimes.indiatimes.com/news/defence/after-pakistan-india-drones-fight-a-new-warfare-straight-out-of-science-fiction-space-war-drone-battles-to-space-dogfights-how-modern-warfare-is-shifting/articleshow/121594037.cms

97   The Economic Times. 2025. "After Drones, a New Warfare Straight out of Science Fiction: Space War, Drone Battles to Space Dogfights – How Modern Warfare is Shifting." The Economic Times, June 3, 2025. https://economictimes.indiatimes.com/news/defence/after-pakistan-india-drones-fight-a-new-warfare-straight-out-of-science-fiction-space-war-drone-battles-to-space-dogfights-how-modern-warfare-is-shifting/articleshow/121594037.cms

98   Ground Control. Satellite-Enabled Drones in Modern Warfare. Ground Control. Last modified October 18, 2024. https://www.groundcontrol.com/blog/satellite-enabled-drones-in-modern-warfare/#:~:text=Satellite%20connectivity%20is%20a%20reliable,time%20video%20footage%20from%20UAVs.

99   Kratos Has a Hypersonic Drone in the Works." TWZ, March 20, 2025. Accessed June 18, 2025. https://www.twz.com/air/kratos-has-a-hypersonic-drone-in-the-works

100   Quantum Computing Drives Smarter Drone Navigation." *SkyPerth*, n.d. Accessed June 18, 2025. https://www.skyperth.com/drones/quantum-computing-drives-smarter-drone-navigation/

101   LTA Drone." *6G Flagship*. Accessed June 18, 2025. https://www.6gflagship.com/lta-drone/

102   Ground Control. Satellite-Enabled Drones in Modern Warfare. Ground Control. Last modified October 18, 2024. https://www.groundcontrol.com/blog/satellite-enabled-drones-in-modern-warfare/#:~:text=Satellite%20connectivity%20is%20a%20reliable,time%20video%20footage%20from%20UAVs.

103   SecQuest. (n.d.). Drone vulnerabilities. Retrieved June 18, 2025, from https://www.secquest.co.uk/white-papers/drone-vulnerabilities

104   Hardbreak. (n.d.). Deauthentication attacks. Retrieved June 18, 2025, from https://www.hardbreak.wiki/network-analysis/protocols/wifi/deauthentication-attacks

105   Weiler, Julia. "IT Security: Security Vulnerabilities Detected in Drones Made by DJI." Ruhr University Bochum, March 2, 2023. https://news.rub.de/english/press-releases/2023-03-02-it-security-security-vulnerabilities-detected-drones-made-dji.

106   Vajiram & Ravi. "India's Drone Warfare Doctrine: Strategic Shifts and Emerging Capabilities." *UPSC Daily Current Affairs*, June 10, 2025. https://vajiramandravi.com/current-affairs/indias-drone-defence-strategy-swarms-threats-future-warfare/.

107   Vajiram & Ravi. "India's Drone Warfare Doctrine: Strategic Shifts and Emerging Capabilities." *UPSC Daily Current Affairs*, June 10, 2025. https://vajiramandravi.com/current-affairs/indias-drone-defence-strategy-swarms-threats-future-warfare/.

108   India's Anti Drone Strategy: Zen Technologies Advocates Hard Kill Solutions in Light of Emerging Threats." *IDRW.org*, n.d. Accessed June 08, 2025. https://idrw.org/indias-anti-drone-strategy-zen-technologies-advocates-hard-kill-solutions-in-light-of-emerging-threats/.

109   Lukmaanias Blog. 2024. "Topic 8: The Ethics of Drone Warfare – Balancing Security and Human Rights." August 24, 2024. Accessed June 18, 2025. https://blog.lukmaanias.com/2024/08/24/topic-8-the-ethics-of-drone-warfare-balancing-security-and-human-rights/.

110   Inside FPV. "Future Trends in Defence Drone Technology." December 13, 2024. https://insidefpv.com/blogs/blogs/future-trends-in-defence-drone-technology.

111   "Ethics of Using Drones in Warfare & Surveillance." *Consensus.app*. Accessed June 18, 2025. https://consensus.app/questions/ethics-using-drones-warfare-surveillance/.

112   Drone Warfare, Politics & Policy." *Number Analytics Blog*, n.d. Accessed June 10, 2025. https://www.numberanalytics.com/blog/drone-warfare-politics-policy

113   Bedi,/ Rahul. *Autonomous Warfare in Operation Sindoor. The Hindu*, May 30, 2025. https://www.thehindu.com/news/national/autonomous-warfare-in-operation-sindoor/

article69633124.ece.

114   Operation Sindoor: India's Drone Startups Shine in Combat, Testing Kamikaze and Surveillance Drones Against Pakistan. *Indian Defence Research Wing*, May 31, 2025. https://www.idrw.org/operation-sindoor-indias-drone-startups-shine-in-combat-testing-kamikaze-and-surveillance-drones-against-pakistan/

115   Operation Sindoor: India's Drone Startups Shine in Combat, Testing Kamikaze and Surveillance Drones Against Pakistan. *Indian Defence Research Wing*, May 31, 2025. https://www.idrw.org/operation-sindoor-indias-drone-startups-shine-in-combat-testing-kamikaze-and-surveillance-drones-against-pakistan/

116   *Operation Sindoor and India's Emerging Drone Industry.* ETV Bharat, May 13, 2025. https://www.etvbharat.com/en/!bharat/operation-sindoor-and-india-emerging-drone-industry-enn25051302634.

7

Winning Without Fighting

*To win one hundred victories in one hundred battles is not the acme of skill.
To subdue the enemy without fighting is the acme of skill.*

—Sun Tzu

Non-Kinetic Warfare

The nature of conflict is undergoing a profound transformation a shift from the traditional clash of steel and explosives to a more subtle, yet equally decisive struggle waged in the unseen realms of the electromagnetic spectrum, cyberspace and the human mind. This new battlespace is defined by non-kinetic warfare, a form of engagement where objectives are achieved without the direct application of physical force. Instead of bombs and bullets, the weapons are signals, code and ideas. This domain is broadly categorized into three interconnected pillars, electronic warfare which targets an adversary's use of the electromagnetic spectrum, cyber warfare which exploits the vast networks of digital information and cognitive warfare which seeks to influence the perceptions and decision making of the target population or its leaders. The future of the battlespace is inextricably linked to the mastery of these non-kinetic tools, where victory may be secured long before the first shot is ever fired or indeed without any shots being fired at all. The rise of these methods reflects a strategic adaptation to a world, where the costs of conventional war are prohibitively high, both in terms of human life and economic devastation and where the lines between peace and conflict are increasingly blurred.

These methods exploit vulnerabilities in an adversary's technological infrastructure and human cognition, offering a potent combination of deniability, cost effectiveness and strategic impact. As traditional battlefields grow more lethal and politically risky, world powers are investing heavily in these silent arsenals, fundamentally altering how conflicts are waged.

Countries are using these silent weapons more and more instead of traditional fighting for several strong reasons:

Reduced Risk: Employing a cyber team or jamming signals is far less riskier than sending soldiers into battle.

Plausible Deniability: It's often hard to prove exactly who launched a cyber attack or a disinformation campaign. Countries can hit an enemy while pretending it wasn't them avoiding direct blame and retaliation.

Cheaper: Developing a new missile costs billions. Developing a powerful cyber weapon or jamming system costs much less relatively.

Hitting Deep: Cyber and cognitive attacks can strike deep inside an enemy's homeland disrupting daily life government functions and the economy without crossing borders with tanks or planes. EW can blind an enemy military from hundreds of miles away.

Avoiding Escalation: Using non-kinetic methods lets countries push back against rivals or achieve limited goals without triggering a massive war that could spiral out of control especially between nuclear powers.

Making the Bang Easier: How Non-Kinetic Aids Kinetic War

Non-kinetic warfare isn't just a replacement for fighting it makes actual fighting much more effective when it happens:

Blinding the Enemy (EW): Jamming radar and communications means enemy forces are deaf and blind. They can't see attacking forces coordinate defenses or call for help. This makes them easy targets.

Paralysing the Enemy (Cyber): Knocking out power grids command systems or logistics software cripples an enemy's ability to move troops supply them or even make decisions. Soldiers might be ready to fight but their trucks won't start their orders don't arrive and their lights are out.

Breaking The Will (Cognitive): Spreading fear confusion and doubt through propaganda and disinformation can destroy enemy morale. Soldiers might lose the will to fight. Civilians might pressure their government to surrender. Support from allies might dry up.

Swift Victory: By combining these effects an attacker can overwhelm an enemy quickly. If the enemy can't see, can't communicate, can't move supplies and loses the will to resist, collapse can happen much faster than through brute force alone.

Electronic Warfare: The Silent Powerhouse

The mastery of the non-kinetic domain acts as a powerful force multiplier for conventional military operations making the actual fighting significantly easier and aiding in the swift capitulation of an enemy. By achieving dominance in the electromagnetic spectrum a military force can effectively blind and deafen its adversary. Enemy air defense radars can be jammed allowing friendly aircraft to operate with impunity. Communications can be disrupted preventing enemy commanders from controlling their forces and leading to confusion and chaos on the battlefield. GPS signals can be spoofed sending enemy units into ambushes or causing their precision weapons to miss their targets. Electronic warfare stands out as the most potent non-kinetic means in this regard due to undermentioned effects:

Ubiquitous: Everything modern militaries do relies on the electromagnetic spectrum radios drones missiles radar surveillance satellites. Attacking this is attacking the nervous system.

Immediate Effect: Jamming or spoofing has instant consequences. A drone drops a bomb loses control instantly. A guided missile veers off course instantly.

Battlefield Dominance: Controlling the spectrum allows your forces to operate freely while hindering the enemy fundamentally shaping the battle space. The Russia Ukraine war powerfully shows how decisive EW can be.

In modern warfare, electronic warfare has become a critical component. With the integration of sophisticated sensors, communication systems and information networks, the electromagnetic spectrum is crowded and contested. EW plays a vital role in gaining the edge in terms of situational awareness and tactical advantage. In the early days of the Russian invasion of Ukraine in 2022, Russian EW disrupted enemy radars and communications, disrupting ground-based air defense systems as well as their own communications.[1] Russian ability to disrupt GPS signals is credited with reducing the success of Ukrainian use of HIMARS and JDAM bombs.[2] According to a report by the Royal United Services Institute on May 19, 2023, Ukraine had lost approximately 10,000 drones per month due to Russian electronic warfare.[3]

Electro-Optical (EO) systems are finding their way into nearly all military applications in the battlefield, such as day and night surveillance and observation, weapon targeting, fire control, tracking, ranging, missile guidance, communication, etc. In the foreseeable future, both opposing sides in a battlefield will be heavily equipped and dependent on a variety of EO systems for successful conduct of their military operations. The side with capability to degrade the opponent's EO systems, will have the winning edge in the battlefield.

There is an overlap between the cyberspace and electromagnetic spectrum and it results in multidiscipline effects. The cyberspace and electromagnetic spectrum create a common operational environment that could be named as the cyber electromagnetic domain. The cyber electromagnetic domain is not meant to equate the terms cyberspace and electromagnetic spectrum, but rather to highlight that there is significant overlap between them and future technological development is likely to increase this convergence.[4]

Also EW is being widely and frequently employed for counter space operations. Global Positioning System (GPS) satellites have proven to be particularly vulnerable to jamming as it blocks users from acquiring useful and accurate positioning, navigation and timing data from those satellites. But the jamming of navigation satellites is primarily restricted to civil GPS signals, as military signals are more robust.[5] Satellites are controlled from ground stations through electronic signals and they pass their data back to ground stations, so attacking those uplink and downlink linkages electronically can render satellites ineffective.[6] The Russian Federation is reported to have completed the development of a laser-based ASAT on the A-60 aircraft, designated 1LK222 Sokol Eshelon. These laser-based systems can both dazzle and blind sensors on satellites. With sufficient power, they are capable of damaging light- or heat-sensitive physical components on satellites.[7]

The application of computer virus concepts has made possible the advent of a new class of electronic warfare. Events of the last few years have demonstrated dramatically that computer viruses are not only feasible but can quickly cause catastrophic disruption of computer systems and networks. Current trends in the development of military electronic systems have significantly increased the vulnerability of these systems to computer virus attack. This has created a new form of electronic warfare consisting in the electronic insertion of computer virus micro code into a victim electronic system through direct or indirect mechanisms.[8]

The need to integrate stealth technology and EW systems into military equipment is anticipated to act as a major demand driver for EW systems, as stealth technology on its own will not be sufficient to combat threats. For example, an aircraft is required to be able to detect radio frequency electromagnetic emissions of enemy aircraft while simultaneously leveraging its own stealth capabilities. Also the proliferation of portable jammers, directed energy weapons, quantum-based technologies and space based assets embedded with Artificial intelligence and Cognitive Radio will define the future of EW technology.[9]

Future EW systems are expected to utilize nano electronics, with Carbon Nanotube (CNT) technology leading the way. A single CNT can perform all functions like antennas, tuners, demodulators, and amplifiers making a complete functional receiver circuit. The nano receiver of the future could have a universal RF processor in which different radio band configurations are set from a vast grid of nano receiver cells, each dedicated for a certain single frequency slot of the radio spectrum. When a different receiver is required, the system is reconfigured or reprogrammed accordingly.[10]

Loitering munitions have come of age and are a significant factor in EW and command and control. The ability to deploy localised precision munitions, particularly in the absence of valuable and vulnerable platforms in range, is changing the battlefield dynamic in much the same way as unmanned aerial vehicle (UAV) proliferation or anti-tank guided missile proliferation. An interesting development is the loitering surface to air missile, known as the 358, Iran has evidently created.[11]

Most of the powerful nations are endeavouring to integrate EW with AI and Autonomy. The development of unmanned systems, including drones and autonomous vehicles integrated with Electronic Warfare capabilities is gaining traction. These systems can be employed for reconnaissance, signal jamming and electronic attacks, providing a versatile and mobile EW platform. Their ability to operate in contested environments while minimising human risk makes them sought after assets on the modern battlefield.

The future of electronic warfare is poised to undergo transformative changes as technological advancements redefine its scope and capabilities. One significant trend is the integration of artificial intelligence and machine learning into EW systems, enabling real-time analysis of vast amounts of electromagnetic spectrum data and enhancing the speed of decision-making in complex environments. This will allow for more efficient detection and jamming of enemy communications and radar systems, making EW operations more adaptive and autonomous.

Cyber Warfare: The Borderless Battlespace

Cyber warfare exploits network vulnerabilities to steal, disrupt, or destroy data and systems. It ranges from disabling critical infrastructure to covert espionage, offering unparalleled anonymity and global reach. Disruptive technologies are increasingly transforming the scope and lethality of Cyber Warfare.

AI-Powered Malware: AI helps attackers find weaknesses faster and create smarter malware. AI also helps defenders spot attacks quicker. Quantum computing (still emerging) could one day break current encryption making secrets very hard to keep. Self-learning viruses like ChatGPT-based phishing tools adapt to bypass security protocols, enabling hyper-personalized deception campaigns.

Quantum Decryption: Future quantum computers could break current encryption, rendering state secrets and financial systems vulnerable. The NSA's "Post-Quantum Encryption" initiative aims to develop unbreakable codes, anticipating this threat.

Peacetime/Shadow War: The Stuxnet worm (discovered around 2010 widely attributed to US Israel cooperation) physically damaged Iran's nuclear centrifuges. This was a landmark cyber attack causing real world destruction.

Russia Ukraine War: Cyber is constant. Russian attacks target Ukrainian power grids government services and banks. Ukraine fights back with its volunteer "IT Army" hitting Russian websites and infrastructure. Russian hackers also spread disruptive malware globally.

> **Israel Iran Conflict:** Cyber attacks are a key tool. Following Stuxnet Iran ramped up its cyber forces. Attacks between them happen often. In June 2025 reports emerged of major cyber attacks disrupting Iranian fuel distribution systems and government communications seen as part of the ongoing shadow war.
>
> **Israel Hamas War 2024:** Cyber skirmishes are frequent with hackers targeting government websites media outlets and critical infrastructure on both sides though often with limited widespread impact compared to kinetic strikes.

Conventional weapons and actions have consequences for the perpetrators, however in the cyber space, any act can range between cyber prank to cyber activism or cybercrime, before entering the realms of a cyber-war. Then is the international opinion and related geopolitical trappings attached to a conventional strike, whereas on the other hand it took years for major media organizations to start pointing towards the US and Israel as the forces behind Stuxnet, and much of the proof came from tacit acknowledgement by American and Israeli government officials themselves. In this new battlefield, there are no rules of engagement, there are no Geneva conventions, there are simply cutting-edge aggressors and vulnerable targets who have not yet realised the doors they've left open. The post-modern battlefield stands to be fundamentally altered by the information technology revolution, at both the strategic and the tactical levels. The increasing breadth and depth of this battlefield and the ever-improving accuracy and destructiveness of even conventional munitions have heightened the importance of C3I matters to the point where dominance in this aspect alone may now yield the able practitioner consistent war-winning advantages. Yet, cyberwar is a much broader idea than attacking an enemy's C3I systems while improving and defending one's own. In Clausewitz's sense, it is characterized by the effort to turn knowledge into capability.[12]

Offensive Cyber Capability

Cyber space like other physical domains of warfare is not a neutral domain, it can be manipulated and controlled in times of no active conflict and even in times of perfect peace. Giving friendly nations, adversaries and non-state actors equal opportunity to do inimical activities with complete deniability.[13] India faces serious threats in all these domains especially cyber from state actors like China and Pakistan and non-state groups. Developing strong offensive cyber capabilities is not optional it's essential for defense. Given this evolving landscape it is imperative that the Indian Armed Forces develop robust offensive cyber warfare capabilities. In the modern strategic environment a purely defensive posture is insufficient.

The ability to deter adversaries and to shape the battlespace to India's advantage requires a credible capacity to hold an enemy's digital infrastructure at risk. This is a field best left to a synergistic partnership between the armed forces and specialized civilian agencies. The armed forces are uniquely positioned to integrate cyber operations with traditional military planning and to execute attacks that have a direct effect on the battlefield. They understand the operational art and can identify the critical enemy systems that if disrupted or destroyed would yield the greatest military advantage. Cyber warfare moves at digital speed. Decisions to counter an attack or launch a disruptive operation against an enemy preparing a strike need to be made in seconds or minutes. The armed forces understand battle needs and timelines far better than civilian agencies often bogged down in slower processes and legal hurdles. They can integrate cyber effects directly with military operations like using cyber to disable an air defense radar just before an airstrike.

However, the cultivation of deep technical expertise the development of cutting-edge tools and the gathering of intelligence on a global scale often thrive in the more agile and specialized environments of civilian intelligence and technology agencies. Civilian agencies are crucial for protecting national infrastructure sharing threat intelligence and setting broader policy. The unique skills in the private sector are also essential. However the actual offensive capability to disrupt or degrade an enemy's military systems in real time during conflict must reside with and be commanded by the armed forces. This ensures speed secrecy and operational effectiveness. Therefore a model where the armed forces provide the operational command and strategic direction while civilian agencies provide the technical talent and intelligence support would create the most effective and formidable national cyber warfare capability. This division of labor ensures that military objectives drive the use of these powerful tools while leveraging the full spectrum of the nation's technological and intellectual resources.

Another key trend is the growing importance of cyber-electronic warfare convergence, where EW systems are increasingly linked with cyber operations. This fusion allows forces to disrupt not just physical systems but also digital networks, creating multi-domain effects that can cripple enemy command, control, and communication capabilities. Quantum technologies, though still in their early stages, are likely to revolutionize secure communications and signal interception in EW by making it possible to detect and counter quantum encryption, while also enabling unhackable communication channels. As discussed in a subsequent section, the rise of directed energy weapons, including high-energy lasers and electromagnetic pulses (EMP), will create new EW tools capable of disrupting or disabling electronic systems from great distances.

With the rapid advancements in technologies in their respective domains, the synergy between cognitive, cyber and electronic warfare will also grow stronger. In future conflicts, the line between influencing human perceptions and disrupting technological systems will become increasingly blurred. Technological innovations in artificial intelligence, machine learning, and electronic domains will enable faster, effectively coordinated campaigns that employ both cognitive and electronic means to disable the enemy's decision-making core.

AI-Driven Cognitive Attacks: AI can be used to analyse electronic disruptions in real-time, crafting cognitive attacks tailored to specific weaknesses in how the adversary responds to EW.

Generative AI: Generative AI is poised to significantly revolutionise cognitive warfare by enabling more sophisticated, scalable, and personalised manipulation of perceptions and behaviours. Generative AI's ability to autonomously create text, images, audio, and video content will empower adversaries to develop and leverage disinformation, deep fakes, and other psychological tools at an unprecedented scale, precision and speed.

Hyper-Realistic Deep fakes: Deep fakes can be used to create false narratives, discredit political campaigns, or incite public unrest by showing leaders or prominent individuals making inflammatory or incriminating statements which may not be taken well by own citizens, friendly nations or the larger international audience.

Automated Disinformation Campaigns: Generative AI can be employed to develop targeted automated disinformation campaigns at a massive scale. AI-generated texts, social media posts, articles, and comments can be tailored to specific audiences and released through bot networks at selected opportune moments. Misinformation and disinformation campaigns can be more seamless and discreet, targeting specific demographics, nationalities, or social groups with hyper-personalised content designed to influence their perceptions or behaviours. These campaigns can be continuously refined by AI systems, based on real-time feedback from info space, ensuring that the cognitive attacks evolve to remain effective and convincing.

Neuro Weapons: These weapons are powered by a basket of disciplines grouped together under the head of Neuro Science Technologies, which employ Directed Energy as well as nano and bio technologies in conjunction with knowledge of cognitive sciences (Nanotechnology, Biotechnology, Information Technology, and Cognitive Science Technologies) to physically degrade the brain's cognitive abilities. However, the employment of Neuro weapons for performance enhancement to own soldiers, as well as to degrade adversary soldiers and civilians, will surely cause an ethical debate. While temporary enhancements, by administering stimulants, may be acceptable, biological enhancements of a permanent nature, such as by employing Clustered Regularly Interspaced Short Palindromic Repeats technology which leads to genetic modifications in humans, will likely be looked at from an ethical perspective as having crossed a red line.[14]

Cognitive Warfare: The Human Domain

Cognitive warfare is aimed at the human mind to manipulate the beliefs, emotions and decisions of the target population, forces and leadership. It combines neuroscience, social media and tailored disinformation to erode trust in institutions, leaders and reality itself. In the connected world of today, social media algorithms spread information (true or false) incredibly fast to huge audiences. Deepfakes create realistic fake videos or audio of people saying or doing things they never did. AI generates a deluge of believable fake text images and videos making disinformation campaigns massive and hard to counter.

Ukraine-Russia Conflict: The Russian "Firehose of Falsehood" propaganda model flooded social, portraying Ukrainian leaders as Nazis and Russia as a liberator. The Russian propaganda has been high volume, multichannel, rapid, continuous and repetitive. It makes no commitment to objective reality or consistency.[15]

Israel-Hezbollah Clash: Israel's "Operation Grim Beeper" used pagers and radios rigged with explosives to assassinate Hezbollah leaders. Preceding this, a sustained cognitive campaign spread distrust within Hezbollah ranks about communication security, amplifying paranoia.[16] Mossad exploited this by leaking compromised personal data of mid-level operatives, accelerating organisational paralysis.[17] With this innovative and unprecedented Pager Attack, Israel not only manipulated the cognitive domain of Hezbollah but also its other adversary's in the region.

Emerging Tools

Neuro-Manipulation: Virtual Reality environments like the Metaverse can effectively alter facial features in political speeches to resemble famous personalities, increasing persuasion by 20% in Stanford trials. Future biometric sensors in VR headsets can monitor physiological responses (pupil dilation, skin conductance) and give real time feedback to customize propaganda and make it more convincing.[18,19]

Neuroweapons: Directed microwave energy (linked to Havana Syndrome incidents) caused brain injuries in US diplomats, demonstrating potential to incapacitate leaders without physical contact. The National Academy of Sciences confirmed pulsed radio frequency energy as the likely culprit, highlighting this strategic threat.[20]

The Electronic, Cyber and Cognitive Interplay

Cyber, Cognitive and Electronic Warfare though driven by distinct technologies are however, interrelated components of modern conflict, all playing vital roles in gaining superiority over an adversary. While cognitive warfare focuses on influencing human perception and decision-making, electronic and cyber warfare targets to disrupt communications, sensor systems and networks, creating conditions favourable for cognitive manipulation. Together, they represent different layers of information operations and non-kinetic warfare that shape the broader battlefield.

Target Domains: Electronic Warfare focuses on controlling or disrupting an adversary's use of the electromagnetic spectrum while Cognitive Warfare focuses on manipulating human minds, perceptions, emotions, and decision-making processes, by leveraging information, disinformation and psychological operations. EW can be seen as a technical enabler of cognitive warfare, creating electromagnetic conditions that make the enemy more vulnerable to psychological and cognitive manipulation. For instance, disrupting an adversary's communications makes it easier to control the narrative and influence how the adversary interprets events.

Disruption and Confusion: Electronic Warfare aims to disorient and confuse enemy forces by jamming communications, radar systems, or GPS signals, thereby degrading their situational awareness. Cognitive Warfare capitalises on this confusion by psychological operations that further erode the adversary's ability to comprehend reality, making decision-making difficult or erroneous. Thus, the confusion created by EW makes the target more susceptible to cognitive manipulation, as disrupted forces are less able to verify information, making them more likely to fall victim to deception and misinformation.

Combined Influence in Decision-Making: Electronic Warfare blocks or distort the enemy's access to accurate data or intelligence, preventing rational decision-making. Cognitive Warfare can then inject misleading or false information into the disrupted channels, guiding adversaries toward making decisions that align with the attacker's objectives. When EW prevents adversaries from obtaining reliable information, cognitive warfare can take advantage of this information vacuum by planting alternative narratives that shape how the enemy thinks or reacts.

Electronic Warfare Amplifies Psychological Effects: Electronic Warfare can be used to support cognitive warfare by targeting specific communication nodes and spreading fear or confusion. For instance, by jamming or spoofing enemy communications, EW can fabricate false orders or communications blackouts, creating a feeling of isolation amongst soldiers or commanders forcing them to believe attack by overwhelming forces. Cognitive Warfare then amplifies the psychological dislocation by disseminating rumours or false information that exacerbates fear, confusion, and stress. The emotional impact caused by disrupted communications and loss of control is a psychological catalyst for cognitive warfare campaigns, deepening the sense of vulnerability and uncertainty.

Information Degradation: Electronic Warfare degrades or disrupts an adversary's ability to receive accurate information through traditional military systems. Cognitive Warfare simultaneously degrades the credibility of the adversary's information sources so that even if the enemy does receive information, they may not trust it. The combined effects of EW and cognitive warfare can lead to a collapse in trust – where an adversary not only loses reliable access to information but also questions the validity of whatever information they receive.

Psychological Impact of Technological Disruption: Electronic Warfare technologies such as spoofing can alter the reality perceived by the enemy by feeding false data. This induces a cognitive dissonance, where the enemy's perceptions of the battlefield are out of sync with reality. Cognitive Warfare takes advantage of this by making it harder to distinguish between true and false information. By attacking the enemy's confidence in their technological systems, cognitive warfare can push adversaries toward making decisions that are increasingly based on false assumptions, leading to strategic and operational miscalculations.

Cyberspace and cyber technologies, have a central role to play in the conduct of state-on-state cognitive operations. The synergetic use of cyber and cognitive operations is often referred to in the literature as Cyber Influence Operations.[21] CIO are a rather complex confluence of cyber and cognitive operations. It is emphasised here that not all CIO require the use of specialist cyber expertise. Simple CIO which do not require high level of cyber expertise may only involve use of cyberspace as a medium for dissemination of preferred narratives through social media platforms. However, the complex ones require cyber operations expertise. The former CIO are referred to as Cyber-enabled Social Influence Operations while the latter are termed as Cyber-enabled Technical Influence Operations (CeTIO).[22]

Cyber Warfare as an Enabler of Cognitive Warfare: Cyber operations provide entry points for cognitive warfare. By attacking, exploiting, or hijacking digital systems, cyber warfare can gain access to media platforms, social networks, or communication channels, which cognitive warfare can then exploit to manipulate perceptions. For instance, a cyber-attack on a news channel's website could allow an adversary to plant fake stories or disinformation that shapes public perception and causes confusion. The Russian hack of the Democratic National Committee in 2016 led to the release of stolen emails, which were used in a disinformation campaign to influence voter perceptions in the US presidential election.[23]

Cognitive Warfare as a Driver for Cyber Attacks: Cognitive warfare strategies can guide the choice of cyber targets. By identifying critical social, political, or cultural vulnerabilities in an adversary, cyber operations can be tailored to exploit those vulnerabilities to spread disinformation, weaken trust, or incite internal divisions. Targeting critical infrastructure like social media platforms or public-facing government websites to spread fake news or deep fakes that manipulate public opinion.

Degrading Information Credibility: Cyber warfare can be used to attack or distort information sources that people trust. When attackers hack and alter official websites, news outlets, or social media accounts, they sow distrust and create chaos by making it harder for people to discern fact from fiction. A cyber operation that takes over a government's emergency alert system could send false warnings, causing panic and undermining trust in the government's ability to provide accurate information.

Creating Disinformation through Cyber Tools: Cyber warfare enables the rapid creation and dissemination of false narratives, amplifying cognitive warfare efforts. Bot networks, deep fake videos, and fake social media profiles can be used to spread disinformation across large audiences, increasing the reach and effectiveness of cognitive manipulation. During the COVID-19 pandemic, state and non-state actors used cyber tools to create and spread conspiracy theories about the virus's origins, treatments, and severity. This effort destabilised public trust in governments and health authorities, furthering cognitive warfare goals of creating division and fear.

Cyber Attacks Leading to Psychological Pressure: Cyber warfare can disrupt essential services causing fear and uncertainty. This leads to a psychological impact, which cognitive warfare then leverages by spreading fear-mongering or panic-inducing disinformation. A cyber-attack on critical infrastructure could shut down power or water supplies in a city. Cognitive warfare tactics, such as spreading rumours of foreign infiltration or government incompetence, could escalate public anxiety and distrust.

Psychological Manipulation through Cyber Infiltration: Hacking into private communications and leaking sensitive information can be used to create internal political conflicts, destabilize leadership, or blackmail key figures. These actions can manipulate emotions such as fear, anger, or frustration, aligning with cognitive warfare goals. Leaked emails during the 2016 US presidential election exploited cyber intrusions to create narratives about corruption, thereby influencing public sentiment and voting behaviour.

Cyber Warfare Weakening Information Integrity: In hybrid warfare scenarios, such as those seen in Ukraine, cyber operations targeting telecommunications systems have disrupted military command-and-control, while cognitive warfare has focused on spreading false information about troop movements and outcomes of battles to confuse both soldiers and civilians.

Cognitive Warfare Exploiting Cyber-Disrupted Communication: Cognitive warfare can take advantage of communication blackouts caused by cyber-attacks to control the narrative. With reliable information sources silenced, adversaries can use social media, propaganda, or fake broadcasts to spread their version of events. During the Arab Spring, governments and non-state actors used cyber tools to shut down social media platforms, but activists and foreign influencers used cognitive tactics to keep the revolution's narrative alive, spreading videos and updates through alternative means.

Cyber Tools for Cognitive Influence: Hacktivist groups or state-sponsored actors can hack into social media platforms or use algorithms to manipulate trending topics, amplify certain posts, or suppress others. This type of cyber manipulation supports cognitive warfare by shaping online discourse and influencing how people interpret unfolding events. The Hamas cognitive campaigns during the ongoing Gaza Conflict involved cyber tools to create bot armies that amplified Israel's overwhelming and indiscriminate use of fire power on social media, supporting cognitive warfare efforts to create social polarisation in the West especially US universities.[24]

Cyber-Supported Data Collection for Targeted Cognitive Warfare: Cyber operations allow adversaries to gather vast amounts of data about individuals and groups. This data can be used in cognitive warfare to tailor disinformation and propaganda to specific demographics, exploiting their biases, fears, and preferences. The Cambridge Analytica scandal involved using personal data harvested from social media to create hyper-targeted political messages designed to influence voters' behaviour – an intersection of cyber-enabled data gathering and cognitive manipulation.[25]

Weakening Public Trust in Institutions: Cyber-attacks on critical systems like electoral infrastructure, banking systems, or healthcare databases can erode public trust in the government's ability to protect them. Cognitive warfare leverages these attacks by spreading rumours or disinformation that further undermine institutional legitimacy. A cyber-attack on election systems could lead to false claims of rigged outcomes, supported by cognitive warfare campaigns that manipulate perceptions of fraud, fostering distrust in democratic institutions. The Indian EVM controversy is a classic case of public trust manipulation.[26]

As military technologies advance, the synergy between cognitive warfare, cyber warfare and electronic warfare will grow stronger. In future conflicts, the line between influencing human perceptions and disrupting technological systems will become increasingly blurred. Advances in artificial intelligence, machine learning, and cyber warfare will enable faster, more effective coordinated campaigns that use both cognitive and electronic means to disable an enemy's decision-making ability.

EW and Cyber operations lay the groundwork for cognitive warfare by disrupting enemy systems, creating opportunities for psychological manipulation. In turn, cognitive warfare takes advantage of the vulnerabilities created by Electronic warfare and cyber operations to influence perceptions, decision-making, and ultimately, the course of a conflict. Together, they form a potent non-kinetic force multiplier, with far-reaching implications for future warfare.

Electronic Perimeter Along the Borders

For almost a century Electronic Warfare (EW) has been playing a crucial role plays a crucial role in border security along disputed borders between adversary nations by providing both offensive and defensive capabilities that shape military strategies and allows swift tactical actions at the functional level. Most of the devices along the borders use the EMS and thus, either get enabled or destroyed by EW capabilities. It not only enables nations to conduct surveillance and intelligence gathering by intercepting communications and radar signals, thus monitoring adversary troop movements and intentions but also allows them to disrupt communications through jamming and spoofing, creating confusion, delaying decision-making and foil infiltration attempts. Additionally, it supports deception operations by mimicking or manipulating signals, which misleads adversaries about real positions or movements along the borders.

Unattended ground sensors, short, medium and long range battle field radars had been the mainstay of border security infrastructure since decades. The advent of electro optical devices, PTZ and IR cameras have further added to the efficacy of the architecture. This architecture on ground further gets reinforced by the aerial assets such as optical devices and radars on aircrafts and helicopters, drones, UAVs, surveillance balloons and tethered platforms.[27] However, this electronic maze remains prone to countermeasures by the adversary and inimical elements. To counter threats to own electronic assets and networks, EW employs electronic countermeasures (ECM) and electronic counter-countermeasures (ECCM) to protect communication and radar systems.

The integration of EW with cyber warfare allows for protection of digital networks against cyber-attacks as well as the protection of electronic systems from intrusions. EW certainly provides tactical advantages along active borders by reducing the effectiveness of enemy systems, providing early warning of adversarial actions and serving as a strategic deterrent by demonstrating technological superiority. It also supports border security personnel by aiding them in maintaining void free surveillance and curb free movement of smugglers, illegal migrants, refugees, infiltrators and terrorists. Overall, EW is essential for maintaining an edge in a contested border environment, influencing conflict outcomes without necessarily resorting to open combat.

The Non-Kinetic Infantryman

Electronic Warfare involves the use of electromagnetic spectrum to disrupt, deceive, or disable enemy communications, radar, and other electronic systems. For infantry, an arm which survives on timely orders, timely execution and timely feedback, EW capabilities are critical for situational awareness, protecting communications and disrupting enemy operations.

Russia's Krasukha-4 is an advanced mobile EW system designed to disrupt airborne (aircraft, helicopters, UAVs, satellites, aerostats) radars, elevated ground radars, X-band guidance radars, GPS signals, RF seekers on missiles and glide bombs, and data links, effectively blinding enemy forces and protecting friendly units.[28]

The US Army's CREW (Counter RCIED Electronic Warfare) system designed to counter improvised explosive devices (IEDs) by jamming their triggering signals, significantly reduces the threat to infantry patrols. The dismounted systems, also called "Manpack" systems, are carried by war fighters to provide protection from RCIEDs.[29]

EW will become increasingly integrated into infantry operations, with soldiers equipped with portable EW devices to counter enemy drones, communications, and sensors. These capabilities will allow infantry to operate more effectively in contested environments where electronic warfare plays a key role.

Cyber warfare extends the battlefield into the digital realm, where infantry units can be both protectors and targets. As military operations become more connected, the ability to defend against and launch cyber-attacks becomes critical. Cyber warfare goes a long way to degrade the enemy forces before their potential is brought to bear on own infantry forces in the battlefield.

Israel's Unit 8200 An elite cyber warfare unit that specializes in intelligence gathering, cyber espionage, and offensive cyber operations. Its operations often directly support infantry missions by disrupting enemy communications and infrastructure. Some of the operations that allegedly get attributed to this unit include the 2005-10 Stuxnet virus attack that disabled Iranian nuclear centrifuges, a 2017 cyberattack on Lebanon's state telecoms company Ogero, the thwarting of an ISIS attack on a civilian airliner travelling from Australia to the United Arab Emirates in 2018 and the recent Pager Blasts in Lebanon.[30]

In future, cyber capabilities will be further integrated into infantry units, enabling them to launch localised cyber-attacks to disable enemy systems, manipulate information, and protect their own networks from hacking attempts. This integration will make cyber warfare an essential component of all future infantry operations.

Mechanised Forces: Seeing the Unseen, Striking the Unreachable

The ever increasing digitization of warfare is also leading to greater connectivity between armoured fighting vehicles and wider elements of military networks. This network-centric approach to warfare is intended to help armoured forces gain more accurate situational awareness and respond more rapidly and effectively to dynamic battlefield conditions. For instance, the NATO's Alliance Ground Surveillance (AGS) system allows for data sharing across different platforms, so that one platform can interact and coordinating tasks, making operations more synergised. The integration of cybersecurity is also a key aspect of these digitalized networks, aimed at protecting connected systems from enemy cyber-attacks, as evident from the US Army's Next-Generation Combat Vehicle (NGCV) program, that focuses on cyber resilience in response to the increasing threats of electronic warfare. Network-centric warfare principles and artificial intelligence are shaping new strategies in modern armoured battles. The Network-Centric Warfare involves the use of advanced communication networks, sensors and data fusion systems to create an "all-connected" and "all-informed" battlefield environment. Meanwhile, AI systems are making it possible to undertake faster and more precise combat decisions and enabling tankmen for joint forces' combined combat efficiency in the battlefield. Theoretically, NCW allows all elements of military force, soldiers, platforms, fixed wings, rotary wings, unmanned aerial vehicles, and command centres, to share, display and take specified actions according to mission protocol in real time, supporting flexible real-time adjustment of combat strategies and operations.

Information superiority leads to operational supremacy. Tanks and other combat vehicles are no longer separate entities, but part of an intelligent network integrated with drones, infantry, artillery, and air support. This will make the firepower more effective, increase protection against threats, and infuse efficiency into logistics and supply chains. For example, the tank detects the enemy's position, strength and direction, and then informs the artillery unit on a shared network. The artillery can immediately engage and neutralise the target.

The integration of Network-Centric Warfare (NCW) with artificial intelligence presents substantial challenges, the intricacy of the systems involved, the necessity for secure and resilient communication networks and the potential risk that adversaries could exploit vulnerabilities in these networks. However, this integration also raises ethical and legal dilemmas concerning the deployment of AI in autonomous weapons systems. Accountability remains a pressing issue, particularly because unintended consequences could arise from such technology. Although we anticipate advancements in machine learning, data analytics and communication technologies, which may enhance the capabilities of armoured forces, the path forward may be challenging. Still the future battlefield will witness broad implementation of AI-driven systems capable of operating autonomously, collaborating with other systems and adapting to real-time environmental changes.

Impact on AFV Design

Integrated Sensors and Communication Systems: Modern AFVs are equipped with advanced sensors, cameras, and communication systems interconnected through a network, enabling real-time data sharing and enhanced situational awareness.

Cyber-Resilient Architecture: As AFVs increasingly depend on interconnected and digital systems, cybersecurity has emerged as a vital consideration in their design. Vehicles are now outfitted with technologies to identify and mitigate cyber threats.

Electronic Warfare Capabilities: AFVs are progressively being integrated with electronic warfare systems capable of jamming enemy communications, disrupting targeting mechanisms and safeguarding friendly forces from electronic assaults.

IoT-Enabled Maintenance: Sensors embedded within AFVs can monitor vehicle health, predict maintenance needs, and report back to centralized systems, ensuring that vehicles are always combat-ready.

K2 Black Panther (South Korea): This advanced main battle tank features networked systems that enable seamless communication with other military assets, enhancing battlefield awareness.[31]

T-15 Armata IFV (Russia): Features advanced electronic warfare systems designed to protect the vehicle from enemy targeting and cyber-attacks.[32]

> *Stryker Dragoon (USA):* This variant of the Stryker vehicle is equipped with EW systems to counter enemy drones and electronic threats on the battlefield.[33]

> *Leopard 2A7+ (Germany):* A highly networked AFV with integrated communication systems that allow it to operate effectively within a broader network-centric warfare environment.[34]

Impact on Tactics

Coordinated Manoeuvre Warfare: Networked armoured fighting vehicles (AFVs) can function as integral components of a unified force, facilitating the exchange of real-time intelligence regarding enemy locations, terrain and potential threats. This significantly enhances operational coordination and enables more intricate, synchronized military manoeuvres.

Predictive Maintenance: IoT-enabled systems facilitate predictive maintenance (a crucial factor), minimizing downtime and guaranteeing that vehicles are combat-ready when necessary. This, in turn, bolsters prolonged operational capabilities over extended durations.

Cyber Defence and Offense: Current tactics encompass utilizing AFVs to both protect against and initiate cyberattacks. However, this introduces a new dimension to armoured conflict, where dominance over the electromagnetic spectrum can decisively influence the outcomes of engagements.

EW-Integrated Operations: AFVs, which are equipped with advanced electronic warfare capabilities, can effectively support broader operational objectives. They do this by neutralizing enemy communications and sensor systems. This action, however, provides friendly forces with a significant advantage in the field. Although challenges may arise, the strategic benefits are clear because they enhance situational awareness and operational effectiveness.

In future wars, being able to see the battlefield and counter the enemy's electronic systems will be key. The ideal AFV will have a multi-spectral sensor suite of radar, infrared, optical and electromagnetic sensors. These sensors, powered by AI will give the vehicle 360-degree situational awareness and be able to detect and identify threats across the electromagnetic spectrum, even in an electronic warfare environment.

The AFV, just like an aircraft, will feature built-in electronic warfare equipment. These systems will be able to jam, spoof, and disrupt enemy communications, radar and targeting. In addition, with AI-powered cyber defences, the AFV will be protected from electronic and cyber threats to ensure the trustworthiness and availability of its mission-critical systems.

EW and New-Age Artillery Weapons and Systems

The rapid development of electronic warfare is reshaping modern military operations, heavily impacting artillery systems. Traditionally artillery has been described by firepower and range, but EW changes the manner artillery would operate and be integrated with other systems and evolves into more sophisticated and responsive weapon platforms.

Counter-Battery Operations and Target Acquisition

EW has brought in a definite change in the manner counter battery operations are being undertaken by the artillery units. Traditionally, some form of counter-battery fire depended upon acquired data from acoustic or radar systems to locate enemy gun positions, but with modern electronic warfare platforms can detect and jam radars and communications networks of enemy artillery, making it more difficult for the enemy to locate or respond to such incoming fire.

For example, Russia's Zoopark-1M counter-battery radar system has combined EW capabilities to pinpoint enemy artillery positions and debilitate them by jamming their radar-based detection systems.[35] AN/TPQ-36/37 Radars provided by NATO allies, detected incoming Russian artillery fire, enabling rapid retaliation. EW did make GPS-guided systems less reliable, pushing Ukraine to adopt alternative targeting methods, such as inertial navigation. Concurrently, you have EW systems being integrated with artillery fire control systems, which induce real-time interference with enemy electronic signals, therefore imposing delays or errors into enemy targeting processes. This makes it hard for them to execute counter-battery fire in an efficient manner.

Communication Disruption and Secure Command and Control

Modern artillery systems heavily rely on secure communications networks for effective command and control (C2). So far, electronic warfare systems capable of disrupting or jamming communications present some of the most serious challenges to artillery units. Countries are developing new artillery systems with enhanced electronic counter-countermeasures (ECCM) to ensure their communication systems can resist jamming and other forms of electronic disruption.

Electronic Deception and False Targeting

One emerging trend in EW-enabled artillery includes the ability to deceive enemy sensors and targeting systems. Electronic deception allows artillery units to create false targets through decoys or manipulation of the electromagnetic signature of actual artillery platforms. Such tactics mislead enemy surveillance and targeting systems, generally wasting valuable ammunition and effort on non-existent or low-value targets.

Electronic Warfare Integration in Precision Fires

The US military has been at the forefront of integrating electronic warfare into artillery doctrine. The M777A2 howitzer is used extensively by the US Army and Marine Corps, it employs advanced digital fire control systems to integrate EW information for the purpose of avoiding enemy jamming. These artillery units operate with the EW teams to locate and suppress the enemy radar and communication systems using signals intelligence (SIGINT). Aiming with HIMARS employs data from the Advanced Field Artillery Tactical Data System (AFATDS), which integrates EW inputs to provide a more thorough picture of the battlefield.[36]

The Russian Borisoglebsk-2 EW system is designed to suppress the enemy's communication and navigation systems, thus hampering the enemy's ability to coordinate counter-artillery fire or maintain operational awareness. It was employed in Ukraine and Syria to disrupt enemy drones and artillery radars, allowing Russian artillery units to operate with relative liberty.[37]

The PLA Rocket Force benefits from the integration of EW systems that enhance its survivability in contested electromagnetic environments. The Chinese military's focus on information based warfare ensures that artillery units are closely linked with EW platforms, allowing them to degrade enemy radars and communication systems before launching precision-guided munitions.

Network-Centric Warfare and Joint Operations

One of the most critical impacts of electronic warfare on artillery is the evolution of network-centric warfare, where artillery units are no longer standalone forces but part of a larger network of electronic warfare, intelligence, and strike capabilities. EW-enabled artillery systems can now operate in sync with drones, satellite systems, and cyber assets to modulate strike capabilities and be more agile and responsive.

The US Army's Multi-Domain Task Forces (MDTF) integrate EW, cyber, and artillery elements into a single operational unit capable of conducting coordinated strikes across multiple domains. Artillery fires are synchronized with EW attacks to blind enemy radars and communication networks, maximizing the impact of artillery strikes while minimizing enemy response time.[38]

Resilience in Contested Environments

Because adversaries have become increasingly adept at using electronic warfare to jam communication and GPS signals, artillery units must also learn the art of operating in contested environments. Modern artillery systems are being designed today with EW defensive features like electronic counter-countermeasures and redundant communication systems, ensuring operations are sustained within a degraded environment.

For instance, US artillery units emphasise on training to work in environments where, indeed, GPS signals might be U/S (unserviceable) or blocked by enemy EW. These units rely on inertial navigation systems (INS) and manual targeting methods as backups to ensure that they can still deliver accurate fire even in the absence of satellite-based data.[39]

Adaptability in Tactics and Speed of Engagement

Developing EW systems that enhance the ability of artillery to respond quickly to the rapid changes of the battlefield still offers great scope for further improvement. Interference with enemy communications and radar networks will, therefore, give windows of opportunity for artillery strikes to be undertaken that will generally give the enemy less than 10 seconds to respond. It also means the artillery unit itself is also enjoying more mobility since it can actually use those gap windows created by EW for a reposition or avoid detection from its enemy.

For example, in Ukraine, the Russian artillery used electronic warfare systems to blind Ukrainian artillery radars, thus allowing the repositioning of the 2S19 Msta-S howitzers without getting spotted and targeted. Combined with mobility and electronic destruction, this gives artillery units a significant tactical advantage in modern warfare.[40]

EW systems are neutralizing enemy sensors and guided munitions, creating a cat-and-mouse game of jamming and countermeasures. Krasukha and Leer-3 Systems disrupted Ukrainian drone communications and GPS signals during the Siege of Mariupol, degrading PGM accuracy.[41]

The Battle of Bakhmut exemplified the convergence of these technologies, wherein, Russia used Lancet drones and Krasnopol shells to attrition Ukrainian forces and in retaliation found Ukraine leveraging Excalibur rounds and counterbattery radars to slow Wagner Group advances. Both sides relied on drones for real-time battlefield awareness, turning the city into a "meat grinder" of technological one-upmanship.[42]

Because electronic warfare will continue to evolve and become more and more potent, artillery must become adept in robust ECCM and secure communication systems so as to ensure successful operations in a contested environment. The integration of EW into artillery systems and operations will be critical in enhancing the ability of the artillery forces to make strikes with precision and stealth against enemy counter-action measures.

Non-Kinetic Turbulence: A New Front in Military Aviation

Electronic Warfare, or EW, is the strategic manipulation of the electromagnetic spectrum to intercept, disrupt, or deceive enemy systems, including radar, communications, and navigation networks. In an age where control of the electromagnetic domain can dictate the outcome of engagements, EW has emerged as a critical enabler across helicopters, UAVs, and autonomous systems, bolstering their survivability and effectiveness in contested environments.

The effectiveness of stealth technology and EW is constantly challenged by advancements in enemy radar, communication, and electronic warfare capabilities. As adversaries develop more sophisticated radar systems, such as those operating at low frequencies, the ability of stealth aircraft to evade detection may be reduced. Similarly, advancements in electronic warfare by adversaries could counteract the effectiveness of EW systems used by military aviation forces.

Helicopters

Advanced EW suites provide a vital layer of self-protection, equipping these platforms with the tools to jam enemy radar and missile guidance systems, thereby thwarting detection and targeting efforts. These systems also enhance situational awareness by integrating electromagnetic sensors that detect and analyse threats in real time, giving pilots a comprehensive understanding of the electronic battlespace. The

Russian Ka-52 Alligator helicopter exemplifies this capability with its Vitebsk EW system, which includes laser warning receivers and infrared jammers. Deployed in conflicts such as Ukraine, the Ka-52 uses these tools to evade air defences, demonstrating how EW empowers helicopters to operate in high-threat scenarios where survivability is paramount.

UAVs

UAVs, meanwhile, leverage EW as both an offensive and defensive tool, carrying specialized payloads to conduct signals intelligence (SIGINT), electronic warfare, and suppression of enemy air defences (SEAD). These capabilities allow UAVs to disrupt enemy communications, degrade radar effectiveness, and neutralize air defence networks, all while requiring robust countermeasures to protect against reciprocal EW threats from adversaries. The US Army's Gray Eagle UAV illustrates this dual role, equipped with EW pods that enable jamming operations and SIGINT collection. Its deployment in contested environments highlights how UAVs are expanding the reach of EW, providing commanders with flexible, persistent options to shape the electromagnetic battlefield.[43]

Autonomous Systems

Autonomous systems take EW to the next level by integrating AI-driven adaptability, allowing them to respond dynamically to shifting electromagnetic conditions. These platforms can execute real-time radar jamming and electronic countermeasures, adjusting their tactics on the fly to counter enemy systems. The US Air Force is actively developing autonomous EW drones to enhance battlefield resilience, pairing AI with advanced EW payloads to create systems capable of independent operation in complex, contested environments. This evolution signals a future where autonomous platforms could dominate the electromagnetic spectrum, offering militaries a decisive edge in electronic warfare.

Cyber Warfare represents a new frontier in military operations, encompassing both offensive and defensive actions targeting computer networks and systems to disrupt, deny, or degrade an adversary's capabilities. As military aviation becomes increasingly reliant on digital infrastructure, the importance of cyber resilience and the potential of cyber as a weapon cannot be overstated, with distinct implications for helicopters, UAVs, and autonomous systems.

Helicopters

Robust cybersecurity measures are essential to safeguard onboard systems from hacking attempts that could compromise navigation, communication, or weapon systems. In contested environments, where adversaries may seek to exploit vulnerabilities in digital networks, these protections ensure operational integrity and mission success. The US Army's Integrated Air and Missile Defence (IAMD) system, for example, incorporates cyber defences into helicopter avionics, providing a shield against electronic intrusions while maintaining secure links to command networks. This integration reflects the growing recognition that helicopters, as critical battlefield assets, must be fortified against cyber threats to remain effective.

UAVs

UAVs face similar challenges but also serve as potent platforms for delivering cyber payloads, such as malware, to disrupt enemy systems. Their reliance on remote control and data links makes them prime targets for cyber attacks, necessitating advanced encryption and anti-hijacking measures to preserve mission integrity. At the same time, UAVs can conduct offensive cyber operations, by hacking into OS of adversary UAVs.[44] Reports suggest that countries and even hackers have employed UAVs to take control or intercept hostile UAVs, showcasing how these platforms can extend cyber warfare into the physical domain, a capability that amplifies their strategic value.[45]

Autonomous Systems

Autonomous systems, by virtue of their independence, demand even more sophisticated cyber defences to protect against tampering or takeover, while also opening the door to AI-driven cyber countermeasures that can autonomously detect and neutralise threats. The US Department of defence is investing heavily in such technologies, developing autonomous platforms capable of defending their own networks and, potentially, launching cyber-attacks with minimal human oversight. This dual role positions autonomous systems at the forefront of cyber warfare, where their ability to operate in isolation could make them both resilient defenders and elusive aggressors in the digital battlespace.

Disrupting the Skies: The Rise of Non-Kinetic Air Defence

The era of air defense dominated by kinetic interceptors striking down missiles and guns downing aircraft is rapidly fading. In its place, a silent revolution is unfolding where victory in the skies is increasingly decided not by explosives but by invisible forces such as electronic jamming that blinds sensors, cyber attacks that paralyse command and control, and cognitive attacks that shatter the confidence of human operators and leadership alike. This shift to non-kinetic warfare has fundamentally transformed air defense from a contest of firepower to a battle for control over the electromagnetic spectrum, cyber space and human perception.[46,47]

Electronic warfare now forms the first and most potent layer of modern air defense. Advanced jamming systems create "bubbles" of electromagnetic denial, rendering enemy radars useless and severing data links between aircraft and ground control. During Russia's invasion of Ukraine, these invisible barriers forced Ukrainian pilots to fly perilously low to avoid detection, while Russian systems like the Krasukha neutralised strike and reconnaissance aircraft, UAVs, low-orbit satellites and surface-to-air missile batteries by overwhelming their targeting radars with noise.[48] Beyond jamming, spoofing techniques send false GPS signals to misdirect missiles or drones off course, as seen when commercial aircraft over Iran suddenly found their navigation systems reporting impossible locations.[49] Cheap decoy drones equipped with electromagnetic emitters now mimic fighter jets, saturating radar screens with false targets and tricking defenders into wasting million-dollar interceptors on worthless objects, a tactic Yemen's Houthis exploited to overwhelm Saudi Arabia's advanced Patriot systems.[50]

Simultaneously, cyber warfare has turned air defense networks vulnerable to cyber attacks. Hackers infiltrate command and control systems to corrupt targeting data, disable fire control software, or induce paralysing false alarms. In the shadow war between Israel and Iran, cyber operations reportedly delayed the activation of Iranian air defenses during drone strikes by corrupting server data, buying precious minutes for attackers.[51] The threat extends beyond the battlefield: malware like Triton, which targeted Saudi industrial safety systems,[52] could be adapted to sabotage factories producing missile components or radar coolants, crippling an adversary's ability to employ its air defenses even before the conflict begins.

Perhaps most insidiously, cognitive warfare targets the human element of warfare i.e. the soldiers and commanders who interpret data and make life or death decisions. Deepfake audio and video fabricate orders from superiors, as Russia demonstrated by releasing a fake recording of Zelensky ordering surrenders of his forces.[53] Radar screens flooded with false targets induce decision fatigue, causing operators to miss real threats amid the noise. Disinformation campaigns on social media erode trust in air defense capabilities, demoralizing troops and undermining public confidence in their military's ability to protect them.

In Ukraine, non-kinetic attacks degraded majority of Soviet-era air defenses early in the war, forcing a shift to decentralized networks using Starlink satellites for jam-resistant communications and AI algorithms to filter electronic interference.[54] The cost asymmetry is equally telling as a $500 GPS jammer can divert a $3 million cruise missile, while a $20,000 drone swarm can exhaust a $10 million Patriot interceptor battery.[55]

Looking ahead, the future of air defense lies in integrating these non-kinetic layers into a seamless AD umbrella. Projects like the NATO's AMDC2 integrate sensors, shooters and EW systems with artificial intelligence, enabling these systems to distinguish decoys from real threats even amid intense jamming.[56] Quantum radars promise immunity to spoofing, while AI advisors will augment human operators, filtering cognitive attacks and prioritizing threats. For nations like India, investing in these capabilities is not optional. Integrating systems like the S-400 with offensive cyber and EW units and AI testbeds is essential to avoid legacy air defenses rendered obsolete not by superior firepower, but by silent, cheap and deniable attacks that conquer the spectrum long before the first aircraft takes off. The age of winning air battles through brute force is over; the future belongs to those who master the invisible forces of disruption.

The Silent Orbit: Non-Kinetic Warfare Redefining the Battle Above

Space, once a sanctuary for peaceful satellites and scientific exploration, has become the ultimate high ground in modern conflict, not through kinetic destruction, but through the invisible, pervasive reach of non-kinetic warfare. Electronic jamming, cyber intrusions and cognitive deception are transforming space technologies from passive enablers into active weapons systems, rendering traditional concepts of space "weaponisation" obsolete. The future of cosmic dominance lies not in missiles shattering satellites, but in silently hijacking their data, blinding their sensors, and poisoning the decisions they inform.

The Electronic Siege: Dominating the Cosmic Spectrum

Electronic warfare has turned the electromagnetic spectrum surrounding Earth into a contested battlefield. GPS jamming and satellite signal spoofing now routinely cripple navigation, communications, and intelligence gathering. In Ukraine, Russian troops deployed truck-mounted Tobol jammers to disrupt Starlink internet critical for drone operations and battlefield coordination, forcing SpaceX engineers to develop countermeasures.[57] Near conflict zones like Egypt, Syria, Israel and the Indo Pak Border, commercial ships and aircraft report sudden, dangerous GPS location shifts, evidence of attacks where spoofed signals place vessels miles off-course.[58] The Chinese SJ-21 which pulled out an unserviceable Beidou satellite out of the orbit could also be maneuvered suspiciously close to other nations' assets, radiating signals to cause electronic sabotage.[59] These invisible assaults achieve what anti-satellite missiles cannot i.e. deniable, reversible and cost-effective disruption. A powerful enough ground jammer can destabilise a spy satellite's functionality, forcing adversaries to fight blind without effective ISR input.

Cyber Warfare: Hacking the Celestial Nervous System

Satellites and their ground stations are now prime cyber targets, turning space infrastructure into a vector for global disruption. The 2022 Viasat hack attributed to Russia, disabled tens of thousands of satellite modems across Europe, halting wind turbines in Germany and disconnecting Ukrainian command centers hours before the invasion.[60] This was not mere espionage, it was a strategic strike on space-enabled warfare. Cyber intruders now target satellite firmware updates, inserting malware like PRESSTEA and CADDYWIPER that lies dormant until conflict erupts.[61] Once activated, it can hijack imaging satellites to deliver false terrain data, disable missile warning systems, or even commandeer thrusters to de-orbit spacecraft.[62] The goal is no longer destruction, it is silent control.

Cognitive Warfare: Weaponising the Cosmic Perspective

Space-derived data, satellite imagery, weather patterns, communications are bound to shapes strategic decisions. Cognitive warfare exploits this by corrupting the interpretation of space intelligence. AI-generated deepfake satellite imagery can fabricate troop buildups or hide real ones. During the 2024 Israel-Hezbollah conflict, actors spread manipulated satellite photos on social media platforms claiming huge losses to each other, demoralising the public as well as frontline units until verified imagery refuted it. In the cognitive domain fear itself becomes a weapon as rumors of imminent ASAT tests or "debris storms" can force adversaries to shield satellites, blinding them during critical operations.

Real-World Convergence: The New Space Battlefield

Russia-Ukraine War: Russia's electronic warfare disrupted Ukraine's access to commercial satellite imagery, while Ukrainian hackers targeted Roscosmos ground stations.[63] SpaceX's Starlink became a non-kinetic lifeline, its rapid software updates countering jamming, a private corporation as decisive actor.[64]

US-China Great Game: In 2018, a sophisticated hacking campaign launched from China hacked into satellite operators, defence contractors and telecommunications companies in the United States and Southeast Asia. The hackers infected computers that controlled the satellites, so that they could have changed the positions of the orbiting devices and disrupted data traffic.[65]

Challenges and Imperatives

Attribution Fog: Was that satellite malfunction due to solar flares, hacking, or sabotage? Deniability enables aggression.

Civilian Collateral: Jamming GPS disrupts global shipping, agriculture, and finance.

India's Critical Path: With China's "Space Silk Road" extending influence, India must

- Integrate Defence Cyber Agency with ISRO to protect civil and military satellites.
- Develop offensive cyber-EW suites for satellite disruption.
- Launch AI-powered satellites to monitor hostile space behavior autonomously.

Space warfare has shed its sci-fi tropes of lasers and kinetic strikes. The true conquest of the cosmos now unfolds in the silent realms of corrupted data, manipulated perceptions and electromagnetic suffocation. Nations that master non-kinetic space dominance will control the ultimate high ground, not by destroying the enemy's eyes in the sky, but by owning what they see, controlling what they say and shaping what they believe. The age of space as a sanctuary is over and the age of controlling space has begun.

The future trajectory of non-kinetic warfare points towards even greater integration and sophistication. The distinctions between electronic, cyber and cognitive warfare will continue to blur as operations seamlessly blend all three domains to create complex synergistic effects. The speed of conflict will accelerate further with AI-powered systems making decisions and executing actions in microseconds creating a hyper-war that humans can only hope to manage and not to control.

This raises profound challenges. The problem of attribution will become even more difficult as AI can be used to create false flags and to mimic the techniques of other actors. Establishing and enforcing international norms and laws for conduct in these domains will be a monumental task as the technology constantly outpaces policy. The ethical dilemmas will also intensify as cognitive warfare techniques become more personalised and persuasive potentially eroding the very notion of free will. Nations will face the constant challenge of defending their critical infrastructure and their societies from these pervasive and persistent threats requiring a whole of nation approach to security that encompasses government industry and individual citizens. The unseen battlefield is the new reality of international conflict and the nations that learn to navigate and to dominate this complex terrain will be the ones to shape the course of the 21st century.

Notes

1   Bronk, Justin, Nick Reynolds, and Jack Watling. 2022. "The Russian Air War and Ukrainian Requirements for Air Defence." https://static.rusi.org/SR-Russian-Air-War-Ukraine-webfinal.pdf.
2   Mizokami, Kyle. 2023. "Why Ukraine's GPS-Guided Bombs Keep Missing Their Targets." Popular Mechanics. April 20, 2023. https://www.popularmechanics.com/military/weapons/a43591694/russian-jamming-gpsguided-bombs/.
3   Jankowicz, Mia. 2023. "Ukraine Is Losing 10,000 Drones a Month to Russian Electronic-Warfare Systems That Send Fake Signals and Screw with Their Navigation, Researchers Say." Business Insider. 2023. https://www.businessinsider.com/ukraine-losing-10000-dronesmonth-russia-electronic-warfare-rusi-report-2023-5.
4   Haig, Zsolt. (2015). Electronic warfare in cyberspace. Security and Defence. 2. 22-35. 10.5604/23008741.1189275. https://www.researchgate.net/publication/288871386_ELECTRONIC_WARFARE_IN_CYBERSPACE/citation/download
5   B. Weeden: "Two points: 1) this is all about civil GPS signals (military signals are much more robust) 2) the DOD could have done more to prevent spoofing of civil GPS, but has not 3)

Galileo, BeiDou & QZSS will all help, but not prevent it completely (see #2)", 18 December 2018, https://twitter.com/brianweeden/status/1074787323357876229.

6   T. Harrison, K. Johnson and T.G. Roberts, Space Threat Assessment 2018, Center for Inter national and Strategic Studies, April 2018, https://csis-prod.s3.amazonaws.com/s3fs-public/publication/180823_Harrison_SpaceThreat Assessment_FULL_WEB.pdf.

7   A. Mathew, "Russia Completes Development of Airborne Anti-satellite Laser Weapon", DefPost, 26 February2018, https://defpost.com/russia-completes-development-airborne-anti-satellite-laserweapon/

8   Bhatta, Prof. (2023). Recent Advances In Electronic Warfare. https://www.researchgate.net/publication/368091732_RECENT_ADVANCES_IN_ELECTRONIC_WARFARE

9   Surya Kumar (SK) Singh. Chief Scientist, eAge Technologies India Pvt Ltd, Technology Trends in Electronic Warfare, December 12, 2023, https://www.linkedin.com/pulse/technology-trends-electronic-warfare-ew-eageit-lo3pe/

10   Rajesh Uppal March 30, 2024 Mastering the Electromagnetic Battlefield: Electronic Warfare technology Trends and Market Dynamics https://idstch.com/technology/electronics/mastering-the-electromagnetic-battlefield-electronic-warfare-technology-trends-and-market-dynamics/

11   Electronic warfare: technology trends By GlobalData Analysts https://defence.nridigital.com/global_defence_technology_jan22/ew_technology_trends

12   John Arquilla and David Ronfeldt, CYBERWAR IS COMING! Naval Postgraduate School, Graduate School of Operational and Information Sciences, Department of Defense Analysis,Monterey,CA,93943

13   General Anil Chauhan, Chief of Defence Staff, Evolving Battlefield Architecture, Page 65, Ready, Relevant and Resurgent A Blueprint for the Transformation of India's Military

14   https://www.usiofindia.org/pdf/20240912144421.pdf, Impact of Technology Enabled Cognitive Operations in Hybrid Warfare, USI Occasional Paper Lieutenant General (Dr) RS Panwar, AVSM, SM, VSM (Retd)

15   Christopher Paul, Miriam Matthews, The Russian "Firehose of Falsehood" Propaganda Model, Jul 11, 2016, RAND https://www.rand.org/pubs/perspectives/PE198.html

16   Jonathan Saul, Steven Scheer and Ari Rabinovitch, Hezbollah pager attack puts spotlight on Israel's cyber warfare Unit 8200, September 20, 2024 Hezbollah pager attack puts spotlight on Israel's cyber warfare Unit 8200 | Reuters

17   Amal Chmouny, The Impact of Israeli Cyber Operations on Hezbollah, Apr 2, 2025 The Impact of Israeli Cyber Operations on Hezbollah

18   Rand Waltzman, Future Challenges to Cognitive Superiority, Irregular warfare centre, May 12, 2025 Future Challenges to Cognitive Superiority – Irregular Warfare Center

19   Christoph Deppe, Cognitive warfare: a conceptual analysis of the NATO ACT 01 November 2024, Frontiers | Cognitive warfare: a conceptual analysis of the NATO ACT cognitive warfare exploratory concept

20   R. McCreight, Neuro-Cognitive Warfare: Inflicting Strategic Impact via Non-Kinetic Threat, 16 Sep 2022 Neuro-Cognitive Warfare: Inflicting Strategic Impact via Non-Kinetic Threat | Small Wars Journal

21   https://www.usiofindia.org/pdf/20240912144421.pdf, Impact of Technology Enabled Cognitive Operations in Hybrid Warfare, USI Occasional Paper Lieutenant General (Dr) RS Panwar, AVSM, SM, VSM (Retd)

22   Panwar, R.S., Cyber Influence Operations - A Battle of Wits and Bits: The Cauldron of Concepts and Terminologies, 13 Oct 2020, Accessed 25 Sep 2024, https://futurewars.rspanwar.net/cyber-influence-operations-a-battle-of-wits-and-bits-the-cauldron-of-concepts-and-terminologies/

23   The Washington Post, https://www.washingtonpost.com/world/national-security/how-the-russians-hacked-the-dnc-and-passed-its-emails-to-wikileaks/2018/07/13/af19a828-86c3-11e8-8553-a3ce89036c78_story.html

24   Wikipedia, List of pro-Palestinian protests on university campuses in the United States in 2024, https://en.wikipedia.org/wiki/List_of_proPalestinian_protests_on_university_campuses_in_the _United_States_in_2024

25   Nicholas Confessore April 4, 2018 Cambridge Analytica and Facebook: The Scandal and the Fallout So Far, https://www.nytimes.com/2018/04/04/us/politics/cambridge-analytica-scandal-fallout.html

26   Ranjit Bhushan In the EVM debate, stalemate between civil society and government 22 Jan 2024, https://www.hindustantimes.com/india-news/in-the-evm-debate-stalemate-between-civil-society-and-government-101705921259672.html

27   High-Tech Border Security: Current and Emerging Trends IEEE Public Safety Technology Initiative https://publicsafety.ieee.org/topics/high-tech-border-security-current-and-emerging-trends

28   Vijainder K Thakur September 20, 2023 Russia's Krasukha EW System Forcing Ukrainian Fighter Jets To Abandon Missions After 'Navigation Fails' https://www.eurasiantimes.com/ ukrainian-fighter-jets-abandon-combat-missions-after-navigation/

29   From Program Executive Office Unmanned and Small Combatants Public Affairs, 27 July 2023, JCREW Counter IED Program Achieves Full Operational Capability https:// www.navy.mil/Press-Office/News-Stories/Article/3469648/jcrew-counter-ied-program-achieves-full-operational-capability/

30   Reuters September 18, 2024 What is Israel's secretive cyber warfare unit 8200? https:// www.reuters.com/world/middle-east/what-is-israels-secretive-cyber-warfare-unit-8200-2024-09-18/

31   Peter Suciu K2 Black Panther: One of the Best Tanks on Earth (Made in South Korea) January 2, 2024https://nationalinterest.org/blog/buzz/k2-black-panther-one-best-tanks-earth-made-south-korea-208305

32   T-15 BMP IFV 11 Jul, 2024 army recognition group https://armyrecognition.com/military-products/army/infantry-fighting-vehicles/tracked-vehicles/t-15-bmp-armata-aifv-armoured-infantry-fighting-vehicle-technical-data-sheet-pictures-video

33   *John Antal* Striking Power for STRYKERs – The US Army Up-Guns its Mobile Striking Power in Europe 7. February 2020 https://euro-sd.com/2020/02/articles/16057/striking-power-for-strykers-the-us-army-up-guns-its-mobile-striking-power-in-europe/

34   Leopard 2A7+ Main Battle Tank Kraus-Maffei Wegmann Germany 14 Sep, 2024 https:// armyrecognition.com/military-products/army/main-battle-tanks/main-battle-tanks/leopard-2a7-germany-uk

35   Global Defense News, Zoopark-1M 1L260, 03 Aug 2024, https://armyrecognition.com/ military-products/army/radars/counter-battery-radars/zoopark-1m-1l260-artillery-counter-battery-radar-system-data

36   Inder Singh Bisht, US Army Seeks HIMARS Fire Control Systems for Ukraine, Taiwan, 29, December 2022, https://thedefensepost.com/2022/12/29/us-army-himars-fire-control-system/ #:~:text=%60%60The%20AFATDS%20provides%20fully%20automated%20support%20for,sur face%20fire%2Dsupport%20systems%20" 2C"%20the%20US%20Army%20states.

37   Borisoglebsk-2, https://en.wikipedia.org/wiki/Borisoglebsk-2#cite_note-18

38   The Army's Multi-Domain Task Force (MDTF), July 10, 2024, https://sgp.fas.org/crs/natsec/ IF11797.pdf

39   Kyle Mizokami, The US Army is Developing Artillery That Has GPS Precision – Without GPS, Apr 10, 2018, https://www.popularmechanics.com/military/weapons/a19735395/the-us-army-is-developing-artillery-that-has-gps-precisionwithout-gps/

40   Global Defense News, Has the Russian 2S19 Msta-SM2 self-propelled howitzer proved its value in Ukraine?, 23 Jan, 2025, https://armyrecognition.com/focus-analysis-conflicts/army/ conflicts-in-the-world/russia-ukraine-war-2022/has-the-russian-2s19-msta-sm2-self-propelled-

  howitzer-proved-its-value-in-ukraine

41 Col Mandeep Singh,Russian Electronic Warfare in Ukraine 2022-2023, July 7, 2023, https://indiandefencereview.com/russian-electronic-warfare-in-ukraine-2022-2023/

42 Bendett, Samuel, and Leonid Nersisyan. 2024. "The Drone War over Ukraine," June. Informa, 168–94. doi:10.4324/9781003454120-8.

43 MQ-1C Gray Eagle ER/MP Unmanned Aircraft System (UAS), USA, March 1 2024, https://www.army-technology.com/projects/mq1c-gray-eagle-uas-us-army/#:~:text=Credit:%20US%20Army.,MQ%2D1C%20Gray%20Eagle%20is%20an%20extended%20range/multipurpose%20(,and%20battle%20damage%20assessment%20missions.

44 Kim Hartmann, Keir Giles, UAV Exploitation: A New Domain for Cyber Power, 2016 8th International Conference on Cyber Conflict, https://ccdcoe.org/uploads/2018/10/Art-13-UAV-Exploitation-A-New-Domain-for-Cyber-Power.pdf

45 Kamkar, Samy. (2013, December) 'SkyJack', Github.com. [Online]. https://github.com/samyk/skyjack.

46 Mark Pomerleau, How Army leaders envision non-kinetic capabilities enabling traditional forces, September 19, 2023, How Army leaders envision non-kinetic capabilities enabling traditional forces, DefenseScoop

47 Ritik singh, The changing face of warfare: kinetic and non-kinetic dimensions, 10 May 2025 The changing face of warfare: kinetic and non-kinetic dimensions - Plutus IAS

48 Analysis: How Russia's Krasukha Electronic Warfare System Disrupts UAVs and Radars in Ukraine, 17 Apr, 2024, https://armyrecognition.com/focus-analysis-conflicts/army/defence-security-industry-technology/analysis-how-russia-s-krasukha-electronic-warfare-system-disrupts-uavs-and-radars-in-ukraine

49 The Dangerous Rise of GPS Attacks, The Radionavigation Labortary, May 2024 https://radionavlab.ae.utexas.edu/spotlight/page/3/#:~:text=March%202024:%20Since%20August%20of,%E2%80%9COn%20Oct.

50 How are Drones Changing Modern Warfare?, Australian Army Research Centre, 01 Aug 2024, How are Drones Changing Modern Warfare? | Australian Army Research Centre (AARC)

51 Reflections of the Israel-Iran Conflict on the Cyber World, June 16, 2025, https://socradar.io/reflections-of-israel-iran-conflict-cyber-world/

52 Martin Giles, Triton is the world's most murderous malware, and it's spreading, 05 March 2019, https://www.technologyreview.com/2019/03/05/103328/cybersecurity-critical-infrastructure-triton-malware/

53 Bobby Allyn, Deepfake video of Zelenskyy could be 'tip of the iceberg' in info war, experts warn, NPR, March 16, 2022 https://www.npr.org/2022/03/16/1087062648/deepfake-video-zelenskyy-experts-war-manipulation-ukraine-russia

54 Mr Duncan McCrory CEng FRAeS, Freeman Air and Space Institute, King's College London, Electronic Warfare in Ukraine, Preliminary Lessons for NATO Air Power Capability Development, 23 March 23 https://www.japcc.org/wp-content/uploads/JAPCC_J36_Art-10_screen.pdf

55 Parth Satam, A $3 Million Missile To Strike A $20,000 Drone? Experts Say Patriot Air Defense Missile No 'Panacea' For Ukraine!, December 24, 2022 https://www.eurasiantimes.com/a-3-million-missile-to-strike-a-20000-drone-experts-say-patriot-air-defense-missile-no-panacea-for-ukraine/

56 NATO AMDC2, NATO Integrated Air and Missile Defence, 13 Feb 2025 https://www.nato.int/cps/en/natohq/topics_8206.htm

57 Amit Chaturvedi, All About Tobol, Russia's Secret Weapon Linked to Jamming of Planes' Signals Tobol electronic warfare system was previously linked to jamming of signals of ships on NATO's eastern flank. World News, Apr 23, 2024 https://www.ndtv.com/world-news/all-about-tobol-russias-secret-weapon-linked-to-jamming-of-planes-signals-5503632

58 Saurabh Sinha,GPS Spoofing reported by civil vessels and aircrafts, Times of India, 20 Mar 205 http://timesofindia.indiatimes.com/articleshow/119266775.cms?utm_source=contento finterest&utm_medium=text&utm_campaign=cppst

59 Theresa Hitchens, China's SJ-21 'tugs' dead satellite out of GEO belt: Trackers, Breaking Defense. January 26, 2022 https://breakingdefense.com/2022/01/chinas-sj-21-tugs-dead-satellite-out-of-geo-belt-trackers/

60 Matt Burgess, A Mysterious Satellite Hack Has Victims Far Beyond Ukraine, Mar 23, 2022 https://www.wired.com/story/viasat-internet-hack-ukraine-russia/

61 Mandiant, The GRU's Disruptive Playbook, July 12, 2023 https://cloud.google.com/blog/topics/threat-intelligence/gru-disruptive-playbook

62 Nate Nelson, How Hackers Can Hijack a Satellite, July 15, 2023 https://www.darkreading.com/cybersecurity-analytics/how-researchers-hijacked-a-satellite

63 Rajiv Thummala, Gregory Falco, Hacktivism Goes Orbital: Investigating NB65's Breach of ROSCOSMOS, Sibley School of Mech. and Aero space Engineering, Cornell University, 15 Feb 2024 https://arxiv.org/html/2402.10324v1

64 Kevin Collier, Starlink internet becomes a lifeline for Ukrainians, NBC news, April 29, 2022, https://www.nbcnews.com/tech/security/elon-musks-starlink-internet-becomes-lifeline-ukrainians-rcna25360

65 Joseph Menn, China-based campaign breached satellite, defence companies - Symantec, June 20, 2018 https://www.reuters.com/article/technology/china-based-campaign-breached-satellite-defence-companies-symantec-idUSKBN1JF2X3/

8

Border Brawls

"Border disputes have always been sour and often bloody. Historically, they have sparked more wars than any other cause except for religion and ethnicity."[1]

—**Michael Ignatieff**

Michael Ignatieff's quote resonates deeply with India's history of border conflicts and wars. India shares borders with multiple countries, each presenting distinct challenges and opportunities. Since India's independence, the Indian Nation has experienced constant threats from its neighbours notably Pakistan and China, which have many a times escalated into armed conflicts. The geo political as well as socio economic circumstances prevailing within the smaller countries around India have thrown up constant security challenges for India, in the form of illegal immigration, refugee crises, smuggling, drug & arms trade, terrorism, insurgency and proxy war. The Indo-Pak border in particular has been a persistent source of conflict since the British carved out Pakistan out of India before leaving in 1947. The unresolved territorial disputes, especially over Jammu and Kashmir have led to multiple wars between the two countries as well as numerous skirmishes and border clashes with dynamics including cross-border terrorism and ceasefire violations. It is the fall out of the endlessly escalating contestation at these borders which led to creation of Bangladesh in 1971, a country with largest porous land border with India.

This border dispute has led to the longest and almost permanent deployment of Indian Defence Forces along its western border with substantial costs to the Nation on account of dealing with Pak aided separatist militant and terrorist movements in Punjab and Kashmir. Not to mention the social impact of these pinpricks over the years in these two states. Similarly, the border between India and China, known as the Line of Actual Control (LAC), has been contentious, with disputes over areas extending right from Aksai Chin in the north to Arunachal Pradesh in the east. The Sino-Indian War of 1962 was a direct consequence of these disputes, and China, with its strategy of 'Salami Slicing' has kept the disputes not only alive but has been calibrating the level of contestation through strategic competition and occasional incursions, in sync with India's growing stature in the world arena.

In fact, present standards of India's defence preparedness, military industrial complex and military technological innovation are largely the result of constant security threats from these two countries. With these two countries posing a collusive threat to Indian territorial integrity, Indian borders could have not been more vulnerable and its Armed Forces more stretched in their efforts to safeguard national security on multiple fronts. Additionally, while India and Bangladesh have had cordial relations since Bangladesh's creation in 1971, spanning approximately 4,096 kilometres, border disputes over enclaves and illegal cross border movement have occasionally flared into localised tensions and conflicts. Illegal immigration of Bangladeshi citizens into India has been a constant source of tension between the two countries, straining resources, altering demographics and contributing to social and political unrest along the borders.

Meanwhile, the India-Nepal border, spanning around 1,751 kilometres, grapples with demarcation issues in areas like Kalapani-Limpiyadhura-Lipulekh, despite strong historical ties and cultural affinities. Along the India-Myanmar border, which stretches approximately 1,643 kilometres, issues of illegal migration, insurgency and drug trafficking dominate the bilateral dynamics. While security cooperation and infrastructure development efforts are underway, the civ war raging in Myanmar has derailed all benign interactions and brought India's Act East Policy to a grinding halt, with the existing military regime having lost all control along the border areas. These conflicts underscore the challenges of managing borders in diverse and geopolitically complex regions like South Asia, emphasizing the profound relevance of Ignatieff's quote in understanding India's security landscape and regional dynamics.

Militancy of yesteryears in Punjab and insurgencies in Jammu and Kashmir and the North Eastern states of India are intricately linked to border dynamics, reflecting complex geopolitical realities and historical grievances. In Jammu and Kashmir, the insurgency has its roots in the limitless ego of the western neighbour manifesting into the unresolved territorial dispute fabricated over the region, exacerbated by competing claims to sovereignty, further aggravated by ethnic, religious and political tensions. The heavily contested Line of Control and the presence of militant groups all along it, forced India to undertake one of the longest and most challenging border fencing project (550 kilometres) in modern times. This project which was a testament to human endeavour conquering treacherous terrain and hostile weather has been infused regularly sprinkling of technology of the times to keep the infiltrators at bay, who themselves keep innovating to defeat Indian efforts in an unending cat and mouse game.

In the North Eastern states, though insurgencies have not actually been fuelled by border disputes, the region's proximity to international borders with countries like China, Myanmar and Bangladesh has facilitated the flow of arms, militants and illicit contraband, and provided safe havens and training areas to inimical elements, complicating efforts to address the underlying ethnic tensions and grievances. Therefore, while it can be assumed that India's internal dynamics such as governance failures and socio-economic disparities play a role in these conflicts, however, it is undeniable that border dynamics significantly have been contributing to the persistence and intensity of insurgencies in Jammu and Kashmir and the North Eastern states of India.

It's not only the land which makes the borders, maritime and airspace are other dimensions of a nation's borders. And the Indian coastline along the Indian Ocean and Arabian Sea, extending over approximately 7,516 kilometres, presents critical maritime security challenges as well as opportunities. India's vast unmanned coastal areas face clear and present danger from threats such as piracy, smuggling, espionage and maritime terrorism, necessitating robust coastal surveillance, maritime domain awareness, and naval capabilities, all of which are a highly technologically and financially demanding pursuit. The Indian coastline serves as a gateway for maritime trade, energy transportation, and strategic connectivity, fostering economic growth and boosts India's geopolitical influence in the Indian Ocean region. Thus, ensuring security of its island territories and its relationships with its sea neighbours are crucial components of India's maritime strategy. The Indian land mass accounts for 1382 islands and with island territories like the Andaman and Nicobar Islands and the Lakshadweep Islands strategically located in the Indian Ocean region, India faces diverse challenges and opportunities in their governance and security. Therefore, the management of these island territories and the waters around them comprises not only ensuring their economic development and environmental protection but also their security in order to preserve their strategic significance in the region.

In creating this protected environment India needs to maintain cordial relations with all the seven neighbours on the adjacent and opposite coasts, namely Pakistan, Maldives, Sri Lanka, Indonesia, Thailand, Myanmar and Bangladesh and promote maritime security, trade and regional stability. Except for Pakistan, rest of the countries look up to India as a regional maritime power and has been undertaking maritime cooperation initiatives such as joint patrolling, information sharing, and capacity building, contributing to enhancing maritime domain awareness and combating common challenges such as piracy, illegal fishing, and maritime terrorism. Indian efforts to safeguard its maritime interests include maritime patrols, coastal radar networks, and naval exercises with regional and international partners. Thus, it becomes imperative for India to upgrade its maritime infrastructure with state of the art technology and systems achieving seamless coverage across the maritime domain and at the same time develop naval platforms to thwart any future threats emanating from the maritime space as a result of geopolitical complexities and contestations in the Indian Ocean Region as well as Indo Pacific Region.

Having different types borders with varied border dynamics driven by continuously evolving bilateral and international relations with the its neighbours presents a complex set of security challenges for India with area specific management solutions required for each set of border. Managing relations with such diverse neighbourhood requires a nuanced approach specially in the face of historical tensions, territorial disputes and divergent strategic interests. Each border presents its own set of security challenges and demands unique mechanisms to address these challenges, thereby straining the existing pool of resources which not only includes the security apparatus and armed forces but also diplomacy, economy and cultural domains. Of course, the regions prone to cross border infiltration and transnational threats have priority claim over the collective pool of resources and need to be addressed first. Developing critical infrastructure such as roads, bridges, communication lines and surveillance systems along these borders is crucial for enhancing border security as well as improving transnational connectivity. Effective border management requires seamless coordination amongst various stakeholders, including security forces, border guarding agencies, custom officials, intelligence agencies and local administration which calls for a secure robust communication system to ensure timely and simultaneous flow of information to all these agencies. Ensuring seamless flow of information is essential for preventing cross border crime, infiltration, facilitating legitimate trade and travel as well as addressing humanitarian issues. Alongside, continuous dialogue and diplomacy to address border disputes and promote mutual trust and regional cooperation, it is imperative to have sound and impregnable borders wherever required. Where no physical intervention is required, discreet surveillance should be of the highest standards so that security doesn't get compromised in the bargain of ensuring diplomatic niceties.

Managing active borders with multiple neighbours inflicts substantial demands on the Indian armed forces and security apparatus. The resources allocation for securing vast borders strains military budgets and logistic capabilities, diverting attention as well as funds from other critical areas such as modernisation and training. Operational challenges including adapting tactics as well as equipment to varied terrain and threat perceptions further increases the complexity and challenges for military units deployed in these hostile conditions. Continuous deployments in challenging border environments with extreme climatic conditions impact personnel health and morale but also the equipment life and reliability. In the diplomatic arena sustained engagement and negotiations consume significant resources and will. Therefore, balancing competing priorities along different borders involve strategic prioritisation and patience by defence and diplomatic planners, often leading to difficult trade-off and strategic dilemmas. It need not be overemphasized that persistent border tensions attract international attention and scrutiny placing diplomatic pressures on India to exercise restraint and pursue peaceful settlements, making it vulnerable to cognitive warfare by adversary nations.

DISRUPTIVE TECHNOLOGIES RELEVANT TO BORDER MANAGEMENT

In recent years, the rapid advancement of technology has transformed various aspects of national security, including the way nations assert territorial claims and manage their borders. This transformation is particularly evident in the realm of border control and border management, where emerging technologies are being rapidly absorbed and increasingly playing a significant role. Unstable borders quite often have broader implications for overall regional stability, affecting trade, investment and cooperation among neighbouring countries, further highlighting the importance of effective border management through round the clock fail safe surveillance, swift reaction to inimical actions and surgical neutralisation of threats. As discussed earlier India, with its complex border dynamics and longstanding conflicts with neighbouring countries, provides a compelling case study for examining the intersection of emerging technologies and border management. This chapter explores the possibility of disruptive innovations in shaping the dynamics of border management by various countries who are spearheading the absorption of the latest technologies into their border security strategies.

Robotics

Robotics is another technology which is transforming the defence sector both in active operations as well as logistics and administration. Simple robots are being deployed for supplementing labour oriented tasks whereas advanced robots are being developed with skills, allowing them to perform tasks that are routine but dangerous and don't require human reasoning, such as bomb disposal, operating in contaminated areas and recce through mined / booby trapped urban settlements. One such programme is the EU-funded ROBORDER, which aims at creating an AI-powered autonomous border surveillance system with unmanned mobile robots in the air, water and ground, capable of operating independently and in swarms. These robots will be fitted with optical, infrared, and thermal cameras, as well as radar and radio frequency sensors to detect criminal activity along the sea and coasts. In this system, cellular phone frequencies are used to triangulate the location of suspects, with cameras used to identify humans, guns, vehicles and other objects.[2]

In the next few years, organisations responsible for border guarding and control will be able to field a semi-autonomous border surveillance system that employs a heterogeneous robotic mix comprising of swarms of aerial, water surface, underwater and ground vehicles integrated seamlessly into a network. Some developers believe that beyond this, there is the opportunity to enhance these types of robotic systems with detection capabilities for early identification of security threats in border and coastal areas, along with marine disasters. The development of such capabilities for border security and law enforcement is likely to go hand in hand with military R&D and the development of improved command and control architectures that allow operators to control or supervise a multitude of unmanned vehicles. The deployment of AI and integrated robotic systems requires robust communication networks and network infrastructure. This capability will require the development of new technical solutions and a substantial investment in the hardware and infrastructure to operate and maintain these robotic and automated.

Once widely accepted and fully operationalised this technology will definitely save on the human cost required to keep the borders safe and secure. However, on the other hand, achieving satisfactory levels of autonomy requires investment in more expensive and advanced robotic systems, though the costs associated with such systems is likely to decrease in the near future due to large scale production. Such systems will require advanced data fusion capabilities and the ability to process multiple streams of sensor data from the individual platforms working in the grid. Extensive testing of the system will be required in different operational environments, with access to relevant contextual and environmental information needed for such testing to ensure the robustness of the system for a dynamic battlefield.

Space Technology

Effective border management is a critical concern for countries such as India which has large disputed borders as it involves ensuring security and integrity of national boundaries while facilitating legitimate trans-border activates. Space technology has emerged as a powerful tool in this domain, offering range of capabilities that can enhance border guarding, border control as well as military operations along contested borders. One of the vital contributions of space technology has been in the realm of surveillance and monitoring. This has been particularly useful in remote or inaccessible areas where traditional ground based platforms either cannot be deployed due to terrain configuration or fail to produce optimum results due to inclement conditions and proper logistic support.

Also, the increasing integration of satellite data with other emerging technologies such as unmanned aerial vehicles and tethered platforms can create a comprehensive and integrated border surveillance system that enhances situational awareness and also facilitates effective decision making by military hierarchy.[3] Satellite imaging has become an indispensable tool for border management, allowing nations to monitor border regions, track troop movements, and assert territorial claims. India's use of satellite imagery in border disputes with countries like China and Pakistan has facilitated accurate mapping of disputed territories and provided valuable intelligence for strategic decision-making. In the future, human analysts will increasingly be supported by AI and over time, the intent is that these AI models will be able to operate autonomously analysing satellite imagery for automated target detection and object recognition.

Deep learning methods will also reduce the need for extensive manual training of algorithms, which will speed up automated object recognition as AI models will be capable of learning on their own to identify characteristics of new objects, areas or targets.[4] The utilization of geospatial data and information emerges as a pivotal tool for defence personnel on the ground. By providing relevant insights, geospatial technology enables better deployment of resources such as patrols, posts, and roadblocks, bolstering intelligence capabilities in border management.[5] However, sometimes the quality of data being used for geospatial analysis is a potential barrier, as high resolution of imagery is key for the effectiveness of analysis. While there has been ever increasing demand for high-resolution data, progress in this area is limited due to data protection requirements and regulatory barriers. From a technical perspective, to access the appropriate geospatial data, relevant infrastructure will need to be put in place that allows end users to make use of GIS software, the Internet and mobile phone networks to automate the collection and sharing of relevant geospatial data.

Satellite based positioning and navigation systems such as the GPS, GLONASS, NavIC and Beidou have been in use for decades now and have enabled movement as well as tracking of personnel, vehicles and equipment not only for the military but larger humanity. These systems such as the Indian Regional Navigation Satellite System (IRNSS) provides precise location tracking, essential for executing complex military manoeuvres and guiding missile strikes with high accuracy whenever required to deter border incursions.[6] The proliferation of military satellites over the years have enabled space based communication systems to support secure and reliable data sharing between security forces allowing for swift response mechanisms along the borders and better situational awareness along the decision pyramid.

Satellites equipped with high resolution imaging and synthetic aperture radar (SAR) have been Instrumental in providing real time surveillance enabling military forces to monitor border regions effectively regardless of weather conditions and terrain. These systems help detect troop movements, infrastructure changes and potential infiltrations particularly in hard to reach forested and mountainous areas. Radar satellites like India's RISAT offer all weather day and night monitoring, significantly improving the ability to guard disputed borders.[7] The comprehensive intelligence picture available to the military planners through space based assets allows them to develop more informed strategies and mobilise forces to dynamically emerging situations along disputed borders.

Space based systems have been the mainstay for secure military communication systems and allowed for communication not only between military formations but between security platforms and networks too. Satellites like GSAT enable secure, encrypted communications in the battlefield where terrestrial networks may have been compromised. This enhances seamless communication between units and formations as well as ensures interoperability between services during joint operations. Moreover, the maritime domain benefits from space-based monitoring systems, which track unauthorized vessel movements and ensure security along coastal borders.[8]

As security beyond visual range is critical aspect around sensitive inaccessible border areas, deployment of space based assets is critical to monitoring border fencing and ensuring integrity of access mechanisms and promptly address any breaches. The use of artificial intelligence has now enabled us to study space-based data and predict the type, frequency and intensity of threats along the borders. In addition, space technology is equally critical to electronic warfare and cybersecurity as satellites support electronic warfare systems that detect and jam enemy communications and provide real-time cybersecurity monitoring to protect military networks.

India's first Electromagnetic Intelligence Gathering Satellite (EMISAT), developed by ISRO, was launched in April 2019 through a Polar Satellite Launch Vehicle (PSLV) from SDSC. The satellite is equipped with DRDO's Electronic Intelligence (ELINT) package *Kautilya*, allowing it to carry out scanning of the electromagnetic spectrum. The capabilities include, but are not limited to, space-based Electronic Intelligence (ELINT) for enhancing situational awareness of the Indian Armed Forces and carrying out direction-finding/location of ground-based radar systems of potential adversary countries.[9]

Beyond the purely military applications of space technology in border security, space technology has much to contribute towards the study of environmental factors that impact border security, specially borders along neighbours where natural calamities and environmental factors directly affect border alignments as well as population displacements. Remote sensing can provide useful data on weather conditions, water levels and condition of vegetation, thereby enabling early warning and prevention mechanisms for impending floods, storms and wildfires which may threaten integrity of national borders.[10]

Blockchain Technology

Blockchain technology given its distributed nature and integrity, holds promise for enhancing transparency, efficiency, and trust in border governance systems. Countries can explore blockchain-based solutions for border management, immigration mechanisms, customs clearance and land registry to mitigate disputes and ensure integrity and accountability of agreements. Blockchain technology offers significant potential in the management of disputed borders by providing a decentralized, transparent, and tamper-proof platform.

A 2020 proof of concept by US department of Customs and Border Protection, focused on the use of blockchain technology to protect intellectual property rights on American imports by giving Customs and Border Protection a secure way to exchange data with manufacturers, retailers, rights holders and importers, ultimately reducing the number of physical examinations.[11] On 08 May 2024, Representative Nancy Mace introduced the "Border Security and Blockchain Technology Act," a pivotal legislation that will harness blockchain technology for border security operations, by mandating the Secretary of Homeland Security to implement a public blockchain-based system that enhances data security and operational efficiency across US borders. This system allows for ensuring the immutability and integrity of sensitive data including biometric data, visa information, and customs documentation and enables real time data input by all relevant government agencies for accurate and timely decision-making.[12]

Blockchain can also be used to develop a passenger identification system for immigration control. The immigration system based on the Hyperledger Fabric will automatically finds the corresponding arrival and departure record into or out of the country and validate entry/exit. Hyperledger Fabric based system is aimed at ensuring a secure, decentralized, immutable seamless border control system to which allows for exact recordkeeping of movement across national boundaries. This system also brings into sync every other port of entry into a unified decentralized system. It also aims to create separate data-paths for international transmission of departure records. The system will also include methods to securely store biometric information to validate/verify passengers automatically.[13] Furthermore, a list of blacklisted or flagged travel documents can be stored and maintained in a smart contract (one of the features of Blockchain technology). This list can be updated regularly and automatically through AI enabled software based systems. Any incremental change made to this list will immediately be visible to all law enforcement agencies and border control points, thus enabling immediate control over the movement of a suspected traveller.[14]

One of its primary applications lies in maintaining secure and immutable records, such as historical border agreements, treaties and land titles. Once the records have been finalised, mutually agreed upon, digitally signed and stored on the blockchain, they cannot be modified unilaterally without informing all the stakeholders, thereby mitigating threats of loss or manipulation. Also, blockchain technology facilitates real-time data sharing between users/countries, enabling all parties involved in the resolution process of a particular border dispute to access the same unaltered data, such as satellite imagery, past documents or maps. This shared access fosters trust, as each side has equal visibility and assurance that the data has not been tampered with.

Smart contracts, can be written to automate and enforce agreements between contesting parties/countries. For example, India can write smart contract based agreements with Pakistan or China giving the terms of a ceasefire or forward line of troop deployment agreement, which would automatically trigger predefined actions if any violations occur, minimizing the need for constant negotiation or third-party enforcement. It would force the nations to stand by their agreement as any default or manipulation would be transparently seen by the parties and can be made public to prove commitment to peace. Furthermore, blockchain can be integrated with IoT devices like surveillance platforms and drones to monitor border areas continuously, with real-time data being securely recorded on the blockchain. This would provide indisputable evidence of activities along the border, reducing accusations of unauthorised incursions.

Blockchain also offers the possibility of decentralised decision-making through transparent representation and feedback systems, allowing all stakeholders, including governments and local groups, to participate in resolving disputes fairly, especially in areas which have been manipulated by proxy actors across national boundaries. This curbs political manipulation and mitigates corruption by ensuring that decisions are transparent and cannot be tampered with. In situations where population along border areas is affected by natural disasters or humanitarian crises, blockchain can promote collaborative efforts between adversary nations without the need for direct interaction.

For example, during massive floods that marred areas of Pakistan Occupied Kashmir and Jammu and Kashmir, information about displaced populations, relief supply chains, and emergency operations could be shared on a blockchain, ensuring transparency while maintaining the necessary distance between adversary governments. This helps in maintaining humanitarian access without compromising security or escalating tensions. However, challenges remain in harnessing the dividends of blockchain technology in border management, due the lack of adequate technological infrastructure, challenges to widespread adoption, and international legal frameworks. However, despite these hurdles, blockchain's potential to enhance trust, transparency, and security makes it a promising tool in the complex world of managing disputed borders.

Analysing Technological Opportunities and Obstacles in the Indian Landscape

Opportunities

As far as border security and border control goes, disruptive technologies offer the forces a plethora of opportunities to enhance their capabilities and effectiveness. The most prominent of these applications are in surveillance and reconnaissance. Unmanned Aerial Vehicles, drones, and satellite imagery has revolutionised ISR through almost real-time monitoring and situational awareness along the borders, boosting the capability of our forces to detect and respond to intrusions swiftly. These technologies not only have expanded the scope of surveillance but have also made the whole process more dynamic, responsive and connected. Furthermore, latest communication technologies aided by satellite based systems are playing a pivotal role in ensuring seamless connectivity between troops and their command centres, thereby decreasing the sense of isolation in difficult terrain.

High-bandwidth networks and satellite communication systems have ensured rapid dissemination of information and coordination, even in remote and challenging deployment areas where traditional communication networks may be unreliable and prone to compromise by the adversary. Advancements in data analytics and AI have empowered the border guarding agencies with predictive analysis capabilities, and now by analysing vast amounts of data from various platforms, AI algorithms can identify patterns and predict potential threats, allowing pre-emptive actions to prevent escalations in border tensions. Additionally, the deployment of remote weapon systems integrated with state of the art sensor networks and smart fences, have bolstered the overall security infrastructure and minimised risks to own troops. These opportunities underscore the potential of disruptive technologies in revolutionising border security and empowering our defence forces with unparalleled capabilities in safeguarding the nation's territorial integrity.

Indian scientists have reportedly been working on all-terrain AI-enabled robots that might be used to patrol international border of the country. The Bengaluru-based Central Research Laboratory of defence PSU Bharat Electronics Limited (BEL) is working on developing a prototype for such a robot by December of 2024. The BEL robot will be equipped with sensors and programmed to communicate with the control centre. The primary objective of the border patrol robots will be to save the lives of security personnel deployed for border surveillance. The estimated cost of these robots will be in the range of Rs 70-80 lakh for small orders. As of now, 80 scientists and engineers in three AI-specific labs – CRLs in Bengaluru and Ghaziabad, and BEL Software Technology Centre (BSTC) in Bengaluru – are working on the AI-enabled patrol robots.[15]

Challenges

India recognizes the challenges in harnessing these advancements, that too despite the appealing opportunities presented by disruptive technologies for better management of border disputes. In remote border areas, the limitations of adequate state of the art technological infrastructure including OEM support and maintenance facilities are one of the foremost obstacles further aggravated by inconsistent power supply and poor network connectivity. Due to the harsh and remote nature of our borders, deploying and maintaining complex equipment is a logistics nightmare, which may result in compromised quality or reliability of border surveillance systems. In addition, on boarding and fielding new technologies demands a heavy cost premium on the defence budget and requires thoughtful prioritisation of resources. Given the recruitment pool of our defence forces at large, the training of specialised personnel to handle hitech equipment and systems takes time and money both as well as complex systems may invite a certain degree of push back from personnel who are not familiar with the new technology being introduced.

Sometimes the procurement cycle of such systems is so protracted that by the time a platform is fielded, the technology itself has become outdated. At the same time, cybersecurity is a growing concern, as adversaries attempt to leverage vulnerabilities inherent in networked systems to impede operations or siphon sensitive data. Some technological capabilities could be especially difficult to deploy in disputed areas over the next few decades though, for regulatory and policy reasons – these obstructions include outdated frameworks and international agreements that would need to be navigated while balancing legal and ethical concerns. Questions of accountability, civilian deaths and the right to privacy are further muddied by calls for greater legislation on the use of autonomous weapons and surveillance systems. These challenges highlight the difficulties of applying emerging technologies to border security and thus underpin the importance of developing comprehensive response strategies.

Leap from Border Management to Warfighting

Disruptive technologies such as AI, space technology, robotics, EW and cyber are increasingly becoming integral to both border management as well as warfighting between adversary nations. These technologies offer significant advantages by enhancing surveillance, decision-making, and combat effectiveness, creating an overlap in their application across both domains. However, while there are commonalities in how these technologies are used, their deployment in an all-out war differs in scale, intensity and objectives.

One of the primary overlapping factors in peacetime employment along borders and during an all-out war is the use of disruptive technologies for surveillance and reconnaissance. AI-powered drones and advanced sensors are employed in border management roles to monitor and secure borders, detect illegal crossings, and gather intelligence on adversary movements. This continuous surveillance prevents border violations and ensures border integrity. However, during active hostilities, these same technologies are utilized for battlefield reconnaissance, providing real-time intelligence and enabling successful kinetic operations. The fundamental purpose i.e. gathering and processing information to gain an advantage over the adversary remains consistent across both scenarios.

Another area of similar employment is in decision-making and command control. AI and big data analytics play a crucial role in both border management and warfighting as they are leveraged to process vast amounts of data to predict threats, optimise responses and enhance well informed decision-making. In the context of border management, these tools help in making informed decisions to prevent conflicts and manage security. In war, however, the decision-making process is far more rapid and critical, often determining the outcome of battles. Secure communication networks are pivotal in both the scenarios, ensuring that command and control structures function effectively. While in border management, these networks coordinate activities between military and law enforcement agencies, during warfighting, they ensure operational coherence under the stress of combat.

While considering the role these technologies play in an all-out war, it come out clearly and emphatically that the scale and intensity of their use in a war is much greater as they get leveraged for kinetic operations. The offensive employment of disruptive technologies manifests in the employment of armed drones, AI enabled weapon platforms, autonomous systems and coordinated cyber-attacks to degrade and destroy enemy systems, as disruptive innovations get deeply integrated with traditional kinetic operations. Cyber and EW operations are often synchronised with physical attacks to cause attrition to enemy capabilities across all domains, whereas, such level of integration may not be typically carried out in border security operations aimed at maintaining stability and preventing escalation.

These emerging technologies not only enhance the effectiveness of combined arms operations across land, sea, air but also across the borderless domains of space, cyberspace and info space. The offensive posture and outcomes of these technologies is a trans dental departure from their defensive deterrence focussed employment for meeting border management objectives. Lastly, the rules of engagement in war allow for a higher tolerance to collateral damage, reflecting the different objectives and ethical considerations compared to border management, where minimising harm and adhering to international norms are paramount.

Therefore, as the nature of warfare evolves by the day, the Indian Army also must adapt by integrating the emerging disruptive technologies that are set to transform the future battlefield. My research largely focusses on providing broad recommendations for the effective absorption and application of such technologies into the warfighting machinery and doctrine of any Army which wants to fight today's war with future capabilities. Accordingly, the chapters that follow will delve into the research development and in-service exploitation of these technologies by the major combat arms, specifically, Infantry, Armoured, Artillery, Aviation, Air Defence and Space. Each of these combat arms plays a distinct and critical role in any army's overall operational effectiveness. In no manner, it means that these emerging disruptive technologies have no role in other arms and services such as Engineers, Air Defence, Signals and those involved in the logistics of war fighting. In fact, some of the most transformational innovations have manifested in these domains, however, including all of them in a single research would have expanded the scope of the research beyond practical limits and diffused the focus from the most relevant aspects based on current realities and strategic significance.

Although Space does not fall traditionally under the direct arena of the Army, and for all military purposes remains the domain of the Air Force, not only in India but in most of the advanced militaries of the world, the increasing militarisation of space and the advent of space-based and anti-space weapons have significant implications for land-based operations. Future conflicts will likely see space as a contested domain, regardless of the nature of specific operations, where control of satellite communications, navigation, and reconnaissance will directly impact the effectiveness of ground forces. Therefore, it becomes prudent to devote adequate time on how the emerging disruptive technologies are transforming the space above the future battlefield and how the integration of space-based capabilities and the development of counter-space strategies will therefore be essential to maintaining operational superiority on the battlefield.

The succeeding chapters will explore these themes in detail, examining how each combat arm can harness the potential of disruptive technologies. The aim is to offer valuable insights that may assist the Indian Army on its path to technological transformation, ensuring that it remains prepared to meet the challenges of future warfare with advanced, technology-driven solutions.

Notes

1   Michael Ignatieff, Blood and Belonging: Journeys into the New Nationalism
2   ROBORDER. Autonomous Swarm of Heterogeneous Robots For Border Surveillance https://cordis.europa.eu/project/id/740593
3   Fatemeh Farajzadeh, Andrew C. Trapp. Optimization for Secure and Humane Border Operations, 28 Jan 2022, https://arxiv.org/abs/2201.10341v3
4   Lockheed Martin automates satellite image analysis. 7 June 2019, https://www.aerospace-technology.com/news/lockheed-martin-satellite-image-analysis/
5   Swetha Rao Harnessing Geospatial Technologies For Border Security 28 May 2024 Indo Pacific Geo Intelligence Forum https://geointelligence.net/blogs/blog-detail.php?id=329&title=harnessing-geospatial-technologies-for-border-security

6   Dr. Ranjit Singh, FIETE Department of Electronics Communication Engineering, Ajay Kumar Garg Engineering College ISRO's NAVIC: Navigation Support to Civil and Military Segments https://www.akgec.ac.in/wp-content/uploads/2023/01/6.1.23_10-Dr_Ranjit_Singh_NAVIC.pdf

7   Brig Arvind Dhananjayan India's Military Satellites: The Armed Forces' Bulwarks For Communication & Surveillance! 17 Apr 2023 https://chanakyaforum.com/indias-military-satellites-the-armed-forces-bulwarks-for-communication-surveillance/

8   *Staff Correspondent* India's Military Communication Satellites: The Backbone of Modern Warfare, May 14, 2023 https://www.iadb.in/2023/05/14/indias-military-communication-satellites-the-backbone-of-modern-warfare/

9   Brig Arvind Dhananjayan India's Military Satellites: The Armed Forces' Bulwarks For Communication & Surveillance! 17 Apr 2023 https://chanakyaforum.com/indias-military-satellites-the-armed-forces-bulwarks-for-communication-surveillance/

10   How Can Space Technology Help With Border Security? New Space Economy https://newspaceeconomy.ca/2023/10/23/how-can-space-technology-help-with-border-security/

11   Natalie Alms, Staff Reporter, Mace sponsors bill to push CBP to use blockchain at the border, May 7, 2024, https://www.nextgov.com/emerging-tech/2024/05/mace-sponsors-bill-push-cbp-use-blockchain-border/396387/

12   *Press Release,* Representative Mace Introduces the "Border Security and Blockchain Technology Act" to Leverage Blockchain in Enhancing Border Security, May 8, 2024, https://mace.house.gov/media/press-releases/representative-mace-introduces-border-security-and-blockchain-technology-act#:~:text=Enhanced%20Data%20Integrity%3A%20Utilizing%20blockchain, accurate%20and %20timely%20decision%2Dmaking.

13   Dhiren Patel, Balakarthikeyan, Veermata Jijabai Technological Institute, Mumbai and Vasu Mistry National Institute of Technology, Surat Border Control and Immigration on Blockchain https://link.springer.com/chapter/10.1007/978-3-319-94478-4_12

14   Arvind Deendayalan, Blockchain is the next big step in automated border control systems, April 17, 2023, https://www.financialexpress.com/business/blockchain/blockchain-is-the-next-big-step-in-automated-border-control-systems/3048806/

15   AI robots to patrol India borders soon, prototype to come in December https://www.businesstoday.in/technology/news/story/ai-robots-to-patrol-india-borders-prototype-to-come-in-december-194617-2019-05-02

9

Infantry: Forward March into the Tech Era

"Wars may be fought with weapons, but they are won by men. It is the spirit of the men who follow and the man who leads that gains the victory."

—General George S. Patton

As the modernisation of Indian armed forces progresses in sync with the rapid evolution of disruptive technologies and their absorption into our forces remain at course, these technologies if leveraged effectively will transform the landscape of the Indian Army's combat arms, with particularly significant implications for the Infantry, Armoured Corps & Mechanised Infantry, and artillery. As these arms form the backbone of our conventional warfighting capability, the infusion of emerging disruptive technologies such as AI, robotics, EW, Cyber warfare, Nano technology and Space based systems into their arsenal is not merely a costly technology demonstration but an inescapable necessity for maintaining battlefield superiority. This battlefield superiority is the fountain head from which the much talked about 'strategic autonomy' flows.

The Infantry, due to territory being at the centre of almost all conflicts, has been always at the forefront of direct engagement. Over the years' lot of scientific pursuits have been undertaken to bring down the human cost of war, which has ushered in the technological transformation enhancing operational effectiveness of the Infantry. Infantry has graduated from it 'Man behind the Machine' dictum to 'Soldier as a System'. Soldier systems are being developed, equipped with endurance suits, advanced connected communications, AI enabled decision aids and augmented reality consoles empowering the infantryman with enhanced situational awareness and real time intelligence. Wearable robotics and exoskeletons have been introduced, improving the soldier's physical capacities and allowing him to carry heavier loads, sustain higher level of operational tempo and remain combat effective for protracted periods through intense battles. These advancements will allow infantry units to operate effectively in diverse and challenging terrains, from urban environments to remote and rugged landscapes, ensuring they remain a decisive force on the battlefield.

The advent of disruptive technologies is poised to revolutionise the traditional concepts of armoured warfare. The integration of autonomous platforms, advanced sensors, AI-driven targeting systems and nano materials into armoured vehicles design and weapons profile will significantly enhance their situational awareness, precision and survivability on the battlefield. Active protection systems (APS) that can detect and neutralise incoming projectiles in real-time will provide a crucial layer of protection, while the development of unmanned ground vehicles (UGVs) could change the dynamics of mechanised warfare by allowing for more flexible and risk-tolerant manoeuvres. The combination of these technologies will enable armoured formations to operate with greater speed and decisiveness, particularly in complex, contested and contaminated battle environments.

As for Artillery, the integration of disruptive technologies is crucial in enhancing its role as a guarantor of precision fire support. Advances in networked targeting capabilities, automated fire control systems, and precision-guided munitions (PGMs) aided by and superior ISR from unmanned, autonomous and space based systems, will allow artillery units to deliver highly accurate and timely firepower with minimum collateral damage. The induction of more and more long-range, high-precision artillery systems, coupled with real-time intelligence and surveillance data, will enable the Artillery batteries to conduct deep strikes with pinpoint accuracy, thereby shaping the battlefield and disrupting enemy operations from afar. Additionally, automated resupply and logistics systems will ensure that artillery units can sustain high rates of fire over extended periods, maintaining their effectiveness during protracted engagements.

As this chapter explores the impact of these disruptive technologies on the Infantry, laying the foundation for subsequent chapters emphasising similar impact Armoured Forces, Artillery, Aviation and space, it will highlight the ways in which these innovations are reshaping their operational doctrines and force structures. The integration of these technologies will not only enhance the effectiveness of these combat arms but also pose challenges that will need to be addressed to fully realise their potential. This chapter therefore, delves into how these technologies will reshape the way the Infantry could operate in the future, with operational effectiveness and present both opportunities and challenges lying ahead.

Since the dawn of organised Infantry warfare, Infantry has been the backbone of military pursuits. From times immemorial, the infantry units have been the primary force responsible for capturing and holding territory, killing the enemy infantry in close combat and doing everything on the frontline that was essential to military victory. Therefore, the evolution of infantry across the generations of warfare, truly reflects the changes in military technology, tactics and the nature of warfare itself. Historically, soldiers were equipped with basic weapons such as spears, swords, and later, came the firearms. The advent of gunpowder and the development of muskets and rifles significantly transformed infantry tactics, leading to the dominance of infantry in the linear tactics of the 18th and 19th centuries. The two World Wars saw further advancements, with the introduction of automatic weapons, mortars, and grenades, which enhanced the firepower and flexibility of infantry units. Whether with a weapon or bare hands, Infantry has been the custodian of the fighting spirit of armed forces, since every other arm has been a modification of this universal prototype of human courage, will and endeavour.

In the next decade or so, the Infantry, traditionally known for its reliance on manpower and basic armaments, will be transformed by a suite of emerging technologies. These technologies will not only enhance the capabilities of individual soldiers but also redefine the very nature of infantry warfare. This chapter examines how advancements in Artificial Intelligence and Machine Learning, robotics, autonomous systems, drones, electronic warfare, directed energy weapons, cyber warfare, the Internet of Things, nanotechnology, and biotechnology will shape the future infantry. Examples of cutting-edge systems and platforms under development, developed or fielded by various countries have been discussed to amplify the future trends.

Wearable Technology and Exoskeletons

According to a technical information paper by the US Defense Centers for Public Health, spine and back injuries accounted for 28.3% of all noncombat wounds among soldiers in the US Army in 2021. The main reason for such casualties among the soldiers was identified to be a result of repetitive or prolonged musculoskeletal loading, such as heavy lifting and other strenuous jobs performed by soldiers.[1] Imagine soldiers from Infantry who have to carry heavy loads in the form of personal gear, weapon and equipment in tough terrain for prolonged times. Not only in war but daily administrative routine on isolated posts sometimes is full of heavy lifting. Wearable technologies and exoskeletons hold tremendous promise in solving this issue by enhancing the physical capabilities of infantry soldiers. These technologies are designed to improve mobility, endurance, and situational awareness, thereby increasing the overall effectiveness of infantry units. In fact exoskeletons have been found to not only improve endurance but also enhance cognitive performance by reducing mental fatigue due to prolonged strenuous work. Wearable technology does not only include exoskeletons but also a range of devices that can be worn by soldiers to monitor their health, enhance communication, and improve battlefield awareness. For example, wearable sensors can track vital signs such as heart rate, body temperature, and hydration levels, providing real-time data to commanders and medics. This information can be crucial in maintaining the health and combat readiness of soldiers in demanding environments.

Exoskeletons are mechanical suits that augment a soldier's strength and endurance, enabling them to carry heavier loads over longer distances with less fatigue. Different from other robots, the operator of an exoskeletons is human who need to make decisions and perform tasks with exoskeletons.[2] China's state owned company, Norinco, which manufactures armoured vehicles and heavy ground munitions, debuted its second-generation military exoskeleton in 2018, a body brace designed which enables infantry soldiers to carry almost 100 pounds of weapons, supplies, and ammunition.[3] The Chinese soldiers PLA Xinjiang Military region were also reported to be using such suits for logistic missions in the high altitude areas of Ngari, Southwest China's Tibet Autonomous Region in the winter months of 2020.[4] A fairly advanced version of such exoskeletons is the AI enabled Lockheed Martin ONYX exoskeleton which is designed to support the lower body, providing additional strength and endurance during activities such as running, climbing, and lifting. It takes about 150 milliseconds for Onyx to respond to a soldier's movement, which is similar to the amount of time taken by human muscles to respond to signals from the brain.[5]

Advanced version of wearable technologies manifests in wearable heads-up displays (HUDs) and augmented reality (AR) goggles which provide soldiers with real-time information about their surroundings, enemy positions, and mission objectives. These devices can overlay critical data onto the soldier's field of view, allowing for faster decision-making and improved situational awareness. The US Army's Integrated Visual Augmentation System (IVAS) is an example of such technology, integrating night vision, thermal imaging, and AR into a single system.[6]

Exoskeletons can also be equipped with additional features, such as integrated power supplies and communication systems, further enhancing the soldier's capabilities. In August 2023, the US Army received the first 20 prototypes of the Integrated Visual Augmentation System 1.2 variant. IVAS is a single platform that features an all-weather fighting goggle and a mixed reality heads-up display that integrates next-generation situational awareness tools and high-resolution simulations to provide Soldiers with improved mobility and lethality, during the day or at night. IVAS provides Soldiers with a single device to fight, rehearse and train.[7] This system has an integral training tool, the Squad Immersive Virtual Trainer, that provides Soldiers with objective-based scenarios and battle drills through holographic and mixed-reality imagery, giving units the flexibility to train their soldiers with minimal resources.[8] This milestone is the latest step in the process of getting the most advanced version of the situational awareness system in the hands of soldiers. As these technologies continue to develop, they are expected to play a crucial role in future infantry operations, particularly in challenging environments such as mountainous terrain, urban areas, and dense forests.

While wearable technology and exoskeletons offer significant benefits, there are also challenges associated with their development and deployment. These include the weight and power requirements of the systems, potential impacts on soldier mobility, and the need for extensive training to operate the equipment effectively. Additionally, the high cost of these technologies may limit their widespread adoption in the short term. Also most of the exoskeletons are being developed to carry out specific tasks rather than developing capabilities which allow them to adapt to dynamics of battlefield environment. For example, adapting to a change from lifting loads to tactical movements during room intervention.[9] Looking ahead, advancements in materials science, battery technology, and AI are expected to overcome many of these challenges, leading to more lightweight, efficient, and user-friendly systems. The future of infantry warfare is likely to see greater integration of wearable technology and exoskeletons, enhancing the physical and cognitive capabilities of soldiers on the battlefield.

Advanced Weaponry and Ammunition

The development of advanced weaponry and ammunition is transforming the firepower and effectiveness of infantry soldier. Smart guns, smart sights, night vision devices, energy weapons, and other innovative armaments are providing soldiers with new tools to engage the enemy with greater precision and lethality.

Smart Guns

Smart guns are firearms equipped with advanced targeting systems, biometric sensors, and electronic control mechanisms that enhance accuracy and safety. Incorporating biometric access, these weapons can be programmed to recognise authorised users, reducing the risk of unauthorised use or accidents. Israel Weapon Industries has unveiled a computerized small arms system that increases strike probability by up to three times. The system, called Arbel, comprises a computer-based platform, an electronic trigger mechanism with sensors, a firing mode, and a removable battery. Arbel uses algorithms and motion sensors to cut out firing inefficiencies from human fatigue, breathing, and impaired motor functions. When the weapon is on Arbel mode, the system monitors shooter movement and trigger status, selecting the ideal moment of fire for greater accuracy, including on moving targets.[10] Additionally, smart guns can be integrated with heads-up displays or augmented reality systems, allowing soldiers to engage targets more effectively in complex environments.

Another example of smart gun technology is the XACT System PGF (Precision Guided Firearm) by a US company Tracking Point, which uses a combination of sensors, laser rangefinders, and ballistic computers to calculate the optimal firing solution for long-range shots.[11] The PGF automatically adjusts for many factors, including Range, Temperature, Barometric pressure, Spin drift, Wind input, Cant and Inclination.[12] This technology significantly improves the accuracy of infantry marksmen, allowing them to engage targets at extended ranges with a high probability of success.

Energy Weapons

Imagine a scenario, sniper firing from a vantage point far in depth from the Line of Control and taking out an infiltrating terrorist at the LC Fence, without being heard, without any flash or recoil and achieving this silent kill without any regards to wind speed and direct, gravity pull on the bullet, or the effect of curvature of earth. Not one of the conventional sniper rifles but only a laser sniper rifle can achieve all the above and much. Only challenge that remains is that of producing a coherent beam of lethal magnitude from a man portable Directed Energy Weapon (DEW). Energy weapons, such as infantry portable DEWs, represent a new frontier in infantry armaments. Unlike traditional firearms, energy weapons do not rely on conventional ammunition, making them highly efficient and capable of continuous operation until the power source exhausts itself.

The US military has been actively researching and developing DEWs for various applications, including counter drone defence and point protection. In September 2019, US Army researchers Adam L. Foltz and James E Burke were issued U.S. Patent 10,408,579 for their directed energy modifications to the blank firing adapter used on M4 rifles. The muzzle attachment contains a piezoelectric generator, powered by firing the blank cartridge, which creates an electromagnetic pulse directed by a horn antenna capable enough to destroy enemy electronics, such as small drones or improvised explosive devices.[13]

For infantry units, portable energy weapons could provide a powerful tool for neutralising threats such as UAVs, drones, incoming projectiles, or enemy personnel. While these technologies are still in the experimental stage and most of its operational use has been still kept under the wraps, they hold significant potential for future infantry operations especially in counterinsurgency operations where fatal force may be avoided by just disabling the target.

Advanced Ammunition

In addition to smart guns and energy weapons, advanced ammunition is also transforming infantry firepower. Precision-guided munitions (PGMs), for example, are being miniaturized for use in small arms and grenade launchers. These munitions use GPS, laser guidance, or other targeting systems to accurately engage targets, reducing the risk of collateral damage and increasing the effectiveness of infantry fire. Northrop Grumman has developed a proximity-fuzed munition that automatically senses when it is within the desired distance of a target, and detonates its warhead. Proximity-fuzed munitions are ideal for sensing and defeating moving targets, especially swarms of small unmanned aerial systems. The munition is enabled with an algorithm which allows it to differentiate between the target and the background clutter and optimist the airburst accordingly.[14]

The XM25 Counter Defilade Target Engagement (CDTE) system, developed by the US Army, is an example of advanced ammunition technology. The XM25 incorporates a laser range-finder and fires airburst projectile that can be programmed to explode at a specific distance, allowing soldiers to engage enemies hiding behind cover.[15] This capability significantly enhances the lethality of infantry units in urban and complex terrain.

The adoption of advanced weaponry and ammunition presents several challenges, including the need for specialised training, logistical support, and maintenance. Additionally, the cost of these technologies may limit their availability to elite units or specific missions. However, as research and development continue, it is expected that these technologies will become more accessible and integrated into the broader infantry arsenal. In the future, the development of more compact, efficient, and versatile weapon systems will likely enhance the capabilities of infantry units.

Communication and Battlefield Management Systems

Effective communication and battlefield management are critical to the success of infantry operations. Often Infantry soldiers are either deployed at isolated posts or fight in a small team in intense battle zones detached from their command centres. Robust communication is their only link with the larger gambit of warfighting, which keeps them abreast of their progress as a small set within the operational scenario. Their effective tactical employment, movement, rest recoup and redeployment all are functions of secure two-way communication. Infantry units rely on a variety of communication systems to maintain contact with each other and with their higher command. These systems include handheld radios, satellite communication devices, and secure data links that enable voice, data, and video communication in real-time. The development of new communication technologies is enhancing the range, reliability, and security of these systems, ensuring that infantry units can operate effectively in even the most challenging environments. Present day advanced communication systems based on satellite links and AI-driven battlefield management tools are revolutionising the way infantry units operate, enabling more efficient coordination, faster decision-making, and improved situational awareness.

One notable advancement in communication technology is the development of the Soldier Radio Waveform (SRW), a secure, software-defined radio system used by the US military. The SRW is a Joint Tactical Radio System (JTRS) networking software that aims to provide voice, data, and video capabilities to small combat units and unmanned systems. SRW is designed to operate as a mobile ad-hoc network (MANET), enabling communication through a self-configuring, infrastructure-less network of mobile nodes.[16]

The objective of BMS is, therefore, to improve the effectiveness of battlefield management by facilitating the collection, processing and dissemination of information in real time, which allows informed decision-making and the coordination of actions by offering both qualitative and quantitative advantages.[17] The new Russian Automated Command and Control System (*Avtomatizirovannyye Sistemy Upravleniya* – ASU) was first deployed at the headquarters of the 6th Combined-Arms Army (CAA) in St. Petersburg, co-located with the OSK HQ, nicknamed as the 'Star Wars HQ'. This system connects all levels of C2, from the strategic down to the tactical, and provide commanders with decision-making options in real time, reducing the time taken by decision cycle by almost three times. The ASU seamlessly function across all types of communication, wired, radio, relay and satellite, along with access to the Russian GLONASS satellite navigation system. It receives information through closed radio and satellite channels and synthesizes a single picture of the battlefield.[18] The advanced version of such ASU's are the AI-driven battlefield management systems that are transforming the way infantry units' plan, execute, and adapt to changing conditions on the battlefield. These systems use AI and machine learning algorithms to analyse vast amounts of data from sensors, drones, and other sources, providing commanders with real-time insights and recommendations.

For example, the US Army's Command Post Computing Environment (CPCE) is a unified software platform that integrates various battlefield management tools into a single system. The goal of CPCE is to eliminate stove-piped legacy systems and provide an integrated, interoperable, cyber-secure, cost-effective computing infrastructure framework for multiple warfighting functions. CPCE provides commanders with a common operational picture, allowing them to monitor the status of friendly and enemy forces, track logistics, and coordinate fire support.[19] The use of AI in these systems enables faster and more accurate decision-making, improving the overall effectiveness of infantry operations. Systems which can be tailored to specific levels of command, varied terrain and varied armed forces with different doctrinal concepts are being developed which allow for countries from a particular political or regional grouping such as NATO to provide uniform Battle Field Management experience to through common protocols and standards. BMSs are part of a broader development trend of Command, Control, Communications, Computers, and Information (C4I) capabilities for battle command in a multi-domain environment.[20]

The integration of advanced communication and battlefield management systems presents challenges related to cybersecurity, interoperability, and the complexity of managing large amounts of data. Ensuring the security of communication networks and preventing cyberattacks is a critical concern, particularly as adversaries develop their own electronic warfare capabilities.

The Infantryman

The absorption of novel technologies into infantry operations will profoundly affect the protection, endurance, mobility and lethality of the infantryman on the battlefield. These technological innovations will not just enhance the individual aspects of soldier's performance but also create synergistic improvements across multiple domains, leading to creation of a more capable and resilient infantry force. Many of these technologies have already made their way into the infantry equipment, weapons and systems thereby initiating a process of technological transformation of the Infantry.

Protection

The primary focus of advancements in protection is to enhance soldier survivability without compromising mobility on operational effectiveness. Meanwhile, advancements in materials science, specifically nanotechnology and advanced composites are enabling production of lighter and tougher body armour. This armour provides the soldiers with the highest level of protection to counter bullets, shrapnel and assure greater survivability against IEDs.

The US Army's Institute for Soldier Nanotechnologies (ISN) is developing nanomaterials that can create ultra-light, flexible body armour. Trials at the Institute for Soldier Nanotechnologies, showed that a polymer patterned in a "lattice-like" structure using nanotechnologies could withstand more force than Kevlar or steel.[21] These materials are have been designed to provide protection equivalent to the traditional armour albeit at a fraction of the weight. In the foreseeable future, soldiers will be wearing smart armour that not only provides physical protection but in collaboration with other wearable technologies shall monitor health indicators, detects chemical and biological threats, and integrate with the soldier's communication systems. This would lead to a significant increase in the soldier's survivability on the battlefield with least diminution of his combat capabilities.

Endurance

Endurance is one of the most critical facet of infantry soldier's effectiveness in battle, as he must operate for extended periods in physically demanding environments. Exoskeletons and advanced biotechnology are at the forefront of increasing soldier's endurance by augmenting his physical capabilities and reducing the strain of carrying heavy loads for protracted periods.

The US Army's Tactical Assault Light Operator Suit (TALOS), first introduced in 2013 was designed to enhance a soldier's strength and stamina. The project is still under development and the suit has not been operationally deployed as more and more technological innovations have been discovered during the development phase itself. Although its individual subsystems, such as the exoskeleton, base layer, visual augmentation system, helmet assembly, armour, power and communications continue to be "refined" in support of other independent applications.[22] The future exoskeleton would be a system of systems and the simultaneous absorption of new technologies into one system may be a never ending process.

Advances in biotechnology, such as gene editing and bio-enhanced nutrition, are being researched to enhance a soldier's endurance by improving muscle efficiency, reducing the need for rest, and accelerating recovery from injuries. Artificial skin could insulate soldiers from environmental extremes, as well as provide frontline treatment for wounds. Uniforms and coatings could contain materials that mimic vegetation to deceive known enemy sensors. Edible vaccines could provide temporary protection against pathogens in exotic locales. Sensor implants could monitor a soldier's health and dispense antidotes to chemical or biological threats, both natural and unnatural.[23] The progress in this field seems to be visibly slow not because of lack of R&D but largely because of ethical and legal constraints on trials, prototypes and operationalising such innovations. Ultimately, Biotechnology could lead to soldiers who require less rest, recover more quickly from injuries, and can maintain peak physical performance for extended durations.

Mobility

Mobility is crucial for infantry soldiers to manoeuvre across diverse terrains and quickly respond to evolving battlefield conditions. Technologies such as advanced materials, lightweight equipment, and autonomous systems are enhancing soldier mobility.

The development of lighter body armour using nanotechnology and advanced composites allows soldiers to move more freely and at greater speeds without sacrificing protection. This is especially important in urban and mountainous warfare, where agility can be the difference between life and death.

Robotic mules like the US Army's Squad Multipurpose Equipment Transport (SMET) are being developed to carry heavy loads, such as ammunition, water, and supplies, thus unburdening soldiers and allowing them to move faster and farther.[24] China too fielded its latest Robotic Dogs fitted with an automatic rifle during the 15-day Golden Dragon 2024 exercise with Cambodia. Two versions of the robot-dog were showcased – one that can fire at the enemy and a lighter dog that can be used in identifying targets. Both have a 4D wide-angle perception system, and just like real-life canines, can move forward and backward, jump, and lie down.[25] India too has inducted similar robotic MULES (Multi-Utility Legged Equipment) for infantry operations in high altitude and inaccessible areas. 100 of them have been contracted for a sum of approximately Rs 300 Crores.[26] With the integration of autonomous systems, soldiers will likely have access to a range of robotic support vehicles that can carry supplies, provide reconnaissance, and even offer medical evacuation, enhancing their operational range and mobility in contested environments.

Lethality

Advanced weapon systems, precision-guided munitions, and networked targeting systems are making a sizeable impact on the lethality of infantry soldiers. These tools allow troops to hit targets from farther away, and with deadlier results. The US Army's Next Generation Squad Weapon (NGSW) program aims to create new rifles and machine guns with high-tech smart sights, laser rangefinders, and ballistic calculators.[27] These smart guns help soldiers shoot accurately in spite of the battle fatigue and environmental conditions. Systems like the Integrated Visual Augmentation System (IVAS) give soldiers AR displays with live targeting info helping them to spot, identify and engage threats faster than they usually have done if unaided by the system. IVAS is a single platform that will improve Soldier sensing, decision making, target acquisition, and target engagement. These capabilities will provide the increased lethality, mobility, and situational awareness necessary to achieve overmatch against our current and future adversaries.[28]

In the years to come, infantry soldiers will carry weapons linked to a wider network of sensors and AI-driven targeting systems allowing them to strike precisely with minimum collateral damage. The use of DEWs and other cutting-edge munitions could also offer new unconventional ways to neutralise enemy forces further boosting infantry soldier's lethality.

The impact of emerging technologies on the protection, endurance, mobility, and lethality of infantry soldiers is profound and multifaceted. These technologies are not only enhancing individual aspects of soldier performance but are also creating an integrated combat system where each technological advancement complements and amplifies the others. As these technologies continue to evolve, the infantry soldier of the future will be more protected, enduring, mobile, and lethal, capable of operating effectively in a wide range of environments and combat scenarios. This transformation will redefine the role of infantry in modern warfare, making them a more versatile and formidable force on the battlefield.

The Queen

In modern warfare, the role of infantry continues to evolve in response to new threats and technological advancements. Today's infantry units are not only equipped with advanced weapons but also supported by sophisticated communication systems, intelligence assets, and precision-guided munitions. The nature of warfare has also shifted, with a greater emphasis on urban combat, counterinsurgency, and asymmetric warfare, where infantry plays a crucial role in complex environments

Robotics and Autonomous Systems

Robotics and autonomous systems are set to revolutionize infantry operations by providing support in logistics, reconnaissance, and direct combat. These systems reduce the physical and cognitive burden on soldiers, allowing them to focus on more complex tasks. Autonomous vehicles can perform tasks such as resupply, casualty evacuation, and even combat roles without exposing human soldiers to risk.

Russia's Uran-9 is a remote-controlled combat vehicle equipped with anti-tank missiles, machine guns, and a 30mm automatic cannon. It's designed to support infantry units by engaging armoured targets and fortified positions. However, like its heavier siblings it has not been widely used in the ongoing Russia-Ukraine War, primarily due to vulnerability of the larger heavier models to tank and artillery fire. There has been a slight deviation in the employment philosophy of robotic systems, as the war has got prolonged, the Russian Army has moved from fielding few and expensive systems towards cheap and plentiful with a staggering number of conventional drones, FPVs and UGVs.[29] These vehicles are relatively small in size and perform mine delivery and 'kamikaze' bombing roles, along with supply and logistics operations. Such UGVs include the Yozhik-R (Hedgehog-R) wheeled UGV supposedly in use in Ukraine, which can carry a 5 kg payload and conduct reconnaissance.[30] Other Yozhik variants in development include a version that can carry two TM-62 anti-tank mines weighing 10 kg each, and a vehicle equipped with a miniature six-round multiple rocket launcher with high explosive anti-tank multi-purpose (HEAT-MP) munitions capable of penetrating heavier armour.[31]

The US Army's Robotic Combat Vehicle (RCV) part of the Next-Generation Combat Vehicle program, is being developed in various sizes (light, medium, heavy) to support infantry with tasks ranging from reconnaissance to direct fire support. The US Army plans to employ RCVs as Scouts and Escorts for manned fighting vehicles to deter ambushes and to guard the flanks of mechanized formations. As of now RCVs are intended to be controlled by operators riding in Next Generation Combat Vehicle (NGCV), but in future improved ground navigation technology and AI might eventually permit a single operator to control multiple RCVs or for RCVs to operate in a more autonomous mode.[32]

It is pretty obvious that at this pace of development and multiple conflicts raging round the globe, in the coming years, autonomous systems will likely become integral to infantry units, performing dangerous and labour-intensive tasks, thereby reducing casualties and increasing operational efficiency. The integration of AI into these systems will enable them to operate independently in complex environments, further enhancing their utility.

Directed Energy Weapons

Directed Energy Weapons, such as lasers and microwave weapons, are emerging as powerful tools for infantry units. DEWs can disable electronics, neutralise drones, and incapacitate enemy personnel without using traditional munitions.

The US Army's Multi-Mission High Energy Laser (MMHEL) is a mobile platform with a solid state laser system, agile beam control system, and supporting laser subsystems, integrated into a combat platform. It aims is to provide a low cost-per-engagement Manoeuvre – Short Range Air Defense (M-SHORAD) prototype system. M-SHORAD protects manoeuvring forces from rocket, artillery and mortar; unmanned aerial system, and fixed and rotary-wing manned aircraft and that too at an average cost per kill of approximately $30.[33] These systems are being miniaturized for potential integration into infantry squads or light vehicles. On similar lines, Russia's Peresvet Laser System is designed to blind or disable enemy satellites and reconnaissance assets, it could also be adapted for ground use to protect infantry units from UAVs and other threats. However, its effectiveness directly depends on environmental conditions: in good weather, it works perfectly, but fog, rain, snow and other adverse weather events can interfere with the passage of the laser beam.[34]

As allied technologies advance and DEWs become more compact and efficient, they will be integrated into infantry units, providing novel means of engaging enemy forces and defending against UAVs and electronic threats. The non-lethal capabilities of DEWs also offer new options for crowd control and non-lethal engagements in urban environments.

Nanotechnology

Nanotechnology has the potential to revolutionise materials, medical treatment and even weapon systems for infantry. Nanomaterials can be used to create lighter, stronger armour, while nano-medicine could provide rapid healing and protection against chemical and biological agents.

The US Army's Institute for Soldier Nanotechnologies focuses on developing advanced materials for soldier protection, including lightweight, flexible armour made from nanomaterials. Defence Materials and Stores Research and Development Establishment (DMSRDE) under India's Defence Research and Development

Organisation (DRDO) is involved in the design and development of lightweight, ultra-high-performance materials and technologies for Personal Protective Gears, advanced fibres and technical textiles for strategic applications e.g. CBRN, High altitude clothing, camouflage & stealth, fire & thermal resistance, ballistic protection and development of new generation nanomaterials for applications in multi-functional textiles / composites, micro / nano-electronics, energy devices, and sensors.[35]

Nano-drug delivery systems could deliver life-saving drugs or vaccines directly to affected cells, providing immediate treatment for wounds or exposure to biological agents. To overcome drawbacks of traditional drug delivery, innovative delivery systems are being designed usually termed as smart drug delivery systems which include nucleic acid-based drug delivery system, cell-based drug delivery system, self-nano emulsifying drug delivery system, self-micro emulsifying drug delivery system, chemical and physical stimuli-based drug delivery system, nanoneedles, patches, ultrasound drug delivery and microchip technology.[36] These advancements will significantly enhance soldier survivability and effectiveness in combat.

Biotechnology

Biotechnology is poised to transform the physical capabilities and resilience of infantry soldiers. Genetic enhancements, bioengineered implants, and advanced prosthetics could create soldiers with enhanced strength, endurance, and recovery abilities.

DARPA's Biological Technologies Office (BTO) is exploring projects such as bioengineered tissue and advanced prosthetics that can be integrated with a soldier's body, potentially restoring lost functions or even enhancing natural capabilities. BTO is inspiring innovation to better prepare warfighters by ensuring peak performance and restoring function to injured warfighters.[37] Its working in varied fields concerning Bio-artificial blood product substitutes that are safe and achieve near parity to natural whole blood functionality, non-invasive peripheral nerve stimulation to promote synaptic plasticity in the brain, Memory restoration and enhancement in humans and e bio-manufacture of critical molecules & materials.[38]

Research is ongoing into gene editing techniques like Clustered Regularly Interspaced Short Palindromic Repeats (CRISPR), which could one day be used to enhance a soldier's physical capabilities, resistance to diseases, or ability to heal from injuries.[39] Genetic modifications will be used to improve physical attributes such as strength, endurance and speed, as well as to increase a soldier's resistance to fatigue, injury and disease. Advanced prosthetics and bioengineered implants could restore lost functions of wounded infantryman or even provide him with capabilities beyond what is naturally possible, such as enhanced vision or integrated communication systems. Additionally, biotechnology could enable the rapid healing of wounds and protection against chemical and biological threats, making soldiers more durable and effective in prolonged violent engagements.

Augmented Reality (AR) and Virtual Reality (VR)

Augmented Reality (AR) and Virtual Reality (VR) are poised to revolutionise infantry warfare by transforming how soldiers train, operate and engage off as well as on the battlefield. These technologies bridge the gap between the physical and digital worlds, offering new dimensions of situational awareness, training and mission planning. This section explores the specific ways in which AR and VR are set to redefine infantry operations in the coming years.

AR and VR technologies are transforming military training by creating immersive and realistic environments where soldiers can practice combat scenarios without the risks associated with live exercises. VR, in particular, allows for the creation of highly detailed virtual battlefields that simulate real-world conditions akin to high end video games, including urban environments, rugged terrains, and complex combat scenarios.

The US Army's Synthetic Training Environment (STE) program leverages VR to create a fully immersive training environment where soldiers can practice individual and collective tasks. The STE can replicate any global terrain, enabling soldiers to train in environments similar to those they may encounter during deployment. The STE presents real dilemmas to soldiers, so that they learn to adapt to unforeseen circumstances and allows cross-domain convergence training.[40] STE will eventually allow Army air and ground warfighting units to rehearse missions together in the same environment from multiple different sites and locations around the world over a network.[41] The British Army's Virtual Reality in Land Training (VRLT) program uses VR to train soldiers in close-quarters combat and urban warfare. VR scenarios can be adapted to reflect current intelligence on enemy tactics and terrain, ensuring that training is relevant and up-to-date. VRLT allows for High Resolution VR Headsets to improve environmental immersion, Mixed Reality to see and interact with physical objects, Avatar customisation replicating realistic facial features and body shapes and After-Action Review Enhancement.[42]

AR enhances situational awareness by overlaying digital information onto the physical environment in real-time. This technology can provide soldiers with crucial data, such as enemy positions, navigation routes, and mission objectives, directly within their field of view. By integrating AR into helmets, visors, or goggles, soldiers can access real-time information without having to divert their attention from the battlefield as mentioned earlier while discussing IVAS. Although initially designed for civilian use, HoloLens 2, the brain of the IVAS has been adapted for military applications, allowing soldiers to view and interact with digital maps, 3D models, and live data feeds during planning and reconnaissance missions.[43]

AR in future would most likely become a standard issue of infantry gear, providing infantry soldiers with real time situational awareness and reducing their cognitive load during close combat and complex operations. Soldiers will be able to access mission-critical information on the move, improving decision-making speed and accuracy in engagements. AR will provide soldiers with visual cues, such as the geographical locations of friendly forces, supply drops, or evacuation routes, directly overlaid on their digital environment. This will reduce the chances of confusion and enhance coordination during dynamic operations. AR can be used to guide soldiers through complex tasks by overlaying instructions or expert advice onto their Heads Up Display (HUD). This can be particularly useful in field repairs, medical emergencies, or navigating unfamiliar terrain. In high-pressure situations, AR could provide soldiers with prompts and techniques to guide them manage stressful situations where they feel isolated and abandoned, which can be critical for maintaining operational effectiveness.

As VR gets superimposed on to AR applications, training simulations will become even more realistic, allowing soldiers to experience and react to a wide range of battlefield scenarios. This realistic training will enhance soldier readiness and decision-making under stress, leading to more effective combat performance. VR will allow soldiers to rehearse missions in a virtual environment that mirrors the actual battlefield. These exercises can include simulated enemy forces, terrain and weather conditions as well as estimated or known enemy capabilities, providing a realistic preview of the mission ahead. Using VR, commanders can immerse themselves in a virtual replica of the battlefield, receiving live feeds from drones, satellites, and ground units. This will enable them to make informed decisions and direct troops with a comprehensive understanding of the operational situation. VR Exposure Therapy can be used by militaries to simulate combat environments to help soldiers acclimate to the stresses of battle. This type of training has been shown to reduce anxiety and improve focus during actual combat.

With the advances in AR and VR technologies, there will be increased seamless collaboration between the commanders and their troops irrespective of the physical separation. This will allow for more flexible and responsive command structure and encourage Directive Style of Command. Beyond tactical applications, AR and VR can play a crucial role in the mental preparation of the soldiers. VR simulations can expose the soldiers to the high stress environments in a controlled setting, helping them build resilience and coping strategies. Also it will allow the commanders to choose their teams accordingly as per their psychological readiness and willingness to do the operation. AR and VR technologies if employed in the right manner are set to initiate fundamental changes in infantry warfare, offering unprecedented capabilities in training, situational awareness, mission planning and stress management. By immersing soldiers in realistic virtual environments and enhancing their real life view of the battlefield with digital overlays, these technologies are set to make infantry operations more effective, coordinated and adaptive. As AR and VR continue to evolve, they will become indispensable tools for modern infantryman, shaping the future of ground combat in profound and far reaching ways.

Impact on Infantry Operations

The introduction of emerging disruptive technologies into infantry operations offers numerous advantages but also presents several challenges that must be carefully considered.

Advantages

Enhanced Physical Capabilities: Wearable technology and exoskeletons significantly improve the physical endurance and mobility of infantry soldiers, allowing them to carry heavier loads and operate in demanding environments for longer periods.

Improved Firepower: Advanced weaponry and ammunition provide infantry units with greater accuracy, range, and lethality, enabling them to engage enemy forces more effectively and with fewer rounds.

Better Communication and Coordination: Advanced communication systems and battlefield management tools enhance the ability of infantry units to coordinate their actions, share information, and respond quickly to changing battlefield conditions.

Increased Resilience: Cybersecurity and electronic warfare measures protect infantry units from cyber threats and electronic attacks, ensuring the integrity of their communication networks and electronic systems.

Disadvantages

High Costs: The development and deployment of advanced technologies are expensive, potentially limiting their availability to only the most well-funded military forces.

Complexity: The integration of multiple advanced systems increases the complexity of operations, requiring extensive training and support to ensure that all components function effectively together.

Vulnerability to Countermeasures: As adversaries develop their own technologies, there is a risk that advanced systems could be disrupted or neutralized by electronic warfare, cyberattacks, or other countermeasures.

Ethical and Legal Concerns: The use of AI, autonomous systems, and advanced weaponry raises questions about accountability, the potential for unintended consequences, and the ethical implications of using these technologies in combat.

Future Trends and Anticipated Progress

Looking to the future, several trends are likely to shape the development and deployment of infantry technologies over the next two decades.

Increased Use of AI and Autonomous Systems: The integration of AI and autonomous systems into infantry operations is expected to continue, with advancements in machine learning, data analytics and robotics driving further improvements in the capabilities of Infantry.

Protection and Mobility: Exoskeletons and wearable technology will become lighter, more efficient and more multifunctional, thereby supplementing the physical capabilities of infantry soldiers, allowing them to be more agile across a varied range of battlefield environments.

Cybersecurity and EW: The development of more advanced cybersecurity and electronic warfare techniques will be crucial in safeguarding infantry units from growing threats to assure that their communication networks and electronic systems are not compromised.

Global Competition: The infusion of advanced technologies into infantry remains a great area of competition between military powerhouses, with countries straining to achieve technological superiority to maintain it on the battlefield.

The Indian Context

The modernization of the infantry must concentrate on equipping our soldiers with the tools and technologies that will essentially enhance their situational awareness, decision making and combat effectiveness on a digital battlefield. The infantry unit of today should be equipped with next-generation wearable technologies, including Augmented reality (AR) headsets and an integrated sensor detector arrays that can provide real-time information on the tactical situation, the enemy's movements and environmental factors. These high-tech systems will enable soldiers to operate with a degree of situational awareness previously reserved to the command centers and thus enhance their responsiveness to dynamic situation on the battlefield. The infantry should integrate unmanned systems, both airborne and ground, which can be used as a force multiplier. The use of lightweight unmanned aerial vehicles (UAVs) for immediate reconnaissance is possible, while robotic systems on the ground can carry out hazardous operations such as bomb disposal or mine clearance, significantly reducing the vulnerability of infantrymen. In addition, secure encrypted communication channels are essential for ensuring uninterrupted flow of information between units laterally and with their higher commanders, even in the face of sophisticated electronic warfare onslaught. Furthermore, redundancy and adaptive countermeasures must be adopted to ensure continuity of communications in a contested electromagnetic environment. Ultimately, the goal is to create an infantry force that is not only technologically empowered but also agile and adaptable, capable of rapidly shifting tactics in response to a constantly changing threat landscape.

Indian Army ground operations fundamentally rely on Infantry units whose progression into modern warfare depends on technological integration to boost soldier survivability, mobility, protection and combat performance. Advancements in exoskeletons and wearable tech emerges as an unprecedented leap that stands to alter soldiers' physical abilities by expanding their combat stamina. Infantrymen equipped with exoskeletons and mechanized suits that boost physical power and stamina can transport heavier loads including advanced weaponry and extra supplies across extended ranges without experiencing fatigue. India's varied landscapes present unique challenges where physical stamina becomes a limiting factor across its rugged Himalayas and dense jungles. These suits made with smart fabrics, enabled by wearable technologies such as helmet and wrist-mounted devices track and deliver real-time health indicators by tracking vital signs such as heart rate and hydration levels while maintaining uninterrupted communication with command centers. The integration system maintains soldier-unit communication while enabling medics to deliver swift injury responses which boosts both battlefield operational performance and soldier well-being. This integration keeps soldiers connected to their units and allows medics to respond promptly to injuries, enhancing both operational efficiency and soldier welfare in the heat of battle.

To maximize lethality while preserving own capabilities, while limiting collateral damage, the Infantry will have to incorporate smart weapons with AI-assisted targeting and intelligent munitions. These advanced systems use battlefield information from various sensors on wind speed, distance, target movement, to generate firing solutions in near real-time and allow soldiers to carry out precision strikes with high accuracy and minimum threat of fratricide. Smart munitions such as guided bullets and grenades have the ability to vary their flight path while in the air to home in on hidden or moving targets, minimizing the chance of collateral damage in urban or populated areas. This capability is invaluable when it comes to conducting counterinsurgency operations or enforcing peacekeeping missions, where distinguishing friend from foe is an enduring challenge. By equipping infantrymen with these tools, the Indian Army can enhance the precision and effectiveness of its ground forces, ensuring they remain decisive in both conventional and irregular warfare.

Looking further ahead, biotechnology enhancements offer a radical yet promising avenue to elevate soldier performance beyond natural human limits.

Certain drugs, genetic modifications or bioengineered implants could improve traits like night vision, muscular endurance or resilience of the brain under stress. For example, retinal upgrades might give soldiers improved low-light vision, an essential edge in night ops, while performance-enhancing drugs could reduce fatigue and sustain troops during extended engagements in challenging environments. While ethical and logistical challenges remain, such advancements could provide the Infantry with a competitive edge, particularly in prolonged conflicts or extreme environments like those found along deserts on our western borders, high-altitude deserts of Ladakh, glaciated peaks on the Siachen or tropical swamps of the North East. The Indian Army must carefully explore these possibilities, balancing innovation with the well-being of its personnel, to ensure its soldiers are equipped to meet the demands of future battlefields.

Training is the bedrock of preparedness for the Infantry, and a new method of training, using virtual reality (VR) and augmented reality (AR) as training simulators, will change how soldiers hone and continuously refresh their skills. VR can easily simulate a mosaic of enemy tactics, environmental hazards and equipment malfunctions, while AR superimposes digital information over real-world environments, improving situational awareness in training. Infantrymen can step into these immersive platforms and experience realistic combat scenarios, from urban firefights to jungle ambushes, without the costs of elaborate training areas or the risk of learning those engagements while on the job training in field. With this technology, commanders can customize scenarios to match the operational parameters such as counter-terrorism, border defence or disaster management, ensuring soldiers are well-prepared for the complexities of modern warfare. By investing in these simulators, the Infantry can maintain a high state of readiness, adapting quickly to evolving threats and mastering the integration of new technologies into their tactical repertoire.

The impact of emerging disruptive technologies on infantry operations will be profound, promising to bring about significant improvements in physical capabilities, firepower, communications and resilience. All these developments will, however, also pose challenges such as operational complexity, cost-related issues, human resource and training challenges and vulnerability to countermeasures. An analysis of advanced militaries clearly reveals a race across the globe to integrate these new technologies into infantry forces, with every nation chasing its own paths with different arrays of priorities. The larger ethical and legal ramifications will have to be encountered as well.

As these technologies continue to flourish, infantry units will become more versatile, lethal and adaptive, capable of operating in a wider set of environments and battlefield scenarios. This transformation will not only improve the lethality and versatility of infantry forces but will also radically change the strategies and tactics employed on the battlefield. The next couple of decades will likely bring incessant innovations; innovations that will redefine the capabilities of infantry forces and guarantee that infantry remains essential for present-day military operations. The challenge will be to balance the rewards that come with these technologies as well as addressing their potential risks and limitations. Thus, ensuring that infantry forces can continue to operate effectively in an increasingly complex and contested battlefield.

Notes

1   Diego Laje, The Rise of the Humanoid: Exoskeletons Revolutionizing Military Readiness MAR 01, 2024 Signal https://www.afcea.org/signal-media/rise-humanoid-exoskeletons-revolutionizing-military-readiness

2   Xinyu Guan, Linhong Ji and Rencheng Wang Development of Exoskeletons and Applications on Rehabilitation, Division of Intelligent and Bio-mimetic Machinery, The State Key Laboratory of Tribology, Tsinghua University, China https://www.matecconferences.org/articles/matecconf/pdf/2016/03/matecconf_icmes2016_02004.pdf

3   Jeffrey Lin and P.W. Singer China's working on the next generation of military exoskeleton, Feb 8, 2018 https://www.popsci.com/china-exoskeleton-next-generation/

4   Chen Zhuo, Global Times, 2020-12-11 PLA border troops in Tibet receive exoskeleton suits: report http://eng.chinamil.com.cn/CHINA_209163/TopStories_209189/9951279.html

5   Bree Watson Power Move, 2019 Forging the future of endurance-boosting technology https://www.ucf.edu/pegasus/power-move-onyx-exoskeleton/

6   Integrated Visual Augmentation System (IVAS) https://www.dote.osd.mil/Portals/97/pub/reports/FY2021/army/2021ivas.pdf?ver=FZDivGDiByhjV9U-NnM9dQ%3D%3D

7   Frederick Shear August 1, 2023 Army accepts prototypes of the most advanced version of IVAS, https://www.army.mil/article/268702/army_accepts_prototypes_of_the_most_advanced_version_of_ivas

8   Frederick Shear August 1, 2023 Army accepts prototypes of the most advanced version of IVAS, https://www.army.mil/article/268702/army_accepts_prototypes_of_the_most_advanced_version_of_ivas

9   Kurt Mudie, Daniel Billing, Alessandro Garofolini, Thomas Karakolis, Michael LaFiandra 17 June 2021 The need for a paradigm shift in the development of military exoskeletons European Journal of Sports Science https://onlinelibrary.wiley.com/doi/10.1080/17461391.2021.1923813#ejscbf01748-bib-0002

10 Inder Singh Bisht, Israeli Firm Unveils Computerized Firing System for Greater Accuracy, April 10, 2024, https://thedefensepost.com/2024/04/10/israeli-computerized-firing-system/

11 TrackingPoint's Xact System™ Precision Guided Firearm Earns "Best of What's New Award" from Popular Science Magazine, Nov 13, 2013 https://www.prnewswire.com/news-releases/trackingpoints-xact-system-precision-guided-firearm-earns-best-of-whats-new-award-from-popular-science-magazine-231740271.html

12 Precision Guided Firearms – Tracking Point https://navyseals.com/2517/precision-guided-firearms-tracking-point/

13 Troy Carter Here's the Army's now-patented EMP rifle attachment for taking out small drones, Jan 16, 2019 https://techlinkcenter.org/news/heres-the-armys-now-patented-emp-rifle-attachment-for-taking-out-small-drones/

14 Krista Alestock Advanced Ammunition: Defeating Impossible Enemy Threats Northrop Grumman https://www.northropgrumman.com/what-we-do/advanced-weapons/advanced-ammunition-defeating-impossible-enemy-threats

15   Paul Scharre Precision-Guided Weapons Come To The Infantry November 11, 2015 https://warontherocks.com/2015/11/precision-guided-weapons-come-to-the-infantry/

16   M.S. Marwick, Project Leader C.M. Kramer E.J. Laprade Institute For Defense Analyses, Analysis of Soldier Radio Waveform Performance in Operational Test May 2015 https://apps.dtic.mil/sti/pdfs/AD1032264.pdf

17   Christian D. Villanueva López 27/05/2024 Battlefield management systems in the Ukrainian war https://www.revistaejercitos.com/en/articulos/los-sistemas-de-gestion-del-campo-de-batalla-en-la-guerra-de-ucrania/

18   Roger McDermott Eurasia Daily Monitor, Volume: 16 Issue: 86 June 11, 2019 Russian Military Introduces New Automated Command-and-Control Systems https://jamestown.org/program/russian-military-introduces-new-automated-command-and-control-systems/

19   U.S. Army Command Post Computing Environment June 29, 2018 https://www.army.mil/article/168119/command_post_computing_environment

20   Tamir Eshel Advancing Battle Management Systems, 21 March 2023 https://euro-sd.com/2023/03/articles/30027/advancing-battle-management-systems/

21   Todd South Jul 27, 2021 Nanotech-built armor could replace Kevlar, steel for soldier protection https://www.armytimes.com/news/your-army/2021/07/26/nanotech-built-armor-could-replace-kevlar-steel-for-soldier-protection/

22   Jared Keller, SOCOM's Iron Man suit is officially dead, Jan 13, 2021, https://taskandpurpose.com/tech-tactics/talos-iron-man-suit-dead/

23   National Research Council (US) Committee on Opportunities in Biotechnology for Future Army Applications. Opportunities in Biotechnology for Future Army Applications. Washington (DC): National Academies Press (US); 2001. 1, Introduction. Available from: https://www.ncbi.nlm.nih.gov/books/NBK207450/

24   C. Todd Lopez, New SMET Will Take Load Off Infantry Soldiers https://www.moore.army.mil/infantry/Magazine/issues/2018/APR-JUN/PDF/4)News3-SMET.pdf

25   Shivani Mago, Meet the Chinese Army's newest recruit – a gun-toting 'robo-dog' with an automatic rifle on its back, 31 May 2024, https://theprint.in/world/meet-the-chinese-armys-newest-recruit-a-gun-toting-robo-dog-with-an-automatic-rifle-on-its-back/2109869/

26   Snehesh Alex Philip, Robo-dogs that can fire, surveil & carry load – Army set to induct its newest soldiers soon, 24 June, 2024, https://theprint.in/defence/robo-dogs-that-can-fire-surveil-carry-load-army-set-to-induct-its-newest-soldiers-soon/2145039/

27   Jared Keller, The Army Has Finally Fielded Its Next Generation Squad Weapons, March 29, 2024, https://www.military.com/daily-news/2024/03/29/army-has-finally-fielded-its-next-generation-squad-weapons.html

28   Integrated Visual Augmentation System Program Executive Office Soldier https://www.peosoldier.army.mil/Equipment/Equipment-Portfolio/Project-Manager-Soldier-Warrior-Portfolio/Integrated-Visual-Augmentation-System/

29   Samuel Bendett, Russian UGV developments influenced by Ukraine War, 19 June 2024, https://euro-sd.com/2024/06/articles/38818/russian-ugv-developments-influenced-by-ukraine-war/#_ftn24

30   Viktor Bodrov, "Robots are rushing to the front line: how machines participate in hostilities" (Роботы рвутся на передовую: как машины участвуют в боевых действиях), Tass.com, April 25, 2024, https://tass.ru/armiya-i-opk/20636347.

31   Ibid

32   The Army's Robotic Combat Vehicle (RCV) Program July 23, 2024 The Congressional Research Service https://crsreports.congress.gov/product/pdf/IF/IF11876

33   Multi-Mission High Energy Laser (MMHEL) U.S. Army Space And Missile Defense Command https://www.smdc.army.mil/Portals/38/Documents/Publications/Fact_Sheets/Archived_Fact_Sheets/MMHEL.pdf

34 Peresvet Russian Mobile Laser System ODIN - OE Data Integration Network (.mil), https://odin.tradoc.army.mil/WEG/Asset/Peresvet_Russian_Mobile_Laser_System

35 Defence Materials and Stores Research and Development Establishment (DMSRDE) https://www.drdo.gov.in/drdo/labs-and-establishments/defence-materials-and-stores-research-and-development-establishment-dmsrde

36 Afreen Sultana, Mina Zare, Vinoy Thomas, T.S. Sampath Kumar, Seeram Ramakrishna Nano-based drug delivery systems: Conventional drug delivery routes, recent developments and future prospects Medicine in Drug Discovery, Vol. 15, September 2022, https://www.sciencedirect.com/science/article/pii/S259009862200015X

37 Biological Technologies Office (BTO) Defense Advanced Research Projects Agency https://www.darpa.mil/about-us/offices/bto

38 Biological Technologies Office (BTO) Defense Advanced Research Projects Agency https://www.darpa.mil/attachments/Breakthrough%20Biological%20Technologies%20for%20Na tional%20Security%20event%20020422.pdf

39 Feng Zhang Questions and Answers about CRISPR Broad Institute https://www.broadinstitute.org/what-broad/areas-focus/project-spotlight/questions-and-answers-about-crispr

40 Jeremiah Rozman, PhD, The Synthetic Training Environment, The Association of the United States Army, December 2020, https://www.ausa.org/sites/default/files/publications/SL-20-6-The-Synthetic-Training-Environment.pdf

41 Mr. Michael M. Novogradac, U.S. Army Operational Test Command, March 5, 2024, Soldiers test new synthetic training environment https://www.army.mil/article/274266/soldiers_test_new_synthetic_training_environment

42 Ministry of Defence and The Rt Hon Stuart Andrew MP, 4 February 2019, British Army tests innovative virtual reality training https://www.gov.uk/government/news/british-army-tests-innovative-virtual-reality-training

43 Deborah Bach U.S. Army to use HoloLens technology in high-tech headsets for soldiers, June 8, 2021, https://news.microsoft.com/source/features/digital-transformation/u-s-army-to-use-hololens-technology-in-high-tech-headsets-for-soldiers/

10

Disruptive Manoeuvres: Mechanised Forces Manoeuvring the Technological Terrain

"The engine is the weapon."

—Heinz Guderian

The title emphasises both the physical and strategic agility of Mechanised Forces as they manoeuvre the evolving battlefield, shaped by emerging disruptive technologies. The title aims to blend the themes of military manoeuvrability with technological adaptation, suggesting that the future of armoured warfare will depend on how effectively these forces can integrate cutting edge technologies into their operational frameworks.

The Mechanised Forces have always played a decisive role in wars and military conflict since World War I. They have been found to be of great use even by non-state actors such as LTTE, ISIS, Al Qaeda and the Taliban. Autocratic regimes have often leveraged its shock, speed and fire power to intimidate domestic populations as was in the case of Tiananmen Square Massacre of 1976.[1] Such varied employment of tanks and armoured fighting vehicles brought revolutionary changes to the character of warfare, affording militaries the capability of disrupting fixed positions and providing for manoeuvrability with better speed and protection than regular infantry or mounted cavalry. Though rudimentary, slow, and fairly prone to mechanical failure in initial operations, armoured vehicles rapidly began to gain a place in the arsenal of modern-day militaries.

Rapid advancements in armoured vehicle technology during the interwar years and throughout World War II produced far more capable and complex tanks, typified by German Panther & Tiger tanks and the American Sherman tank. Armoured vehicles became part of a blitzkrieg approach which advocated for speed, surprise and firepower to deliver an overwhelming attack on the enemy.[2] After the War, refinements continued to enhance mechanised forces' capabilities in terms of firepower, mobility, and protection.

During the Cold War, the arms race between NATO and the Warsaw Pact contributed to many advances, which collectively improved and refined main battle tank designs and associated mechanised operations. Newer models of main battle tanks like Soviet T72 and American M1 Abrams were developed with heavy firepower, improved mobility and advanced armour representing a symbol of a nation's military potential and deterrent capabilities. The commencement of the Russia Ukraine War in 2022 has led to constant discussions around the vulnerability and survivability of mechanised columns in modern battlefield environment and the possibility of lighter forces supported by ISTAR capabilities and indirect fires defeating mechanised forces during high intensity warfighting.

The last few decades have also witnessed transformative changes in the inventory as well as role of mechanised forces arising from emerging technologies and unique threats. Asymmetrical warfare, fighting in urban and built up areas and the proliferation of sophisticated anti tank weapons pose new challenges to mechanised forces, while efforts are underway to improve their survivability, firepower and manoeuvrability.

However, heavy armoured forces, through their substantial combat power, ensure that a force can remain mobile while in direct contact with enemy forces, and as such heavy armour still has a valuable role to play on the battlefields of the future.[3] Although Russia armoured forces could the achieve the obvious easily as they were not the main component of Kremlin's war strategy, the Kremlin's main effort hinged on its Special Services, which were supposed to destabilise Ukraine and disorganise its system of state and military administration. The military invasion was not supposed to meet serious organised resistance, other than in Donbas.[4] The setbacks suffered by the Russian military in the battles of Kyiv, Kharkiv and Kherson were primarily the consequences of an initial miscalculation in the planning of the invasion, and the employment of forces improperly structured and commanded for conventional warfighting.

Following the unsuccessful coup de main, attempts were made to correct the operational deficiencies, but because of the losses taken in the opening phase of the war, the Russian military had still been unable to accomplish its tasks. However, by 2023, Ukraine was facing a different beast entirely as Russia had completely evolved its strategy. It had changed from a blitzkrieg strategy to one that involved protracted warfare. It had placed its main bet on the assumption that Ukraine's Western partners, especially the United States, would eventually get tired of supporting Ukraine, allowing Russia to finally gain a decisive edge on the battlefield.[5] Several innovations rapidly proliferated the mechanised formations of both sides, as extensive use of anti tank missiles, attack heptrs and cheap drones reigned explosives onto the AFVs.

Despite the subtle opposition against tanks within the uniformed community, tanks are still a vital asset in modern war fighting, regardless of the claims that their role has diminished or not needed at all. The fighting in Nagorno-Karabakh, as well as the recent war in Ukraine, demonstrated the importance of tanks and mechanised forces at large. The Russian military has suffered heavy losses to its mechanised forces, from which many assume the relative ineffectiveness of tanks. However, there are three things that led to the staggering Russian losses. Firstly, Russian tank crews and tank commanders made false assumptions regarding the nature of combat because they didn't have sufficient warning to prepare adequately. They assumed that they would not need to fight with their tanks because they expected to face little resistance from the Ukrainians. Thus, their tactical employment at the front line was inadequate. Secondly, poor strategy made logistical preparedness much worse with the combat groups running out of fuel and supplies frequently. Lastly, the Russian military did not employ tanks supported by enough infantry. The infantry protects the tanks from anti tank weapons, land mines and improvised explosives devices and the tanks protect the infantry from the enemy armoured vehicles and infantry in fixed positions. Russia failed at this. While antitank weapons play a role, the main killer of tanks in the Ukrainian war has been artillery. The Russian military could not apply all its forces in an integrated manner, despite Russian military training focusing on combined arms employment. In the end, the most common cause of a Russian loss of a tank was crew abandonment. This was often caused by the lack of fuel. Many recovered Russian tanks have had little or no damage.[6]

The tank a product of Second Mechanization Revolution, has been the cornerstone of land warfare for nearly 100 years. Since World War II, tanks have been optimized to engage enemy defences, buildings, vehicles but primarily tanks, and, secondly, to withstand hollow charges. The main armament of all the tanks has been high-velocity cannon mounted on tracked behemoth of around 50 tons with a 1,500-horsepower engine. But increasingly in the modern battlefield tanks are not firing at each other in direct line of sight close combat. They detect each other or by other platforms much earlier and engage other tanks with guided missiles which have a better hit probability. New age munitions in the form of precision guided missiles and EW suites enabled with DEWs will make the main gun redundant in next few years. Similarly, the tanks armour has reached the limits of its ability to withstand precision, tandem hollow-charge, fire-and-forget munitions aimed at the top of the tank. The retreat of the Israeli army's 401st Brigade Merkava Mark 4 tanks from the Saluki Wadi in 2006 in the face of second-generation Russian Kornet anti-tank missiles fired by Hezbollah, the total destruction of the Armenian army in the war against Azerbaijan (2020), the stranded Russian convoy on its way to Kiev, and the rout of the Russian armoured battalion during river crossing in Donbass, indicate towards the inability of present armour to withstand new munitions.[7]

In fact, active defence systems are fitted on heavily armoured fighting vehicles as an after-market accessory, while they are to replace the heavy armour, whose utility has dropped in any case. Present fighting vehicles with active defence systems are, in fact heavier and slower, compromising speed for safety. Heavy armour is not preferable for armour fighting vehicles, just like the 350-400mm steel armour of the past became increasingly unnecessary for the warships of the present; instead, it has become a liability. The futuristic active defence systems, smart sighting systems and EW suites must be laying strong foundation for development of lighter and faster tanks needing just the absolute necessary armour protection against small arms, shrapnel and explosions. Such a transformation will allow for the anticipated decrease in weight, corresponding diminution in engine size and weight, and redirection of designs towards electronically guided defensive and offensive systems.[8]

The Israeli Defense Force, in 2019, revealed the prototypes of the Carmel tank. Carmel is a light tank not armed with a single big cannon designed to take out tanks, but with missiles, a light calibre automatic gun and a reduced crew of two soldiers. It is generously equipped with latest EW suites, AI enabled fire control system and it symbolises the future of tanks and AFVs.[9]

Regardless of the fast evolving character of global conflicts, mechanised forces still remain a cornerstone of modern military strategy. They provide essential capability to engage and destroy enemy forces, secure key terrain and protect friendly forces. Mechansied forces offer a unique combination of mobility, firepower and protection that is critical in both conventional and hybrid war scenarios. In contemporary conflicts, the versatility of mechanised forces is evident in their roles across various theatres of operation. Armoured fighting vehicles have been providing the needed firepower and protection for the accomplishment of strategic objectives in variety of theatres, from combined arms operations to peacekeeping and counterinsurgency.

However, infusion of technologies such as cyber warfare, EW and autonomous weapon platforms have introduced new complexities into modern warfighting rendering mechanised forces increasing vulnerable in highly connected battle space. These new developments create a compelling case for leveraging these new technologies into mechanised forces to keep them effective, threat ready and relevant. Ongoing modernisation of mechanised forces is not just confined to upgrading existing platforms but also creating novel platforms which involve the inclusion of disruptive technologies that would transform the nature of employment these forces in future conflicts.

The Transformation of Armoured Fighting Vehicles (AFVs)

The evolving landscape of technology is facilitating advancements in AFVs. These developments will offer greater operational flexibility on the modern battlefield.

Furthermore, electrification and hybridization in the automotive sector has had an impact on the AFV manufacturing processes, providing the impetus for manufacturers to find ways to improve the fuel efficiency and emissions of the vehicle. An increased interest in soldier safety has driven manufacturers to include improved survivability such as lightweight stealth armour, active protection systems and improved communications and electronic warfare capabilities. AFVs are now built to be modular and flexible to meet operational requirements. Disruptive industry platforms and munitions from defence industry and innovative start-ups are challenging traditional theories and creating renewed competition in the AFV market that is driving product innovation and unique systems.[10]

Advanced Materials and Enhanced Protection

The latest AFVs have benefitted significantly from advancements in materials science, especially the introduction of advanced lightweight, high strength materials including ceramics, advanced alloys and composites. The use of these materials has assured better protection and mobility by reducing the overall weight or "footprint" of the vehicle. The modular armour concept gives operational flexibility, since the vehicle is able attach as much additional armour as dictated by the mission requirements. Some integrated design programs like the US Army's Future Combat Systems, developed to replace the M1 Abrams tank and M2 Bradley infantry fighting vehicle, have emphasized on the modular designs to allow the platforms to be adapted to varying threats and environments.

The FCS program was characterised as a high-risk venture due to the advanced technologies involved and the challenge of networking all of the FCS subsystems together so that FCS-equipped units could function as intended.[11] The program got shelved in 2009, mainly due to delayed timelines, complexity of integrating emerging technologies and budgetary constraints.[12] As the threat scenarios towards armoured vehicles continue to evolve, so does the technology that works to protect them. Traditional armour, while resilient against many types of conventional weapons, is increasingly being supplemented by advanced armour materials and active protection systems to counter newer threats such as missiles, drones, IEDs and energy weapons.

Modern composite armours were specifically developed to afford increased protection against shaped charges and kinetic energy penetrators. The recent innovations in the armour technology such as Chobam and reactive armour have focused on improving the balance between protection, weight and mobility. Nowadays modern composite armours are generally multi-layered multi-material, made from ceramics, metals and polymers, which dissipate the energy of the incoming projectile and prevent penetration.

Electromagnetic Armour or Electric Armour, is another emerging technology that employs high energy electric fields to disrupt or neutralise incoming threats. This type of armour can reliably protect against ATGMs or missiles that are dependent on explosive charges to penetrate the traditional armour. This type of technology generates a magnetic field around the vehicle so that the warhead detonates before it reaches the armour, and thus reduces its efficiency. The main benefit of electromagnetic armour is weight efficiency. Traditional explosive reactive armour (ERA) increases the weight of the AFV by 10 to 20 tons, whereas electromagnetic armour can achieve effectiveness with significantly less weight, potentially only a few tons, thereby reducing the load on the vehicle and allowing for other uses of its weight capacity and engine power.[13]

Active Protection Systems is one of the most revolutionary systems in the development of the protective capabilities of the armoured vehicles. These systems detect, track and neutralise incoming threats before they strike the vehicle, usually by means of mix of radars, sensors such as explosive charges, and directed energy weapons. The Trophy system, from Israel, is one of the better-known APS systems with a lot of history having been successfully deployed on tanks and armoured vehicles through various conflicts. The Trophy system employs infrared and radar inputs to detect incoming rounds, and then fires a munition to destroy or neutralise the incoming threat in mid-air. This system has proven effective against a variety of threats, including RPGs and ATGMs, and is credited with saving lives in combat.[14] Other countries, such as Russia and the United States, have also developed their own APS technology. Russia's Arena and Afganit systems[15] provide capabilities very much similar to the Trophy, with variations in operational capabilities when it comes to the countermeasures and detection systems. The Afganit active protection system is organic to the T-14 tank and T-15 heavy infantry fighting vehicle.[16]

While APS do truly represent a strong case for the enhancement of survivability in armoured vehicles, it comes with its own set of challenges. Complexity and cost are a major issue for consideration, besides collateral damage attributed to the use of countermeasure and also the vulnerability of APS to electronic warfare that may shut down, disable, or otherwise disturb its sensors and degrade its target acquisition. The future of APS as part of armour technology is likely to include greater integration with other systems on the battlefield, such as UAVs and UGVs, providing a layered defence and complete situational awareness. The development of new materials might also usher in the era of lighter, highly effective armour materials to enhance the mobility and protection of the armoured vehicles.

Impact on AFV Design

Lightweight and High-Strength Materials: Development has enabled composite materials, ceramics, and advanced-alloys which allow for lighter but equally protective armour. Modern advancements allow easier mobility while reducing fuel consumption, almost without sacrificing protection.

Modular Armour Systems: Modular armour provides flexibility and customisation as per mission mode, more plates fitted in high-threat zones, while lighter configurations can be used for low intensity conflicts.

M1 Abrams SEP V3 (USA): The latest variant of the M1 Abrams tank, touted as the best tank on earth, incorporates composite armour and modular systems for increased protection and flexibility.[17]

Rheinmetall Lynx KF41 (Germany): This IFV features modular armour and a flexible design, allowing it to be configured for different roles on the battlefield. The Lynx also features a modular architecture with inherent growth potential; a high power-to-weight ratio for maximum agility across all terrain types; scalable lethal and non-lethal effects for the full spectrum of operations in open and urban environments; a modular and adaptable survivability system with low visual, thermal, and acoustic signature levels; state-of-the-art command ability with a full suite of situational awareness and battlefield information systems; and significantly improved habitability, with low noise and vibration profiles as well as full air conditioning and a next-generation human-machine interface.[18]

Impact on Tactics

Mobility and Agility: Lightweight AFVs with advanced, extremely high-quality protection will easily go through various terrains of almost all sorts and kinds, enabling rapid deployment along with repositioning in the battlefield.

Adaptable Defence: The ability to customize armour based on mission-specific threats allows for more versatile and effective tactics, tailored to the operational environment.

Hybrid Propulsion and Mobility Innovations

The transition toward hybrid-electric propulsion systems represents a significant advancement that affords AFVs the capability to function quietly. This not only diminishes their thermal and acoustic signatures, but also, bolsters their stealth abilities. The British Army's Ajax reconnaissance vehicle, incorporates these technologies to enhance both strategic mobility and survivability.[19] In addition, adaptive suspension systems in programs such as DARPA's Ground X-Vehicle Technologies (GXV-T) support vehicle maneuverability and terrain adaptability, thus allowing for rapid response and mobility on a variety of landscapes, including urban environments and rough terrain.[20] Although these innovations are promising, they face implementation challenges because of the complexities involved.

Impact on AFV Design and Tactics

Hybrid-Electric Propulsion: Hybrid electric propulsion is a new frontier in the design of AFVs. Hybrid engines will considerably fuel lower consumption and produce lesser noise enhancing both range and stealth.

Adaptive Suspension and Mobility Systems: Advanced suspension systems allow AFVs to navigate rough terrain with greater ease, improving their ability to traverse challenging environments.

Ajax Family of Vehicles developed by UK uses advanced suspension and hybrid propulsion to enhance mobility and reduce noise, making them more effective in a variety of combat scenarios. The Puma Infantry Fighting Vehicle from Germany has an advanced powertrain and suspension system, which are specifically engineered for superior mobility across diverse environments.[21] However, the effectiveness of these innovations is contingent upon their integration in combat scenarios, as they must adapt to the complexities of modern warfare.

Modular Design and Multi-Role Versatility

The idea of modularity goes beyond just weapons; it incorporates the total design of the vehicle. Modern Armed Fighting Vehicles, such as the British Ajax and the German Boxer, are designed to perform multiple missions via rapidly interchangeable mission modules. Hence, a single platform can be reconfigured for either recce, direct action, medical evacuation or command and control modes while responding to the operational needs of the mission, while avoiding the cumbersome logistics of multiple specialized vehicles. It does also create complexity, but in most cases the benefits outweigh the disadvantages.

Impact on AFV Design

Scalable Systems: Vehicle systems are becoming progressively more scalable, thus allowing for upgrades or alterations in capabilities as new technologies arise or mission requirements change.

Modular Payloads: Armoured Fighting Vehicles are currently being developed with modular payload bays that can be customized for a variety of roles. This means that one platform can serve multiple roles.

Boxer MRAV (Germany/Netherlands): Known for its modular design, the Boxer can be quickly reconfigured for various missions, from troop transport to battlefield command and control.[22]

Armata Universal Combat Platform (Russia): A modular platform that can be adapted for different types of vehicles, including tanks, IFVs, and self-propelled guns, depending on the mission.[23]

Impact on Tactics

Mission-Specific Configurations: The modular nature of contemporary AFVs enables commanders to rapidly reconfigure these vehicles for particular missions. This provides greater flexibility in both planning and executing operations.

Scalable Force Structures: With scalable systems, tactical units can be adapted to the scale of the conflict. It can involve either a small, specialized task force or a larger, combined arms operation. Scalable force structures are highly beneficial for countries with varied terrain involving different battle environments ranging from deserts to glaciated landscapes.

Advancements in Munitions and Ammunition

Alongside advances in vehicle technology, there has been a slew of innovation in the munitions and ammunition used by AFVs, all of which significantly enhance the effect these platforms can create on the battlefield. New munitions are increasingly developed to defeat advanced armour systems or maximize damage against positions that are fortified or concealed. Programmable ammunition, such as the U.S. Army's XM1113[24] and the use of airburst capabilities for shells like those fired from the XM25,[25] allows for rounds to be detonated at exact distances thereby engaging enemy behind cover or in trenches. Also, smart munitions which can adjust flight path to hit moving targets or change targets after launch, similarly increases the lethality of AFVs. Technologies like these increase the hit probability against fast moving/distant targets and improve the overall economy by reducing the amount of ammunition needed to neutralize threats.

This evolution in munitions matches the advancements in AFV platforms; so these vehicles can deliver precise firepower in any combat scenario. Smart ammunition and precision guided munitions (PGMs) have increased the firepower of mechanised forces. They allow for more accurate targeting and increased lethality. While they're more efficient they also reduce collateral damage and the need for multiple shots to get a kill. That's especially important in modern warfare where precision beats volume.

Smart ammunition means munitions with various guidance systems, sensors or other tech that can change their trajectory and behaviour in flight. This includes a whole range of projectiles, guided tank rounds and anti-tank missiles that can track and engage moving targets. The main advantage of smart ammunition is the probability of hit against targets in complex or cluttered environments. For example, the M829E4, the latest generation of kinetic energy penetrators used by the US Army's M1 Abrams tanks – has advanced guidance and targeting systems that make it more accurate and lethal against modern armour.[26] The Israeli LAHAT missile can be launched from a tank's main gun, it uses laser guidance to engage targets at range, including beyond line of sight.[27] But effectiveness of such tech can be affected by environmental factors and while they improve targeting precision they also add complexity to deployment.

The primary obstacles linked to smart ammunition and Precision Guided Munitions (PGMs) encompass their expense and the risk of countermeasures that might hinder their guidance systems. As adversaries enhance their electronic warfare capabilities to disrupt GPS and laser-guided munitions, the efficacy of these weapons could consequently be compromised. However, ongoing research initiatives are concentrated on addressing these issues, particularly through the creation of alternative guidance systems that are less vulnerable to jamming or various forms of interference. Notwithstanding the challenges, the future of smart ammunition and PGMs appears promising and is likely to witness significant advancements in miniaturisation. This would enable a greater number of rounds to be transported on each vehicle, alongside improvements in both accuracy and lethality.

The Ideal AFV

Tomorrow's battlefield will be nothing like today's. With AI, ML, robotics, advanced electronics and potentially nukes the very nature of armoured warfare is being turned on its head. These technologies aren't just building on what we have but are changing the design, functionality and tactics of Armoured Fighting Vehicles (AFVs).

Modular Design: The Key to Flexibility

In the fast pace of future warfare adaptability is key. The future AFV will be built on a modular and adaptable design to evolve and meet the changing demands of the battlefield. At the heart of this design is nanotechnology enhanced armour, a game changer that's strong and flexible. This armour is not only light but self-healing so the vehicle can stay in action even after being damaged.

This modular approach goes beyond armour. The reconfigurable structure of the AFV can change from one combat role to another. For example, a base configuration can be a main battle tank but with the addition or removal of modules it can become a recce vehicle, command and control centre or an infantry fighting vehicle. This level of flexibility means the AFV can be configured for specific missions, environments and threats.

Autonomous and AI Enabled

As AI and ML continue to evolve, they will increasingly be embedded within AFVs and affect the way they operate. AFVs will be able to function, navigate, acquire, and engage with human inputs serving only as a minor enhancement to the capabilities of the vehicle. AI will do the data processing in real time and the vehicle will identify and prioritise threats and make decisions on the most likely best course of action.

ML will also allow the AFV to learn from experience. They will get better by reviewing and analysing previous engagements, adapting their tactics and strategies to successfully deal with potential threats. This constructive learning will be priceless in an environment where enemy forces constantly seek, develop and exploit new ways to conduct attacks.

Robotic Integration and Swarm Tactics: Multiplying Force with Autonomous Systems

Above all, armoured warfare in the future will require far more than AFVs. An advanced and potent AFV will act as the networked node for UGVs and UAVs. The AFV will be accompanied by an array of autonomous and robotic wingmen in combat formations, which will provide the needed support to the armoured squadron with tasks ranging from reconnaissance, mine clearance, close combat engagements to supply and logistics support or anything else imaginable, at the same time allowing the AFV to focus on its primary mission.

The concept of swarm tactics is already a reality. AI will enable the AFV to coordinate a swarm of autonomous systems, executing complex multi directional attacks that overwhelm enemy defences. These swarms will accrue the AFV a significant tactical advantage by executing the mission with a degree of synchronization and adaptability that would be almost impossible to be countered by human operated systems.

Next Generation Weaponry: Precision, Power, and Lethality

Add the firepower of directed energy lasers or rail guns and AI enabled missile munitions and it is pretty clear that a future AFV will deliver more force further and faster than any other ground system. High energy lasers and microwave systems will allow the vehicle to engage and kill at the speed of light and will offer a precise and cost effective means of threat engagement against variety of aerial threats. This capability would also be engineered to work in tandem with kinetic weapons, offering a range of kinetic options against high value targets.

The AFV will also be able to launch tactical nuclear weapons in a worst-case extreme scenario. These low yield, targeted nuclear projectiles would offer a way to neutralise the most difficult obstructions or provide deep strike weaponry during consequential wartime scenarios with minimum collateral damage

Survivability and Stealth: Evading Detection, Withstanding the Unthinkable

Survivability would be a key characteristic of the ideal AFV. With the addition of stealth, even if the enemy knows that an AFV is in the area, they will not be able to track or target it. Also, radar-absorbent materials, heat signature reduction systems, and active camouflage will provide AFVs the ability to operate in austere environments with lower chances of being detected. Those stealth capabilities could then be coupled with a variety of active and passive defense systems, including smoke screens, electronic countermeasures, and anti-missile systems, all operating through AI interfaces to enable fast and effective responses to incoming threats.

In addition, the vehicle's survivability will be strengthened by its cyber defense package. As AFVs increasingly rely on digital systems, they will become prime targets for cyberattack. Autonomous fighting vehicles or otherwise, will employ AI-based cyber defense to protect the AFVs electronics and networks from unauthorized access or irreversible damage, allowing the vehicle to continuously perform and operate correctly, even during advanced and aggressive electronic warfare.

Sustainable and Efficient Power Systems: Endurance for the Long Haul

In today's complex operational environment, the most capable AFV will require efficient and sustainable Power Systems. A hybrid powertrain's aspirational end state is one that uses traditional fuels (a gas or diesel engine) along with advanced batteries or fuel cells that provide a consistent level of energy efficiency and optimal power output. This hybrid powertrain will lower the vehicle's thermal and acoustic signature and increase the operational range of the vehicle. The vehicle would also have energy harvesting capability (solar panels or kinetic energy recovery systems) to supplement recharging of the battery and powering electronics. The self-sustaining nature of the power system will be valuable in prolonged operations where the existing supply lines may be stretched.

Reinforcing Armour: Emerging Ecosystem

Tanks and combat troop carrying vehicles are not the only platforms that constitute mechanised forces, there is an entire array of other platforms in the mechanised ecosystem. While creating a capable mechanised fighting force to prevail in a diverse and challenging environment, supporting platforms and systems are also pivotal. These systems augment the core strike capabilities of the tanks and troop carrying armoured fighting vehicles, through a range of key functions including reconnaissance, fire support, air defence and logistical support.

Armoured Reconnaissance Vehicles (ARVs) are wheeled/tracked combat vehicles deployed for reconnaissance missions ahead the mechanised fighting groups, to gather enemy intelligence and provide timely early warning to friendly forces of imminent enemy action. These vehicles are equipped with advanced sensors and robust communication systems. The optical, infrared and radar based advanced sensor suites enhance threat detection across varied terrains, as is the case with German Wiesel ARV. Multisource data fusion and analysis technologies help processing data acquired from several sensors creates real time situational awareness for the operator and accelerate target identification. Stealth and mobility technologies such as low radar absorbent materials and high quality suspension systems mitigate the prospect of these ARVs being spotted by radar and thermal imaging sights whilst retaining manoeuvrability.

It can evidently be argued that UAVs have become increasingly relevant in both modern overt and covert operations. Equipped with high resolution cameras and sensors like electro-optical/infrared and synthetic aperture radar, as seen in the American RQ-11 Raven,[28] UAVs can navigate and engage targets autonomously and make quick decisions, while secure communication and data links permit communication between the operational nodes and command centre, ensuring dynamic modification to their range, mission, target and effect.

Engineer Vehicles assist mechanised operations through a range of technologies, such as heavy equipment, dozers, and mine clearing systems. An example would be the M60 Armoured Vehicle Launched Bridge (AVLB),which can deliver bridging capabilities to assist in crossing obstacles.[29] Advanced mine clearing systems focus on safety and effectiveness in detecting and clearing mines. There modular design makes them highly versatile, enabling quick adaptation to various combat support roles.

Armoured Recovery Vehicles are critical for battlefield recovery and repair. Equipped with recovery and repair tools, such as winches and cranes, vehicles like the British Challenger ARRV assist in recovering and on the spot repairing of damaged AFVs.[30] Built with enhanced armour and protection, these vehicles have matching mobility enabling them to keep pace with the high tempo of mechanised operations.

Self-propelled artillery systems are equipped with advanced fire control systems like GPS and laser guided targeting to improve precision. The German PzH 2000, for instance, uses automated fire control for rapid and precise strikes.[31] Automated ammunition handling systems enhance efficiency, while survivability enhancements like armour and active protection systems safeguard the crew during operations.

Multiple Launch Rocket Systems (MLRS) feature precision-guided munitions for long-range accuracy, like the American M270 MLRS, which can launch various munitions. Rapid reloading systems reduce downtime, increasing operational tempo.[32]

Self-Propelled Air Defence Vehicles are equipped with advanced radar and tracking systems for detecting aerial threats at extended ranges, like the Russian Pantsir-S1.[33] They use a combination of surface-to-air missiles and autocannons for comprehensive air defence, as seen in the American M6 Linebacker.[34]

Mobile Command and Control Vehicles are designed for battlefield coordination, utilizing secure communication networks for encrypted and reliable communication. The German GTK Boxer Command Post integrates real-time data processing to enhance command capabilities during operations.[35]

Armoured Logistics Vehicles prioritize protected cargo transport for essential supplies in combat zones. The American HEMTT provides armoured transport with efficient load management systems, ensuring safe and timely delivery of supplies.[36]

Mobile Medical Evacuation Vehicles are equipped with medical equipment integration, providing life-support systems for on-site care and evacuation. The German AEV Puma is an example, offering medical stations and secure evacuation capabilities for effective casualty management.

Electronic Warfare Vehicles disrupt enemy communications with jamming and interception equipment, such as that found on the American M7 Bradley Fire Support Team.[37] They also feature signal intelligence systems to intercept and analyze enemy communications, providing a tactical advantage.

Cyber Defence Systems focus on secure communication protocols and real-time threat detection to protect operational technology from cyber-attacks, enhancing battlefield cyber resilience.

Amphibious Assault Vehicles like the American AAV7A1 are designed with amphibious propulsion systems to transition between water and land smoothly.[38] These vehicles are equipped with enhanced armour and protection for survivability during amphibious operations.

Amphibious Combat Vehicles feature advanced navigation systems for manoeuvrability in aquatic environments, as seen in the South Korean K200. These vehicles are combat-ready to support troops during amphibious operations.[39]

Tactical UAVs like the Israeli Skylark are known for compact and agile designs suited for diverse terrains. They provide real-time data transmission to ground forces, offering timely intelligence during operations.[40]

Strategic UAVs, such as the American Global Hawk, is a high altitude long endurance (HALE) UAV, allowing for extended surveillance missions. It uses high-resolution synthetic aperture radar (SAR) and electro-optical/infrared (EO/IR) sensors with long loiter times for detailed monitoring over vast areas.[41]

Integration of all these platforms with latest technologies will deliver a lethal and potent mechanised fighting force. Incorporating cutting-edge technologies, each platform, from reconnaissance vehicles to logistics to air defence systems will enable multi domain dominance and ensure operational success in a variety of battle scenarios over all types of terrain. Future developments are likely to see these capabilities improve as R&D matures, enhancing the mechanised force's ability to respond to emerging threats and challenges.

Future Ammunition and Munitions for Armoured Platforms

New technologies are enabling radical changes to the design and operational concept of armament systems. Technological innovations in ammunition, munitions, and armaments to raise the lethality and precision of armoured platforms such as tanks and mechanized infantry vehicles have become increasingly complex and hi tech. These emerging technologies are fundamentally transforming the design and effectiveness of armament systems, ensuring that military forces remain adept at countering evolving threats on the modern battlefield.

Probably the most innovative advancement in the military world is the advent of smart and precision-guided munitions. Where traditional ammunition had relied on explosive force and kinetic energy, smart munitions through the integration of advanced guidance systems have huge relevance in enhancing accuracy and effectiveness. They are capable of extremely long-range precise strikes due to the laser, infrared, or GPS guidance, which minimises collateral damage and greatly enhances the effect of the projectile. The US made Excalibur system GPS enabled artillery munitions, achieving unprecedented pinpoint accuracy and minimal collateral damage. Similarly, the German LARS (Langstrecken-Artillerie-Raketen-System) employs laser-guided rockets to enhance targeting precision in artillery fire.[42]

Electromagnetic and Directed Energy Weapons are pushing the boundaries of conventional armament systems. Presently, due to the large size and considerable power consumption of such weapons have not allowed them to be integrated on to the AFVs, however, militaries are constantly working towards achieving this. The Active Protection Systems on the AFVs, though defensive is an example of such platforms. Directed energy weapons, including high-energy lasers and microwave weapons, offer a futuristic solution to counter both airborne and ground-based threats. High-energy lasers can disable or destroy enemy projectiles and drones at the speed of light, while microwave weapons can disrupt electronic systems and incapacitate enemy vehicles. The US Navy's Laser Weapon System (LaWS) has demonstrated the capability to neutralise small boat threats with precision, showcasing the potential of directed energy in modern warfare.[43] In a similar vein, the Chinese military has been developing the "Silent Hunter" laser weapon system, which aims to neutralise drones and low-flying aircraft using high-energy lasers.[44]

Hypervelocity Projectiles (HVPs) are another area of rapid advancement, characterised by their extremely high speeds and kinetic energy. The idea is to construct a HVP and use electromagnetic railgun technology to penetrate modern armour and defensive systems at a relatively low cost. This technology promises to enhance the penetrative power of projectiles, particularly to be effective against heavily armoured targets. The US Navy's railgun program which later got shelved, demonstrated the capability of launching projectiles at speeds exceeding Mach 6, thus increasing chances of penetrating enemy naval armour and coastal defences.[45] Similarly, the Chinese "Electromagnetic Railgun" project is advancing the development of high-speed projectiles designed to penetrate enemy armour and defensive systems.[46]

Modular and Multi-Role Ammunition systems are being designed to enhance flexibility and adaptability on the battlefield. In these systems, ammunition types can be rapidly reconfigured and equipped to deal with multiple situations, and target variety of targets, be it armoured vehicles, aircraft, or infantry. As is the case with the modular artillery systems that offer a range of shell types depending on the tactical requirements, such as high-explosive, incendiary, and smoke rounds. The German PzH 2000 self-propelled howitzer exemplifies this concept with its capability to fire array of munitions from a single platform. The US Army's M109A7 Paladin also utilizes modular ammunition types to provide versatile fire support across different combat situations.[47]

Advanced Material Technologies are revolutionizing ammunition and munitions design by enhancing performance and durability. Composite materials and advanced alloys in projectile construction have enhanced penetrative capabilities as well as achieved weight reduction. Improved propellant technologies like solid-state and advanced composite propellants are achieving greater ranges and accuracy with safety and reliability. Nanotechnology and advanced coatings further enhances performance and longevity by ensuring that ammunition can withstand not only the rigors of combat but environmental conditions as well. For example, the 125mm ammunition of the Russian T-90 includes advanced composite materials for penetration and performance improvement.

Automatic Ammunition Handling Systems will be changing the landscape for reloading and management of munitions on-board armoured platforms. These systems are designed to reduce crew workload and increase the efficiency of operations via automatic loading, storage, and distribution of ammunition. Increases in robotic loaders and smart ammunition management systems that monitor and optimize ammunition levels in real-time have made it possible for a platform to remain combat-ready with low human intervention. An automated ammunition handling system on the South Korean K2 Black Panther tank dramatically reduces crew workload and increases operational efficiency.[48]

The integration of emerging technologies into the development of ammunition, munitions, and weapon systems is setting new standards for firepower, precision, and defence in armoured platforms and mechanized infantry vehicles. By enhancing accuracy, lethality, and improving defensive capabilities, advancements in munitions and allied systems ensure that armoured forces are better equipped to face the rigour of battle. With research and development outside of the conventional bounds, the future of armoured combat promises to be defined by much more sophisticated and effective armament systems.

Armoured Fighting Vehicles for India's Diverse Terrains

Terrain and environment are not the only factors that the Indian Army must account for in its choice of armoured platforms. The Indian Nation is subjected to a variety of threats including an overt nuclear threat from Pakistan and a critically subtle but nonetheless very real military threat from China and a situation of such threats be collusive either by destiny or design. Notwithstanding, the geopolitical challenge, India's military operations must contend with one of the most varied and challenging landscapes in the world, ranging from vast scorching deserts on the west to dense jungles in the east, broken riverine areas, icy glaciers, high-altitude regions, and extensive coastal zones.

Given the varied nature of operational environment in India, the designing the ideal Armoured Fighting Vehicles requires an appreciation of two different types of vehicles, the main battle tank and the troop carrying AFV essential for mechanized infantry. In fact, a consideration of the feasible platforms isn't complete without the consideration of the main battle tank, rather two main battle tanks, one each for the plains and the high altitude areas. It should also be kept in mind that the considerations shouldn't be limited to just the main battle tank but must include the specifications of the entire mechanised platforms in the combined arms ecosystem.

The open plains and deserts along the western front with Pakistan are tailor made for heavy tanks which have superior armour and firepower. These tanks will not only deliver lethal firepower but are nearly impenetrable on the battlefield. Heavy tanks are pivotal for large scale conflicts where devastating firepower and heavy armour turn the tide of battle. While the open terrain is best suited for employment of heavy tanks, the rugged high altitude landscape along the northern and north eastern borders with China dictate the need for light tanks for mobility in steep unfriendly mountainous terrain where heavy tanks will struggle. Light tanks are most suitable for quick manoeuvre, rapid deployment and operations in confined spaces, making them ideal for the challenging conditions of the Himalayan terrain. Thus, Heavy Tanks against Pakistan and Light Tanks against China form a critical aspect of military calculus of India's unique geopolitical threats.

The ideal AFVs for India must be multi-role, multi-use, highly adaptable platforms suitable for strategic mobility for swift deployment across the country's diverse and challenging terrains. The Army must be prepared for manoeuvre warfare. A high degree of mobility is required in all types of terrain. The infantry will have to be given matching mobility as per the dictates of the terrain and what is of the utmost importance is that infantry is given suitable protection, this is a great necessity in view of the lethality paradigm in the battlespace having changed. Therefore, whether it is a main battle tank designed for direct combat or a troop-carrying AFV tasked with safely transporting infantry, these vehicles must leverage the latest in military technology to ensure that India's armed forces remain capable of defending the nation's borders and interests, no matter where the conflict may arise. By integrating modular design, AI, advanced sensors, robotic systems, and sustainable power solutions, these AFVs will set a new benchmark for mechanised warfare in the region.

Varying types of terrain on our borders has been a restricting factor that comes into play whilst working out Indian Army's requirements. As far as the aspect of wheeled versus tracked ICVs is concerned, the issue has been considered and Indian Army chose a delicate mix of both. The RFI for a wheeled ICV has been processed. The Future Ready Combat Vehicle (FRCV) is on course and as per current indications should be available by the end of the next decade.[49] New and emerging technologies have been duly factored in, however, deeply related to it is the HR issue of getting the most suitable qualified manpower to leverage state of the art weapons, platforms and systems.[50]

Indian Army will have to study the recent global conflicts and the ongoing ones which have thrown up plethora of lessons on not only the technological mix required by an effective mechanised force but also the structure leveraging that technology. Therefore, deliberations in sync with our threat perception and areas of interest must be carried out to decide upon structures and organisations that keep our mechanised forces relevant in the face of multi front multi terrain employment. We need to be more lean and mean or we are actually more lean than required? May be we need more Armoured and Mechanised Regiments but with smaller tank troops and squadrons. We definitely want to have nimble formations which have strategic mobility but at the same time have minimal signature.

Advantages and Disadvantages of Emerging Technologies

The introduction of emerging disruptive technologies into armoured forces brings with it a range of advantages and disadvantages, which must be carefully weighed to maximize operational effectiveness.

Advantages

Enhanced Lethality: PGMs and smart munitions have improved the accuracy and lethality, enhancing single shot kill probability of modern AFVs.

Greater Flexibility: The advancements in modular AFVs and their integration with UGVs and UAVs allows for more versatile operations, with the option to deploy autonomous systems in high risk environments.

Increased Survivability: New Active Protection Suites significantly enhance the protection of armoured vehicles against a range of modern missiles and munitions, reducing casualties and increasing survivability of the platform.

Improved Situational Awareness: Modern communication systems and Battle Management System enhances the ability of mechanised forces to gather, process and share information in real time, leading to better decision making and fighting a coordinated battle.

Disadvantages

Complexity: The integration of multiple advanced systems increases the complexity of systems as well as resultant operations, requiring specialised training and technical support to ensure that all components function effectively.

Cost: The development and deployment of advanced technologies is time consuming & expensive and maintaining & upgrading these systems often puts strain on military budgets and national economies.

Vulnerability to Countermeasures: Highly integrated platforms with smart electronic and AI enabled systems are efficient but at the same time highly vulnerable to enemy action. They face a permanent risk of being disrupted or neutralised by electronic warfare, cyberattacks or other countermeasures.

Ethical and Legal Concerns: As discussed earlier, the use of AI and autonomous systems faces questions about accountability, the potential of unintended consequences and the ethical implications of using machines to make life and death decisions.

Future Trends and Anticipated Progress

Looking into the future, several trends are likely to shape the development and deployment of mechanised forces over the next two decades.

Enhanced Protection: Advances in light weight, resilient, stealth armour materials and APS will continue, with a focus on more effective protection without compromising mobility.

Network Integration: The integration of mechanised forces with its own elements as well as into a fully networked battle space will be the new character of mechanised operations, with real time information sharing and coordinated operations becoming the mainstay.

Increased Autonomy: The continuous development of UGVs and autonomous AI enabled platforms will usher in greater autonomy in mechanised operations, with vehicles capable of operating independently or swarms in concert with larger military objectives.

Global Competition: The competition between military powers will continue to fuel development of advanced armoured platforms, with ongoing efforts to achieve technological superiority.

Ethical and Legal Considerations: With the omnipresence of AI and autonomous systems, ethical and legal frameworks will need to evolve to address the challenges associated with their employment, especially in lethal and fatal combat.

The Indian Context

For the armoured and mechanised corps, the evolution ideally needs to occur on many fronts to ensure that conventional, heavy armour will transition to a networked, sensor heavy, agile force. Armoured platforms ideally need to be outfitted with state of the art detector suites that should include thermal imaging, LiDAR and synthetic aperture radars, all feeding into a combined digital grid, on which each operator will receive a common output for their designated role in an operation. The integration of sensors will enhance situational awareness within an increasingly complex and fluid battle space, as well as enable the rapid identification and tracking of adversary's movement in real time. Mechanised forces will leverage real time data fusion technologies to coordinate their movement with other combat arms, and in turn establish a coordinated defensive line.

In addition, deploying UGVs for reconnaissance, road and mine field clearance, and remote turret operations can reduce risk to human operators and increase operational tempo. UGVs with advanced machine learning systems can conduct a wide range of operations and missions, including reconnaissance, targeting and actual combat. Autonomous systems can negotiate severe terrain such as rugged desert and mountains, ignore treacherous urban environments, identify enemy positions and threats and return fire with striking accuracy and minimal collateral damage. This not only protects servicemen from hostile environments but provides mechanised columns the capacity to sustain operational tempo, leveraging technology to overcome and outmanoeuvre opposition from conventional battles to asymmetric wars.

To mitigate the threats like hypersonic munitions and advanced antitank systems, the Indian army must invest in advanced metallurgy, particularly in nano material composite systems or reactive armour systems. Nanotech has the potential to create armour that is light weight and extremely durable utilizing the molecular behaviour of material to develop alloys that can withstand extreme kinetic energy.

Reactive armour, on the other hand, actively counteracts with the approaching projectiles by exploding outwards in order to disrupt or detonate the projectile and absorb their energy before they penetrate to the core of the vehicle. Such innovations ensure that tank and other armoured vehicles can survive on increasingly lethal battlefields, protect the crew, and retain the Corps' ability to project power even against technologically advanced adversaries.

With mechanised formations more dependent on heavily interconnected electronic systems for navigation, target identification and communication, cyber security has become a key priority. Although the integration of electronic and networked capabilities offer significant operational advantages, they also introduce vulnerabilities that the adversary could exploit through cyberattacks. Once its control systems have been compromised, a hacked vehicle may be rendered unusable and ineffective or, in worst case, turned against a friendly force.

To avert the risk described above, the armoured corps should establish a robust cyber security protocol that encompasses encrypted communications, intrusion detection systems and continuous software updates to mitigate risks. Such measures ensure the integrity of own digital infrastructure, ensuring that the vehicles remain under friendly control and active even in the face of sophisticated cyber threats.

In the end, exploring energy weapons introduces a futuristic and realistic advancement for the Mechanised Forces defensive capability. Directed energy weapons, like a high powered laser, can be mounted on an armoured vehicle providing a point defense system from an increasing number of aerial threats, including drones, missiles and low flying aircraft. Laser weapons use instantaneous energy beams in contrast to conventional kinetic energy projectiles and are able to destroy with instant lethality and precision, adding immensely to the tempo of operations, limited solely by power generation capability. In the face of the increasing number of cheap, drone swarms that pose a significant challenge to conventional mechanised columns, this capability is particularly valuable. The mechanised forces can enhance their survivability and versatility by integrating energy weapons, thus creating a decisive edge in a contested environment where conventional countermeasures are likely to be exhausted.

The integration of emerging technologies is ushering in a major change in Armoured Fighting Vehicle (AFVs) construction and tactics, resulting in more adaptable, resilient, and effective combat vehicles to meet the demands of modern warfare. The tactical implications of these technologies are profound, enabling more dynamic and flexible approaches to combat. From autonomous systems that are powered by artificial intelligence to hybrid propulsion systems or advanced satellite communications integrated into a network-centric warfare system, the armoured warfare of tomorrow will certainly be far more lethal and smart, ensuring that military commanding them may also command a decisive edge over the forces on the other end of the barrel.

Notes

1   Tiananmen Square, US Government Office of the Historian, 1989, https://history.state.gov/milestones/1989-1992/tiananmen-square

2   Maj Thomas B. Gukeisen School of Advanced Military Studies United States Army Command and General Staff College Fort Leavenworth, Kansas The Operational Art of Blitzkrieg: Its Strengths and Weaknesses in Systems Perspective, 26 May 2005, https://apps.dtic.mil/sti/pdfs/ADA435929.pdf

3   https://static.rusi.org/heavy-armoured-forces-in-future-warfare-occasional-paper-december-23.pdf

4   The authors published an analysis of these operations in 2023. See Jack Watling, Oleksandr V Danylyuk and Nick Reynolds, 'Preliminary Lessons from Russia's Unconventional Operations During the RussoUkrainian War, February 2022–February 2023', RUSI, 29 March 2023.

5   https://static.rusi.org/lessons-learned-ukraine-offensive-2022-23.pdf

6   https://hdiac.org/technical-inquiries/notable/technological-lessons-learned-from-the-conflict-between-russia-and-ukraine/

7   Azar Gat1 July 20, 2023 The Future of the Tank and the Land Battlefield https://www.inss.org.il/wp-content/uploads/2023/07/special-publication-200723.pdf

8   Ibid

9   Udi Etsion 05.08.19 Israel's Defense Ministry Demonstrates New Autonomous Armored Vehicles https://www.calcalistech.com/ctech/articles/0,7340,L-3767703,00.html

10 https://www.linkedin.com/pulse/emerging-trends-armoured-fighting-vehicles-afv-market-global-op1cf/

11 Andrew Feickert Specialist in Military Ground Forces, The Army's Future Combat System (FCS): Background and Issues for Congress, August 3, 2009, https://dml.armywarcollege.edu/wp-content/uploads/2023/01/CRS-FCS-Issues-for-Congress-2009.pdf

12 Christopher G. Pernin, Elliot Axelband, Jeffrey A. Drezner, Brian B. Dille, John Gordon IV, Bruce J. Held, K. Scott McMahon, Walter L. Perry, Christopher Rizzi, Akhil R. Shah, Peter A. Wilson, Jerry M. Sollinger Lessons from the Army's Future Combat Systems Program The RAND Corporation https://www.rand.org/content/dam/rand/pubs/monographs/2012/RAND_MG1206.sum.pdf

13 Christian Baghai, The Future of Tank Warfare: Electromagnetic Armor, Feb 26, 2024, https:/

/christianbaghai.medium.com/the-future-of-tank-warfare-electromagnetic-armor-578db90e18a8

14   TROPHY APS, Active Protection System Revolutionizing Ground Maneuver Operations, Apr 11, 2024, RAFAEL, https://www.rafael.co.il/blog/trophy-aps/

15   Russian T-90M Tanks Integrate New Arena-M Active Protection System - Can Now Shoot Down Incoming Missiles, Military Watch Magazine, August 23, 2024, https://militarywatch magazine.com/article/t90m-arenam-protection-missiles

16   Analysis Russian Afganit active protection system is able to intercept uranium tank ammunition TASS 11012163, Feb 6, 2018, https://armyrecognition.com/focus-analysis-conflicts/army/ defence-security-industry-technology/analysis-russian-afganit-active-protection-system-is-able-to-intercept-uranium-tank-ammunition-tass-11012163

17   Maya Carlin The M1 Abrams SepV3 Might Be the Best Tank on Earth Right Now https:// nationalinterest.org/blog/buzz/m1-abrams-sepv3-might-be-best-tank-earth-right-now-210083

18   Lynx KF41: Modular IFV for Full Spectrum of Operations, Aug 10, 2020, https:// www.czdefence.com/article/lynx-kf41-modular-ifv-for-full-spectrum-of-operations

19   British Army achieves first firing-on-the-move test with Ajax infantry fighting vehicle, 9 Aug, 2024, https://armyrecognition.com/news/army-news/army-news-2024/british-army-achieves-first-firing-on-the-move-test-with-ajax-infantry-fighting-vehicle

20   Ground X-Vehicle Technologies (GXV-T) Defense Advanced Research Projects Agency https:/ /www.darpa.mil/program/ground-x-vehicle-technologies

21   N.R.P. Schützenpanzer Puma: Germany's deadly new Infantry Fighting Vehicle, June 26, 2015, https://defencyclopedia.com/2015/06/26/schutzpanzer-puma-germanys-deadly-new-infantry-fighting-vehicle/Schützenpanzer Puma: Germany's deadly new Infantry Fighting Vehicle

22   Boxer 8x8 MRAV, Army Recognition Group, 24 Sep, 2024, https://armyrecognition.com/ military-products/army/armoured-personnel-carriers/wheeled-vehicles/germany-boxer-uk

23   Armata Universal Combat Platform https://en.wikipedia.org/wiki/Armata_Universal_Combat _Platform

24   Audra Calloway, Picatinny Arsenal Public Affairs, Army developing safer, extended range rocket-assisted artillery round, August 25, 2016, https://www.army.mil/article/174013/army_ developing_safer_extended_range_rocket_assisted_artillery_round

25   XM25 Counter Defilade Target Engagement System https://www.military.com/equipment/ xm25-counter-defilade-target-engagement-system

26   M829A4 (formerly M829E4) Armor Piercing, Fin Stabilized, Discarding Sabot – Tracer (APFSDS-T) FY15 Army Programs, https://www.dote.osd.mil/Portals/97/pub/reports/FY2015/ army/2015m829a4.pdf?ver=2019-08-22-105950-793

27   LAser Homing Attack Missile (LAHAT) https://www.army-technology.com/projects/laser-homing-attack-missile/?cf-view

28   RQ-11 Raven Unmanned Aerial Vehicle https://www.army-technology.com/projects/rq-11-raven/

29   Ukrainian army reportedly received M60 AVLB mobile bridges, Army Recognition Group, 30 Jan, 2024, https://armyrecognition.com/news/army-news/army-news-2024/ukrainian-armed-forces-have-received-m60-avlb-mobile-bridges

30   The British CRARRV Armoured Recovery Vehicle TankNutDave.com https://tanknutdave.com/ the-challenger-armoured-repair-and-recovery-vehicle/

31   PzH 2000 Combat Proven Artillery Superiority https://www.knds.de/fileadmin/user_upload/ broschueren_2024/KNDS_B_Ansicht_PzH2000_EN.pdf

32   MLRS M270 The combat proven, high volume, long-range precision fires launcher supporting combined Joint All-Domain Operations https://www.lockheedmartin.com/en-us/products/ m270.html

33   96K6 Pantsir-S1 (SA-22 Greyhound) Russian Short-Range Air Defense Gun/Missile System https://odin.tradoc.army.mil/WEG/List

34 M6 Bradley Linebacker https://weaponsystems.net/system/1261-M6+Bradley+Linebacker

35 BW – GTK BOXER A1 FÜFZG http://tank-masters.de/?page_id=1439

36 Heavy Expanded Mobility Tactical Truck A4 HEMTT A4 https://oshkoshdefense.com/vehicles/heavy-tactical-vehicles/hemtt/

37 BFIST M7A3 M7 FIST 26 Jul, 2024 Army Recognition Group https://armyrecognition.com/military-products/army/artillery-vehicles-and-weapons/radar-vehicles/bfist-bradley-fire-support-team-m7a3-m7-fist-technical-data-sheet-specifications-pictures-video-10201175

38 Assault Amphibious Vehicle https://defenseinnovationmarketplace.dtic.mil/wp-content/uploads/2018/05/2018_ATIP_7_01_AAV.pdf

39 Kapil Kajal South Korea deploys upgraded K200 variants | 2022-12-16 https://www.janes.com/osint-insights/defence-news/land/south-korea-deploys-upgraded-k200-variants

40 Skylark 1 https://www.israeli-weapons.com/weapons/aircraft/uav/skylark/Skylark.html

41 RQ-4 Global Hawk, Air Force https://www.af.mil/About-Us/Fact-Sheets/Display/Article/104516/

42 LARS https://weaponsystems.net/system/141-LARS

43 Navy Shipboard Lasers: Background and Issues for Congress August 6, 2024 Congressional Research Service https://sgp.fas.org/crs/weapons/R44175.pdf

44 China Shows Off Silent Hunter 33kW Laser, Drones and Tanks March 13, 2017 The Defense Systems Information Analysis Center (DSIAC) U.S. Department of Defense https://dsiac.org/articles/china-shows-off-silent-hunter-33kw-laser-drones-and-tanks/

45 Ashish Dangwal, Discarded by US Navy in 2021, Japan Keen to Delve into US Research to Master Electromagnetic Railgun Tech, May 22, 2024 https://www.eurasiantimes.com/discarded-by-us-navy-in-2021-japan-keen/

46 Gabriel Honrada, China firing railguns while US blows hot and cold, May 22, 2024, https://asiatimes.com/2024/05/china-firing-railguns-while-us-blows-hot-and-cold/

47 Paladin M109A7 155mm Artillery System, USA https://www.army-technology.com/projects/paladin-m109a7-155mm-artillery-system/

48 South Korean K2 Black Panther Tank https://fighting-vehicles.com/tanks/k2-black-panther-tank/

49 Lt Gen A B Shivane PVSM, AVSM, VSM Future Ready Combat Vehicle and Light Tanks Defence Research and Studies, Mar 23, 2024, https://dras.in/future-ready-combat-vehicle-and-light-tanks/

50 New Age Warfare, Emerging Technologies & Armoured Vehicles, Seminar Report, *South Asia Defence and Strategic Review,* December 9, 2019, https://www.defstrat.com/magazine_articles/new-age-warfare-emerging-technologies-armoured-vehicles/

11

Shells to Smart Systems: Artillery's Quest for Precision and Power

"Artillery adds dignity to what would otherwise be an ugly brawl."
—**Frederick the Great**

Artillery has been a decisive force in military operations for centuries, providing the firepower necessary to support infantry and armoured units, disrupt enemy formations, and destroy critical infrastructure. The role of artillery has evolved significantly over time, with advancements in technology leading to increased accuracy, range, and lethality. In modern warfare, artillery continues to play a critical role, whether in traditional battlefield engagements, counterinsurgency operations, or urban combat.

As warfare becomes more complex and technologically advanced, the importance of artillery is only expected to grow. Emerging disruptive technologies are set to revolutionize the capabilities of artillery units, making them more effective and adaptable to the demands of future conflicts. This chapter explores these technologies and their impact on the future of artillery warfare.

The world of artillery warfare is on the cusp of a profound transformation, driven by the relentless march of disruptive technologies. As emerging innovations in artificial intelligence, robotics, hypersonics, and directed energy reshape the landscape, the very nature of artillery systems and their employment is being fundamentally altered.[1]

At the heart of this revolution lies the growing influence of artificial intelligence and machine learning. These technologies are enabling a new generation of autonomous and semi-autonomous systems, capable of targeting, tracking, and engaging enemy forces with unprecedented speed and precision. The integration of AI-powered algorithms into artillery systems is enhancing their decision-making capabilities, allowing for more effective and efficient use of firepower. The increased proliferation of these intelligent systems in years to come, is poised to redefine the role of human operators, who will increasingly serve on the loop providing decision-support rather than direct controllers.

Robotics, too, is transforming the face of artillery warfare. The development of unmanned ground vehicles and loitering munitions is introducing a new level of mobility, flexibility, and lethality to artillery forces.[2] These robotic systems can operate in hazardous environments, extend the reach of artillery, and provide enhanced reconnaissance and targeting capabilities. The integration of advanced sensors, navigation systems, and autonomous control further amplifies the impact of these robotic platforms, blurring the lines between traditional artillery and emerging robotic warfare.

Hypersonic weapons, with their unparalleled speed and manoeuvrability, are also poised to revolutionize artillery warfare. These next-generation projectiles can penetrate advanced air defense systems and strike targets with devastating precision, challenging conventional notions of artillery engagement. (Brady & Goethals, 2019) The development of such high-speed, long-range artillery systems is redefining the strategic and operational calculus of modern warfare, forcing a rethinking of doctrine, tactics, and force posture.

As these diverse technological innovations converge, the future of artillery warfare is being reshaped in profound ways. The integration of AI, robotics and hypersonics is ushering in a new era of artillery capabilities, redefining the very nature of land-based firepower. These disruptive technologies are not only enhancing the lethality and precision of artillery but also introducing new strategic and operational considerations that will require a fundamental rethinking of doctrine, training, and force structure.

BREAKING FORTRESSES AND MINDSETS: ARTILLERY'S PSYCHOLOGICAL AND STRATEGIC IMPACT

The Russia-Ukraine war has emerged as a laboratory for modern artillery warfare, showcasing how emerging disruptive technologies – such as precision-guided munitions (PGMs), drones, artificial intelligence (AI), and electronic warfare (EW) – are reshaping battlefield dynamics. Though Russia initially delivered massed fires and restricted the use of PGMs but as the war got protracted, both nations started innovating to gain tactical advantages, turning artillery duels into a high-stakes contest of technological adaptation.[3]

From Mass Barrages to Surgical Strikes

Artillery has long dominated the Russia-Ukraine conflict, but the shift to Precision-Guided Munitions (PGMs) has reduced reliance on indiscriminate shelling, favoring precision instead. Russia deployed the Krasnopol-M2 Laser-Guided Shells extensively during the Battle of Sievierodonetsk.[4] Paired with drone-corrected targeting, Krasnopol shells enabled Russian forces to dismantle Ukrainian fortifications with minimal waste of ammunition. While not traditional artillery, Iskander Ballistic Missiles have been used for precision strikes on Ukrainian logistics hubs, during the battles in Donbas region.[5]

Excalibur GPS-Guided Shells supplied by the US, allowed Ukraine to conduct pinpoint counterbattery strikes. Excalibur shells neutralized Russian artillery positions hidden in urban areas, slowing Moscow's advance.[6] Though a rocket artillery platform, HIMARS' GPS-guided missiles disrupted Russian supply lines in Kherson (2022), demonstrating how PGMs can shape operational outcomes.[7] PGMs have forced both sides to prioritize concealment and mobility, reducing static artillery deployments.

Eyes in the Sky

Drones have democratised reconnaissance and strike capabilities, offering real-time targeting data and low-cost attack options.[8] Orlan-10 Surveillance Drones provided real-time coordinates for Russian artillery during the Siege of Mariupol, enabling relentless bombardments of Ukrainian positions.[9]

On the other hand, Ukraine modified Commercial Drones (DJI Mavic) for military use. These drones identified Russian positions during the Kharkiv Counteroffensive, directing devastating HIMARS strikes. Drones have made artillery more responsive but also increased vulnerability, as both sides prioritize drone-jamming technologies. DJIs had become so widespread and useful because of their relatively low cost and simplicity that Russia's top military leaders have named them a "true symbol of modern warfare," noting they have elevated the importance of artillery to heights not seen since World War.[10]

Accelerating the Sensor to Shooter Cycle

AI is streamlining artillery targeting, reducing decision-making time from hours to minutes. During the Battle of Donetsk, Russia tested AI-Driven Target Recognition systems that analyzed satellite imagery and drone feeds to prioritize high-value targets, such as command centers.

Ukrainian forces exploited Machine Learning for Predictive Targeting by using AI algorithms to anticipate Russian artillery movements in Zaporizhzhia, enabling preemptive strikes. While still nascent, AI is shifting artillery from reactive to proactive operations.

ARTILLERY IN THE AGE OF ASYMMETRY

In recent decades, artillery has evolved from traditional tube-based systems into highly networked, precision-guided platforms. Artillery remains a cornerstone of modern warfare, but its application in asymmetric conflicts has shifted from mass bombardment to precision strikes, enabled by technological advancements. Modern conflicts in the Middle East have increasingly featured asymmetric warfare, where state-of-the-art military systems confront irregular tactics and decentralized organizations. Artillery, long a mainstay of conventional warfare, has adapted to this environment. Both state and non-state actors have leveraged innovations to enhance targeting accuracy, speed of response, and operational flexibility. The Israel-Hamas-Hezbollah conflict and the Syrian Civil War provide compelling case studies of how state and non-state actors have adapted artillery to achieve tactical and strategic objectives.

The Israel–Hamas conflict of 2023 saw extensive artillery exchanges, particularly during Israeli operations against Hamas positions in Gaza.

- **Hamas Rocket and Mortar Attacks:** In the initial phase, Hamas launched over 3,000 rockets and mortars toward Israeli cities, including Ashkelon and Tel Aviv. These attacks overwhelmed Israel's Iron Dome missile defense system in specific instances, forcing the Israel Defense Forces (IDF) to rely on counter-battery artillery strikes.[11]
- **IDF Artillery Response:** The IDF's M109 155mm self-propelled howitzers, in coordination with airstrikes, targeted Hamas launch sites, underground tunnel networks, and command centers. Israeli artillery units used precision-guided shells, including the EXCALIBUR GPS-guided 155mm artillery rounds, to strike Hamas positions with minimal collateral damage.[12]
- **Technological Adaptation:** Drones provided real-time target acquisition, allowing for rapid artillery adjustments. Hamas countered this by using mobile launch platforms, which they moved quickly after each salvo to evade detection.

On the sidelines, a renewed conflict erupted between Israel and Hezbollah along the northern border. This conflict brought forward next-generation artillery systems and highlighted a rapid evolution in tactics by both sides. The intense artillery exchanges demonstrated a balance between high-precision strikes and rapid repositioning, with both sides learning from each engagement to refine their operational doctrines.

- **Mobile and Concealed Positions:** Hezbollah improved camouflage and mobility of its artillery, using a combination of older systems with enhanced tactics to evade detection.

- **Hybrid Firepower:** Aside from conventional rocket systems, Hezbollah introduced modified MLRS platforms capable of launching a mix of guided and unguided munitions, complicating Israel's targeting process.
- **Precision Engagements:** Israeli forces integrated advanced digital fire control networks with next-generation 155mm howitzers. These systems enabled rapid identification and engagement of Hezbollah mobile artillery units.
- **Counter-Battery Enhancements:** Utilizing upgraded radars and real-time data links, Israeli units effectively tracked and targeted shifting Hezbollah rocket launchers. A notable artillery duel near the town of Marjayoun forced Hezbollah to frequently relocate its assets.[13]

Meanwhile, the Syrian Civil War entered a new phase marked by urban artillery engagements in contested cities such as Homs and Idlib. Both regime and rebel forces adapted their artillery strategies to the complexities of densely populated environments. Both sides increasingly relied on real-time satellite imagery and drone surveillance to optimize their artillery fire, though the regime's superior integration of electronic warfare tools often provided them with a tactical advantage in counter-battery operations.

Rebel forces continued to utilize improvised mortar systems, adapting them with locally sourced modifications to improve range and accuracy. In the narrow alleys of Homs, rebels leveraged the urban terrain for cover, launching short-range artillery attacks before rapidly dispersing. The Syrian regime, bolstered by its allies, began deploying upgraded versions of the BM-21 Grad and 2S19 Msta howitzers equipped with digital fire control systems and supported by use of first-person view suicide drones. However, the experimental autonomous artillery drones capable of adjusting firing solutions on the fly field-tested by the Kataib Shaheen Brigade of the Hayat Tahrir al-Shamto overwhelmed the regime forces and continued to maintain a tactical edge.[14]

While state actors leverage advanced systems to maintain a technological edge, non-state adversaries persistently adapt with innovative tactics and hybrid artillery solutions. The evolving interplay between precision strikes, counter-battery measures, and rapid mobility suggests that artillery will remain at the forefront of future military engagements, continually shaped by advancements in automation, AI, and real-time intelligence.

THE RISE OF ROBOTIC ARTILLERY

With technology continuing to change the battlefield paradigms, one area on the threshold of a radical overhaul is artillery. Artillery, which has traditionally depended on human crews for targeting, loading, and firing, is becoming a more automated and autonomous field of warfare. The increasing advent of robotic artillery would further enhance the accuracy, speed, and survivability of the engagements, but they also bring new challenges for military organisations as they integrate these scientifically advanced platforms into established force structures. Subsequent paras discuss some of the robotic artillery platforms, along with the advantages they confer tactically, and assesses the difficulties in integrating them into modern armies.

Autonomous Self-Propelled Howitzers

Such artillery systems illustrate an evolution of traditional platforms such as the German PzH 2000 or the US Paladin, wherein automation improves performance. These autonomous howitzers can load, aim, and fire without human intervention, greatly increasing the rate of fire and accuracy. The next generation of these systems may even be entirely autonomous, capable of independent operations on the battlefield and able to execute intelligent decisions almost on-the-spot.[15]

Unmanned Ground Vehicles (UGVs) with Artillery Capabilities

The UGVs with artillery systems can be categorised as another emerging class in robotic artillery. These remotely piloted vehicles can be armed with mortars, rockets, or even larger caliber guns and operate autonomously. Russia's Uran-9 proves that UGVs can operate as mobile fire support units, acting independently or in hybrid modes with human task forces to launch precision strikes.[16]

Mobile Rocket Systems

Robotic Multiple Launch Rocket Systems (MLRS), designed to quickly fire volleys of rockets over long distances, are therefore optimally placed to deliver a saturation fire to support advancing troops or disrupt enemy formations. Autonomous or semi-autonomous versions, e.g. HIMARS, could even be fielded to work in synchronisation with reconnaissance drones and AI-enhanced fire control systems, maximizing battlefield utility.[17]

Robotic Artillery Tactical Advantages

The advantage of institutionalising development robotic artillery is considerable and offers an evolution in the very methods artillery will be utilised in future combat.

Speed and Precision

One of the great advantages of robotic artillery is that such systems can process data and engage with the targets significantly faster than human operators. These systems, powered by artificial intelligence, will calculate and adjust targeting data in real-time; this will greatly enhance precision and response time. In doing so, they will reduce the likelihood of collateral damage to almost negligible levels, which is vital in such complex urban environments or close to civilians.

Reduced Risk to Humans

Robotic systems increase survivability and efficiency of the human crew on the battlefield, especially in high-risk zones. This minimises the threat to personnel, particularly in forward positions or under fire from counter-battery units. Robotic artillery systems can efficiently operate in extreme environments as well as in inactive deployments or No War No Peace scenarios, whether in super high-altitude areas, deserts, or nuclear- or chemical-agent environments, wherein the endurance and survivability of human operators cannot be guaranteed.

Continuous Action

Robotic artillery systems will facilitate undistracted operation for as long as required without rest or rotation of personnel. This presents advantage in prolonged engagements where fire support must continue to be rendered. Moreover, they would be able to spread both in remote areas and independently, providing artillery support to units fighting far from supply lines.

Logistical Efficiency

Robotic artillery indeed offers more efficient logistics and coordination. Using reconnaissance drones, intelligence assets, and sensor-to-shooter systems, robotic artillery can respond to threats more effectively. In a coordinated system, data from multiple sources can be used to direct precision strikes from robotic platforms, ensuring that artillery is always in the right place at the right time.

Challenges of Integrating Robotic Artillery into Force Structures

However, as much as these advantages sound good, integrating robotic artillery with existing military formations presents several challenges.

Interoperability with Legacy Systems

Many present-day militaries are deeply dependent on legacy artillery systems that were designed decades ago. Often, such systems lack the digital infrastructure needed to communicate with autonomous platforms, thereby making it difficult to integrate these new robotic systems. The costs and HR considerations involved in the upgrading or replacement of legacy systems may further delay the adoption of robotic artillery.

Cybersecurity Risks

The growing degree of automation and AI introduces untried vulnerabilities in the shape of a cyber threat. Autonomous systems may be susceptible to hacking, either through capability or means that create misunderstood direction concerning firing, or unauthorised activation, or even shutting down systems. The integrity of robotic artillery systems from hacker interventions will remain one of the most important aspects for military planners to consider.

Ethical and Legal Concerns

Robotic artillery especially the autonomous systems also come under the umbrella of societal considerations both ethical and legal. The concept of autonomous systems applying lethal force without human intervention is controversial, and thus, international laws surrounding the use of autonomous weapons remain in their infancy. Parties to military conflict, therefore, must operate within the legal and ethical frameworks agreed upon by the societies within which they are located.[18]

Command and Control Complexity

Successful command and control (C2) is of utmost importance for successful integration of robotic artillery. Militaries will have to develop new tactics and doctrines that will adequately count on the presence of autonomous systems. Coordination that emerges between human forces and robotic platforms require solid communication lines, backed up by a seamless C2 framework to ensure that human commanders are always in control while capable of leveraging the robotic systems.

Advanced Targeting and Fire Control Systems

The advanced targeting and fire control systems are changing the way artillery units hit targets, giving them the ability to do so with greater accuracy, speed, and flexibility. Using AI with real-time data generated from multiple sensors and data fusion, these systems will give artillery operators the precision and effectiveness for using the accurate firepower with coordination.

Modern targeting and fire control systems are, however, very common, and AI is at work here as a real time processor of information, gathering data from several sources such as drones, satellites and land-based sensors, enabling rapid fabricating and drawing conclusions within seconds of events occurring.

As an example, the US Army's Advanced Field Artillery Tactical Data System (AFATDS) is a digital fire control system that integrates AI and data fusion technologies to coordinate artillery fire across multiple platforms. AFATDS can process targeting data from several different types of sensors and uses a set of rules to prioritize these targets based on the objectives of the mission. AFATDS can automate the coordination of fire missions, thereby reducing the time from target identification to delivering precision and effective fire support.[19]

Careful consideration exists regarding the acceptance of advanced targeting and fire controls, which are based on some features of complexity, interoperability, and secure communication meters. Therefore, developments around such key fighting military assets as UAVs-drones will ensure a far-reaching impact of these systems on the battlefield.

The future looks busy with more AI, machine learning, and data analytics improvements aligned with targeting and fire control-related systems. Such improvements allow more predictability and adaptability in targeting applications, making artillery units far better able to engage in contested and dynamic environments.

HYPERSONIC WEAPONS AND ARTILLERY: A NEW ERA OF SPEED AND REACH

Hypersonic weapon systems are a game-changer in warfare, with serious implications for the future of artillery. Hypersonics are defined as weapons that fly faster than Mach 5, through the atmosphere, and drastically alter the way long-range engagements and strategic deterrence are done, and what artillery means.

China's DF-17 hypersonic glide vehicle, for instance, is designed to evade detection and interception by leveraging its speed and manoeuvrability, posing a threat to key assets in scenarios such as a future China-Taiwan conflict. The US has been developing a counter to the Chinese with systems such as Glide Breaker to intercept hypersonic threats during their glide phase and space-based sensors such as the Hypersonic and Ballistic Tracking Space Sensor (HBTSS) for tracking these weapons worldwide. The rise of hypersonic weapons demands innovation in detection and interception, pushing air defense systems to adapt rapidly to this high-speed, elusive threat.

These hypersonic weapons are set to redefine artillery limitations with respect to range and response time. While conventional artillery systems can only fire projectiles at limited angles and target velocities, weapons flying on hypersonic trajectories are set to ring forth distance bands into unimaginable speeds and hit targets previously considered evasive from conventional artillery. Therefore, this would truly mean:

- **Strategic Depth Strikes:** Hypersonic artillery can fire upon critical enemy infrastructures like command-and-control centers or force concentrations several miles behind enemy lines, disrupting their potential to wage war effectively.
- **Rapid Global Strike Capability:** The speed of hypersonic weapons allows for a rapid response against emerging global threats, produced deterrent by an assured, rapid, and catastrophic retaliation.
- **Compression of Decision-Making Cycles:** The shorter time of flight for hypersonic weapons compresses the decision-making cycles for both

attackers and defenders and requires faster information processing, target acquisition, and engagement reactions.

Shifting the Paradigm of Artillery Development

The integration of hypersonic technologies is instilling a paradigm shift in artillery development toward:

- **Development of Advanced Propulsion Systems:** The development and fielding of such hypersonic artillery systems would require substantially advanced propulsion technology, such as scramjet engines, ramjets, and variational advanced propulsion technologies that allow the projection of hypersonic speeds.
- **Thermal Management and Materials Science:** The extremes of temperature and aerodynamic forces expected to be experienced by hypersonic flight projects necessitate improvised materials and thermal management solutions to ensure weapon integrity and accuracy.
- **Guidance and Control Systems:** Precision and reliability in guidance systems are essential for smoothly manoeuvring hypersonic projectiles through the atmosphere, exploiting atmospheric drag, and achieving the utmost accuracy over long ranges.

Strategic Implications and the Future of Artillery Warfare

With the emergence of different hypersonic weapons, we can hypothesize about major strategic implications ranging from:

- **Destabilizing Arms Races:** The pursuit of hypersonic-backed capabilities may result in fomenting regional and global arms races through nations that aim to build up countermeasures and keep strategic parity.
- **Violation of Strategic Stability:** The speed and surprise offered by hypersonic weaponry could impair traditional deterrence strategies, rendering them susceptible to miscalculations and escalation.
- **New Doctrinal and Tactical Approaches:** The unique properties of hypersonic weapons will lead to new doctrines, tactics, and force structures to exploit their potentials and counteract their uses by adversaries.

The integration of hypersonic technology into artillery heralds the dawn of a new chapter of artillery warfare; as the technology matures and proliferates, it will continue to revolutionize the face of conflict, requiring a total rethinking of strategic thought, posture, and in many ways artillery warfare itself.

THE NANO EDGE IN ARTILLERY

Nanotechnology is now gaining ground in its transformative role in development of military platforms, including artillery weapons and systems, as result of the manipulation of matter at the molecular or atomic scale. Meant to improve materiel, energy conservation, and performance characteristics, nanotechnology has revolutionized how artillery is designed, operated, and employed on the battlefield.

The effect of nanotechnology on artillery design

Nanotechnology has the potential to bring tremendous improvements to gun systems by making them lighter, stronger, and more durable. This would allow for the development of more mobile artillery platforms, extended-range munitions, and the increased survivability of weapon systems and those operating them.

Lightweight and Stronger Materials

Traditionally, artillery systems like the tanks, have been heavy, limiting their mobility, especially in difficult terrains. Nanomaterials, such as carbon nanotubes and nanocomposites, provide extremely high strength-to-weight ratios that can dramatically reduce the weight of Artillery barrels, chassis, and munitions without compromising their durability. Countries are leveraging the use of nanomaterials in the production of lighter artillery platforms and leading to faster deployment and redeployment of gun systems, which is critical in modern day dynamic battlefield environment.[20]

Nanotechnology-enhanced armours for artillery platforms can also resist enemy fire and counter-bombardment strikes better than before. The case in point is Russia, where coatings and armour developed based on nanotechnology for their self-propelled artillery systems aim at reducing these platforms' susceptibility to enemy detection and precision-guided munition attacks.

Extended Range and Precision

Recent research in nanotechnology has been directed towards developing systems that can improve the range and precision of artillery munitions. Alterations to the chemical composition of conventional propellants and explosives could be made by manipulating factors at the nanoscale, making these systems more efficient and powerful.[21] These would produce more energy, enabling the shells to travel further and at higher speeds. Such development and application of nano materiels is being pursued by strategic rivals like China, Russia and the United States aiming at range extension and improved battlefield efficacy of their artillery systems.

Nanotechnology-enabled guidance systems and sensors integrated into the artillery shells will provide far greater accuracy, rendering massed artillery fire unnecessary, allowing for precise strikes instead. This trend coincides with the general consensus within the field of artillery toward the development of precision-guided munitions.

Self-Healing Materials

Another area of interest is that of self-healing materials, which could augment the durability of current artillery systems in a combat environment. However, with the introduction of nanoscale self-repairing mechanisms within the artillery components, such as barrels and chassis, allowing artillery systems to require less maintenance, thus enhancing their operational life.

Nano-Enhanced Munitions

The U.S. Army is one of the leading organizations implementing nanotechnology to improve artillery systems. In particular, one of the important projects involves the use of nano-enhanced propellants to increase the range and accuracy of standard artillery shells. Research into nano-thermites, energetic materials with a much higher rate of reactivity due to their nanoscale composition, offers the prospect for a revolution in artillery munitions with its possible faster burn rates and greater energy outputs.[22] Russia's military modernization efforts include the incorporation of nanotechnology in its artillery systems. For almost a decade countries had been experimenting with nanotechnology-based stealth coatings to reduce the radar and thermal signatures of its artillery systems, making them harder to detect and neutralize.

Furthermore, research in nano-energetic materials aims to increase the destructive power of artillery shells. The additional explosive yield can be achieved thereby in a relatively efficient manner, since these materials render possible the making of the munitions smaller and lighter. Thus, the mobility and logistics of existing artillery systems and their effectiveness can be enhanced further.

Moreover, China and India both have been working towards developing and deploying lightweight artillery systems by leveraging nano materiels. This would enhance the manoeuvrability of artillery units in across the high-altitude border regions.

Increased Lethality and Precision

With improved designs, enhanced ranges and precision-guided munitions, artillery guns guarantee striking long-precise blows against vital targets whilst expending fewer rounds. This surgical precision, which assists in reducing collateral damage, is most crucial in an environment where urban and civilian casualties are increasingly questioned.

SPACE TECHNOLOGY IN REVOLUTIONIZING ARTILLERY WEAPONS AND SYSTEMS

Space technology is transforming many aspects of military operations, and artillery warfare is no exception. With advancements in satellite communications, space-based reconnaissance, precision navigation, and real-time data sharing, the integration of space technology into artillery systems has given rise to more accurate, responsive, and lethal firepower. Various countries are moving fast to operationalise space-based systems into their artillery platforms, creating new paradigms in artillery warfare.

Satellite Navigation and Precision Strikes

One of the most important contributions of space technology to artillery is how navigation satellite systems like GPS (Global Positioning System), GLONASS (Russia), and Beidou (China) have improved the accuracy of artillery fire. While old artillery relied on range tables and manual refinements, satellite navigation allows PGMs and rockets to strike with an accuracy of 1 meter.

This advancement has significantly extended the range and precision of artillery systems. For example, the US Army had developed the Excalibur artillery shell, which can leverage GPS guidance to hit targets as far away as 40 kilometers with a circular error of probability (CEP) of only a few meters. This development remarkably reduces collateral damage and enhances artillery effectiveness when operating in urban or populated areas.[23]

Countries continue to advance in their efforts to modernize and equip their artillery systems for both target acquisition and precision targeting in real-time. The Russian artillery, for its part, relies heavily on GLONASS for greater accuracy of its long-range rocket artillery, such as the Tornado-S, which is able to fire GPS-guided rockets over a distance of nearly 120 kilometers.[24]

Considerable efforts are on by the South Asian arch rivals, India and Pakistan in this direction. With India having already launched its own GPS system called the NavIC, short for Navigation with Indian Constellation, also known as Indian Regional Navigational Satellite System (IRNSS) developed by the Indian Space Research Organisation (ISRO) and Pakistan trying to develop one so that its dependence on Beidou is minimised.[25]

Real-Time Targeting and Reconnaissance

Space-based reconnaissance satellites provide real-time imagery and data for the detection of enemy positions and for directing artillery fire. They allow artillery commanders an unprecedented view of the battlefield, providing multi dimensional capabilities in battlefield awareness from high-resolution optical and synthetic aperture radar (SAR) satellites for monitoring own as well as enemy movements.

Most advanced militaries use space-based assets for battlefield surveillance and target acquisition. The satellites in real time provide coordinates of enemy strongholds permitting artillery units to respond to changing scenarios on the battlefield rapidly. Space-based ISR (intelligence, surveillance, and reconnaissance) platforms reduce the need for human reconnaissance on the ground, minimizing the risks associated with recce and surveillance missions.

China is also now investing heavily in its space-based ISR capabilities to support its ground forces. The PLA relies on satellite intelligence to direct the fire of its strategic artillery systems, thereby enhancing the overall efficiency and effectiveness of its artillery operations. This becomes further enhanced with the integration of satellite imagery into the artillery fire control systems, hence, improving the speed and precision of its artillery units.[26]

Enhanced Communications and Networking

Satellite communications completely transform command and control in artillery operations. Space-based communication systems offer artillery units working in remote or contested areas continuous, long range, and secure channels. This facilitates real-time coordination of artillery units with other military services like air force or naval forces, hence assuring that artillery fires are synchronized with broader operational goals.

These countries, like the United States use the WGS (Wideband Global SATCOM) system to provide reliable communication links between their artillery units and headquarters. This system allows for seamless data transfer and real-time updates on battlefield progress, ensuring that artillery units are able to respond to threats and opportunities quickly and efficiently. The integration of satellite communications also enhances the ability of artillery to operate in decentralized formations, making them more agile and less vulnerable to enemy counter-attacks.[27]

The United States has a long history of being a pioneer in the use of space technology to augment artillery systems. A key example is the Excalibur GPS-guided artillery shell that relies on satellite navigation to do precision shooting over long distances. The US Army's Extended Range Cannon Artillery program aims to substantially extend the range of its artillery systems and is heavily reliant on space-based data for target acquisition and fire control.[28]

Russia's artillery modernisation efforts heavily rely on the integration of its GLONASS satellite navigation system into its long-range rocket artillery. The Tornado-S multiple launch rocket system (MLRS), an evolution of the BM-30 Smerch, is equipped with GLONASS guided rockets capable of precision strikes exceeding 120 kilometers. GLONASS gives Russian artillery units a critical advantage to strike accurately on targets located at long distances.[29]

Aside from GLONASS, Russia is using satellite ISR platforms for real-time targeting and post-strike evaluations. By feeding satellite information back into artillery fire control systems, Russia is boosting the accuracy and lethality of its artillery platforms such as the 2S35 Koalitsiya-SV self-propelled howitzer.[30]

Integrated Multi-Domain Operations

Space technology integrates artillery into multi-domain operations, where land, air, sea, cyber, and space assets operate in tandem. Artillery units are now a part of a much larger network that includes UAVs, satellites, and electronic warfare systems. This integration allows artillery to respond to data gathered from multiple domains, giving it yet another aspect of flexibility and efficiency.

For example, the US military's concept of Joint All-Domain Command and Control (JADC2) propounds integration of artillery seamlessly into a larger and networked battlespace. Space-based communications and ISR platforms allow artillery units to engage targets identified by other air or space assets, enabling coordination of fires across all domains.[31]

Challenges and Adaptation

Despite the numerous attributes that space technology may yield, incorporating space technology into the artillery system also comes with challenges. A major consideration is keeping satellites safe from and insulated from forms of electronic warfare or anti-satellite operations put up by foes. As space becomes a contested domain, nations will need to develop countermeasures and redundancies to ensure that their artillery units can continue to rely on satellite-based systems.

Artillery units will also need to adapt to the increasing reliance on space technology by integrating satellite-based data into their fire control systems. This will require new training and operational procedures, as well as closer coordination with space and cyber units to ensure uninterrupted access to space-based assets.

Towards a Universal Artillery Design

The development of artillery systems tailored to specific terrains can be quite expensive, for each environment lays unique requirements on the design, mobility, and durability of the platform.

Modularity in Design

One of the ways cost can be reduced is to develop modular artillery systems which are adaptable to the requirements of different terrains. These modular platforms can have interchangeable components such as chassis, barrels, fire control systems, which can allow military developers to modify the same base system for operations in desert, mountain, or plains.

For instance, the Caesar 155mm howitzer is an example of such a versatile system that can operate in both desert and plain environments. A modular design allows for easy adaptation, such as adding enhanced cooling systems for desert operations or ruggedizing the chassis for mountain terrain.[32]

Multirole Platforms

Investing in multirole artillery platforms that can be easily adjusted for multiple environments can help reduce the need for separate systems. Platforms that prioritize mobility and adaptability, such as wheeled artillery systems, can perform well across a variety of terrains.

The NORA B-52 self-propelled howitzer, used by Serbia and other nations, demonstrates a level of versatility across different terrains due to its wheeled configuration, which is suited for both varied terrain and regions.[33] Similarly, the US Excalibur GPS-guided artillery shell can be used on multiple platforms and is capable of precision strike across various terrains, which significantly reduces the need for multiple specialized systems for a specific terrain.

Different terrains, require tailored artillery systems so as to get the most out of them in battle. However, militaries can reduce costs by adopting modular, multirole artillery platforms with advanced fire control systems and standardized components. This approach allows for a balance between operational flexibility and financial sustainability, ensuring that modern armies are prepared to engage in a variety of terrains without incurring excessive costs for area-specific systems.

After all, the success of artillery design in future will greatly hinge on being adaptable platforms supporting effectiveness across many terrains through precision-guided munitions and advanced targeting techniques that relieve the burden of developing numerous systems.

ARTILLERY AMMUNITIONS

The landscape of artillery warfare is undergoing a rapid and profound transformation, driven by technological innovations and hard-won lessons from recent conflicts. As we deliberate on the latest innovations transforming artillery munitions, we are in for a bouquet of extended ranges, breathtaking lethality, and unprecedented precision that promises to reshape the very nature of modern warfare.

At the forefront of these developments is a dramatic increase in production capacity across major military powers. The United States, recognizing the urgent importance of artillery in the modern-day war, is planning an ambitious challenge to double the 155mm shell ammunition production. From a modest 14,400 shells per month, it has ramped to a remarkable 40,000 shells each month-a staggering 178% increase. But this is only the beginning. With the aim of reaching 100,000 in total by 2025, the US has promised investment of $5.3 billion in boosting its own domestic capacity.[34] New facilities have been added, including Universal Artillery Projectile Lines in Mesquite, Texas, signaling long-term commitment to artillery supremacy.[35]

Europe has also begun its own initiative on the enhancement of artillery ammunition production. The European Union Act in Support of Ammunition Production (ASAP) goal is to see production attain 2 million shells annually by 2025. Similarly, nations have made considerable strides. Germany has signed an •8.5 billion deal involving Rheinmetall for 155mm shells, with France and Sweden on track to double their munitions and explosives loading capacity by 2025. These concerted efforts across the Atlantic reflect a growing recognition of artillery's pivotal role in modern warfare and the need for substantial stockpiles to sustain high-intensity conflicts.[36]

In tandem with the growth in production capacity, there is a growing focus on enhancing the range of artillery ordnance. The United States is developing the Extended Range Artillery Munition (ERAM), aiming to significantly increase the reach of existing artillery systems. South Korea has unveiled a new long-range 155mm shell capable of striking targets up to 60km away, a 30% increase over previous capabilities.[37]

Perhaps most impressive is Norway's contribution. Nammo's ramjet-powered 155 mm projectile, which holds the potential to extend artillery range to a staggering 150 km. These advancements in range are not merely about hitting targets farther away; they represent a fundamental shift in the role of artillery on the battlefield, allowing for deeper strikes into enemy territory and the potential to disrupt supply lines and command structures more effectively than ever.[38]

Precision has become the watchword of modern artillery, with militaries worldwide investing heavily in the development of "smart" artillery shells. The integration of course-correcting fuzes and the proliferation of Excalibur-like precision rounds across multiple countries signify a new era of artillery employment. These precision-guided munitions offer a multitude of benefits: they reduce collateral damage, allow for more efficient use of ammunition stocks, and provide the ability to engage high-value targets with fewer rounds. The implications of this increased precision are far-reaching, potentially altering tactics, reducing logistical burdens, and minimizing the risk of civilian casualties in conflict zones.

Advancements in propellant technology are also playing a crucial role in enhancing artillery capabilities. Rheinmetall's development of the DM92 propellant for L52 barrels and the creation of "Supercharge" extended range charges, which offer a 10% range increase for L39 cannons, exemplify the ongoing efforts to squeeze every last kilometer of range out of existing artillery platforms.[39] These propellant advancements, combined with precision guidance systems and increased production capacity, are creating a perfect storm of artillery lethality and effectiveness.

Furthermore, the combination of automated loading systems and increased production capacity is enabling a faster tempo of operations. Artillery units can now sustain high-intensity fire missions for longer periods, potentially overwhelming enemy defenses through sheer volume and persistence of fire. This capability, when coupled with advanced target acquisition systems and real-time battlefield intelligence, creates a formidable force multiplier on the modern battlefield.

However, these advancements are not without their challenges and considerations. The logistics and supply chain implications of increased artillery activity are significant. The demand for raw materials, particularly explosives and propellants, is skyrocketing, necessitating robust and resilient supply chains to maintain high-intensity operations. Cost considerations also come into play, as precision-guided munitions are significantly more expensive than traditional rounds. Military planners must grapple with the delicate balance between quantity and quality in ammunition stockpiles, ensuring they have the right mix of munitions to address a variety of battlefield scenarios.

Moreover, as artillery capabilities advance, so do counter-measures. The development of improved air defense systems to counter long-range artillery and the potential for electronic warfare to disrupt precision-guided munitions present ongoing challenges. Military strategists must continuously adapt their tactics and technologies to stay ahead in this ever-evolving arms race.

THE IMPACT OF EMERGING DISRUPTIVE TECHNOLOGIES ON ARTILLERY DOCTRINE AND STRATEGY

Emerging disruptive technologies are revolutionizing warfare across all domains, and artillery is no exception. Once defined by sheer firepower and strategic positioning, modern artillery is increasingly shaped by innovations such as artificial intelligence (AI), autonomous systems, drones, and advanced sensors. These advancements are altering not only how artillery systems operate but also the doctrines that guide their use and the skills required of future artillery crews. Additionally, the ethical implications of these technologies are beginning to raise new questions about the future of warfare.

Massed Firepower Is Obsolete; Time for Precision Strikes

Traditional artillery doctrine called for massively concentrated firepower to overwhelm enemy positions and suppress large swaths of terrain. The emergence of precision-guided munitions (PGMs) has brought about a gradual shift toward greater flexibility and precision in artillery doctrine. There is no need to douse an area with shells to destroy a target; rather, targeted precision means increased effectiveness, as it restricts collateral damage and allows better ammunition management. The change thus reduces the logistical burden on artillery units and allows for greater war stamina, especially in contested or resource-scarce environments.

Integration into Network-Centric Warfare

One of the most obvious roles played by the disruptive technology is the integration of artillery into the realm of network-centric warfare. Increasingly, artillery units have been integrated into a wider sensor network, including UAVs, satellites, and ground-based radars. Data is shared in near-real time, aiding artillery in rapid and more accurate engagement of targets. The sensor-to-shooter loop has been shortened and made more efficient, as artillery platforms almost instantaneously receive actionable data from variety of sensors and can effectively act on it.

The evolution in doctrine forces artillery to operate as a most coordinated component of a multi-domain battle; an amalgamation of ground forces, air units, cyber operations, and space assets. The responsibilities of artillery are now even greater, and the forces are responding to aerial and space-based reconnaissance much before they get inputs from forward Artillery Observers.

Decentralised and Agile Artillery Operations

The employment of artillery is increasingly becoming decentralized because of the induction of agile, mobile, robotic and autonomous systems aided by multi domain surveillance architecture. Rather than static gun positions-a losing proposition vexed by counter bombardment, modern artillery systems can redeploy quickly, fire, and then relocate. The pathological aspect of this 'shoot and scoot' is growingly evident in a conflict setting where the adversary superior direction-finding systems and radars as well as integrated fire control system to direct swift counter bombardment within a matter of minutes.

The introduction of UGVs and automatic howitzers further stimulates this shift toward agility. Autonomous artillery units can now be deployed to destroy targets with minimal crews, enabling the artillery to conduct operations in remote, heavily contested, or hazardous environments. The doctrine of artillery, hence, is witnessing a transformation from deployment of large, centralised batteries towards smaller, agile, dispersed units that can fight a cohesive and integrated multi domain battle while maintaining smaller, dispersed stockpiles to reduce vulnerability to ammunition chains.

AI-Driven Fire Control and Decision-Making

The introduction of AI into fire control systems is transforming the decision-making process in artillery operations. AI is capable of processing a vast amount of battlefield data, identifying the most critical targets, and calculating the best possible firing solutions at speeds rapidly surpassing human operators.

However, this change is going to require strategic transformations. As the battlefield intelligence and command systems become increasingly automated, commanders of artillery units will have to strike a balance between speed and accuracy on one hand with human oversight on the other. Doctrines must shift toward an understanding of AI's limitations, ensuring that human operators remain in control whenever necessary, particularly where collateral damage may ensue and appropriate rules of engagement are unambiguous.

Cost Imposition Strategies

The efficient use of artillery technologies has enabled weaker forces to impose disproportionate financial and logistical burdens on adversaries, a tactic exemplified in both the Russia-Ukraine and Israel-Gaza conflicts. In Ukraine, the strategic pairing of low-cost commercial drones (e.g. DJI Mavic) with expensive precision-guided munitions (PGMs) like Excalibur shells forces Russia into a resource-draining arms race. compelling Moscow to expend far greater resources on countermeasures, including advanced electronic warfare systems and costly air defense interceptors.

Similarly, in Gaza, Hamas's use of rudimentary, mass-produced rockets pressurises Israel to deploy its USD 50,000 per-missile Iron Dome interceptors, creating an unsustainable economic imbalance.[40] Even with Iron Dome's 90% interception rate, the sheer volume of Hamas rocket salvos ensured that Israel expends millions of dollars daily during escalations, while Hamas replenishes its arsenal at a fraction of the cost. These strategies exploit the asymmetry of technological investment, leveraging affordability and scalability to exhaust adversaries' financial and industrial capacities.

The lesson for modern militaries is clear: hybrid arsenals with low-cost, mass-produced systems are essential to sustain prolonged conflicts while crippling opponents' ability to keep pace. This approach, seen in the US Army's shift to 155mm "smart-fuze" shells, balances precision with affordability, ensuring artillery can sustain prolonged conflicts without bankrupting defense budgets.

Urban Warfare Challenges

The complexities of urban warfare have starkly exposed the limitations and ethical quandaries of modern artillery, particularly in dense, civilian-populated environments. In Gaza, Israel's reliance on munitions like the Spice-2000 and M982 Excalibur shells is constrained by the enclave's labyrinthine urban terrain, where Hamas embeds rocket launchers and command centers within residential areas, schools, and hospitals. To navigate this challenge, Israel has turned to AI-powered "precision siege" tactics, using systems like Fire Weaver to fuse data from drones, satellites, and ground sensors to isolate high-value targets. However, even with advanced algorithms, collateral damage remains pervasive, as seen in the 2023 conflict where strikes on Hamas infrastructure resulted in significant civilian casualties.

Conversely, Russia's 2022 siege of Mariupol demonstrated how PGMs can accelerate urban conquest but at catastrophic humanitarian cost. By combining guided Krasnopol shells with indiscriminate "area effect" weapons like thermobaric rockets, Russian forces flattened the city, killing thousands of civilians and drawing global condemnation for violations of international law.[41]

These cases underscore a grim reality: while precision technologies reduce collateral damage in theory, urban battlegrounds – with their intermingling of combatants and non-combatants – inevitably test the limits of ethical warfare. Militaries now face dual pressures to refine AI-driven targeting for greater accuracy while grappling with the moral imperative to avoid civilian harm, a balance that remains elusive in asymmetric urban conflicts.

EW is the New High Ground

The Russia-Ukraine war has underscored EW as a decisive factor in modern artillery operations. Russian systems like the Krasukha-4 and Leer-3 have been used to jam Ukrainian drones, GPS signals and communications, degrading the accuracy of precision-guided munitions such as Excalibur shells and HIMARS rockets.[42] Conversely, Ukraine's integration of NATO-supplied counter-drone systems and frequency-hopping radios highlights the growing emphasis on EW resilience.[43]

In Gaza, Israel's suppression of Hamas's rocket launch coordination through spectrum warfare demonstrates how EW can neutralize asymmetric threats. Militaries now prioritize EW dominance to protect their own artillery's targeting capabilities while crippling adversaries' command-and-control networks. This shift has elevated EW from a supporting tool to a core component of artillery doctrine, mandating investments in adaptive jamming systems, encrypted communications, and AI-driven spectrum analysis to outmanoeuvre opponents in the invisible battlespace.

Artillery as Part of Multi-Domain Strategy

The modern battlefield is multi-domain. Ground, air, sea, cyber, and space are interconnected operations. Artillery is increasingly viewed as one instrument of a greater multi-domain strategy, wherein its traditional utility of administering fire support is now extended to disrupting enemy communication, suppressing electronic warfare systems, and targeting airborne or naval units with long-range precision strikes.

This shift in doctrine places a requirement on artillery units to be flexible and adaptable-so that they can operate in many environments and across perhaps very dissimilar missions. Artillery units must now be ready to attack in all hostile theatres, working in cooperation with air and naval forces, cyber teams as well as space-based assets to provide coordinated, integrated firepower mechanics. The kinds of disruptive technologies coming up, and how such technologies are going to affect artillery doctrine and strategy, is a major shift from the conventional and huge firepower to more precise, agile, and networked operations. Technologies like AI, automation, precision-guided munitions, and drones are slowly getting tied into the hearts of artillery systems, coupling autonomous fire detection with decision-making for systems.

Artillery is now a critical component in a multi-domain battle strategy; their systems can deliver targeted strikes with impressive speed and accuracy. These advancements nonetheless also demand a modification in command and control, human oversight of AI systems, and strategic flexibility in the light of evolving threats.

Skills Needed by Future Artillery Crew

The technology that drives artillery systems is not refashioning the role of artillery systems; they are also redefining the technical and operational skills of the personnel who will be manning them in the future. The traditional role of artillery personnel, loading shells, calculating firing solutions manually, maintaining equipment, has now been supplemented and, in some cases, replaced by new technical and operational competencies developed over the years.

Technical proficiency

The artillery crews of the future will be skilled in the planning and operation of advanced software and automated systems. AI and autonomous systems used with artillery operations will force soldiers to learn programming, identification of problems with a system, and optimally setting the numerous systems. This will likely require extensive training in data analysis, machine learning, and robotics, skills that are far removed from the traditional form of artillery training.

Integration with Multi-Domain Operations

Modern artillery increasingly takes its place within a networked battlefield, where such domains as the air, land, sea, and cyberspace are intrinsically related to each other. Future artillery crews will require skills in integration with intelligence, surveillance, and reconnaissance (ISR) systems and coordination with other units like air support and cyber teams. Working in this multi-domain battlespace will be essential in allowing artillery units to provide timely and effective support to larger military operations.

Cybersecurity Awareness

Since artillery systems are becoming heavily dependent on digital technology and data networks, the danger of cyber-attack is increasing. The increasing reliance on digital technology and network-based command and control in artillery systems enhances the risk of cyber threats. Crews should undergo necessary invulnerability training in the area of cybersecurity so that they can protect their respective platforms against computer hacking, electromagnetic jamming, or electronic warfare. The consequence of such cyberattacks on control systems of artillery weapon systems performing fire operations or vectoring autonomous units could be devastating in real operational settings, hence the need for artillery personnel to keep vigilant.

Comfortable with Decision-Making in Automated Systems

Autonomy is incrementally but profoundly transforming artillery decision-making, by integrating sensor fusion, AI, networking and data-driven algorithms into traditionally human-led fire support systems. Degrees of autonomy are already reshaping several layers of the artillery kill chain by transforming target identification, prioritisation & acquisition, automated fire control, counter battery engagements and predictive maintenance & resupply. This would mean more technical knowledge but also likely higher intuition in moments of quick decisions.

Emerging disruptive technologies are shaping the doctrines and strategies of modern artillery-artillery, bringing in a whole new mix of capabilities and challenges. AI, automation, unmanned systems, and precise guided munitions are changing the way artillery fights, allowing them to do so in an unprecedentedly quick, precise, and cohesive manner. However, these advances will also place deductive reasoning into some of the traditional military decisions, from technical proficiency on the part of artillery crews through cybersecurity awareness to the ethical decision-making.

The Indian Context

Artillery systems must be reformed from static, conventional platforms into dynamic, networked assets that provide accurate, rapid, and sustained fire power, in support of contemporary multi-domain battlefield. To achieve this, it is of paramount importance that digital fire control systems should be upgraded to incorporate advanced algorithms and machine learning, allowing for real-time adjustments based on dynamic battlefield data. Artillery units must be part of an integrated sensor based network that allows for connectivity among ground, aerial, and even space-based assets. These sensors must possess the necessary capabilities to provide an integrated operational picture that maximizes the ability to plan and modify fires, ensure target discrimination, leading to accurate sustained strikes by artillery. Future artillery systems must include long-range precision-guided munitions, to be able to strike targets located deep behind enemy lines, while limiting collateral damage. Increase in the inventory of self-propelled artillery platforms is also imperative to improve their survivability against counter-battery fire. This enhanced mobility, combined with digital connectivity, will enable the artillery units to rapidly redeploy, strike, and then relocate to avoid detection and neutralization by enemy forces. By modernizing fire control, enhancing sensor integration, and improving mobility, the artillery arm will become a pivotal element in both offensive and defensive operations, capable of delivering sustained and accurate firepower in support of intense operations.

Artillery continues to be a decisive force in shaping the battlefield, and its development must focus on achieving precision, speed, and flexibility to deal with capable enemies. The use of precision-guided munitions is a cornerstone of Artillery's transformation, with unguided general purpose artillery shells becoming increasingly ineffective against mobile or hardened targets. GPS and laser guided artillery rounds, which can change their trajectories in flight, represent a level of accuracy never seen before, enabling a gunner to destroy high-value targets, such as an enemy command post or tank column using a single round. This kind of precision gives the gunner a narrower window of opportunity to destroy a high value target, reduces ammunition expenditure and minimizes collateral damage, and increases operational efficiency, especially in situations where civilians or friendly units might be nearby. Such munitions for the Indian Army will guarantee that artillery remains a flexible, reliable, and capable asset that delivers decisive results, across a spectrum of theatres with minimal wastage.

Efficiency in ammunition handling is another critical area, and automated loading systems can enhance the Artillery's effectiveness tremendously. Robotic systems for loading shells into artillery pieces remove the physical strain of the crew, who are often susceptible to fatigue and injury after engaging the enemy for an extended time. Machines can work with speed and precision, loading rounds into howitzers or rocket launchers at uniform rates, resulting in a capability to sustain a higher rate of fire. The ability to provide sustained fire is especially advantageous in high-intensity conflicts, where repeated barrages are needed to suppress enemy movement or breach defences. By limiting the crew's exposure to dangerous activities, and reducing fatigue, automated systems make operations safer and efficient, while allowing crews to maintain continuous pressure on the enemy.

To effectively strike the enemy in depth, the Artillery must aim toward long-range systems, specifically those which can deliver hypersonic projectiles. Hypersonic artillery munitions once a reality, will travel at speeds greater than Mach 5 and will significantly decrease the time between launch and impact, allowing for targeting of far-off enemy targets, such as supply lines or air fields, before the adversary can have the opportunity to react or reposition assets. Such an extended reach and swift response is vital for India, given its vast borders and the need to project power beyond immediate frontlines, whether against conventional forces or insurgent bases in neighbouring countries. Developing or procuring such systems would provide the Indian Army with a strategic advantage, allowing it to disrupt enemy operations at unprecedented ranges and with minimal warning, thereby shifting the balance in its favor during critical phases of conflict.

The survivability of artillery units is dependent on them being able to effectively operate against enemy counter bombardment. The key to counter-battery artillery systems is counter-battery radar. These radar systems are designed to detect incoming artillery rounds, or rockets, immediately identify their point of origin in a matter of seconds and give precise coordinates for retaliation. This detection and destruction cycle is so quick that the target artillery or rocket batteries are neutralized before they can relocate, subsequently protecting friendly forces and infrastructure from sustained bombardment. For the Indian Army, which faces adversaries who themselves rely heavily on artillery on the northern and western borders, developing effective counter-battery systems will help to maintain the operational capabilities of its own artillery while at the same time degrading or denying the enemy his fire support capabilities. Integrating these radars to a networked command and control system will provide artillery units a seamless capability to detect and respond whilst also providing resilience in contested environments.

As militaries around the globe continue to make sense of these disruptive technologies, deep consideration needs to be made for the ethical implications of autonomous and AI-enabled systems, so the future of artillery is both useful and accountable. Integrating these technologies into the artillery doctrine, in other words, is going to require technical innovation, but also clerical strategic planning, and ethical oversight for seeing the realization of their promises while keeping their threats in check.

The rush by armies all over the world to adopt and to adapt to these technologies is fundamentally transforming the very nature of land warfare. Tomorrow's artillery will be able to reach out further, striking with better precision and sustaining operations longer than we have ever seen. It carries with it of course new logistics, tactical, and counter-tactical challenges which will now begin to factor into their consideration. As artillery acts against the arms race. As the artillery arms race continues, it is clear that the ability to effectively harness these new capabilities will be a critical factor in determining the outcome of future conflicts. The artillery, long known as the "King of Battle," is not just maintaining its crown, it is redefining the very kingdom over which it reigns.

Notes

1   "Future Trends in Artillery Technology: Innovations and Impacts," Total Military Insight, accessed February 25, 2025, https://totalmilitaryinsight.com/future-trends-in-artillery-technology/

2   Bode, I. and Watts, T. F. A. Loitering munitions and unpredictability: Autonomy in weapon systems and challenges to human control. Center for War Studies, 2023.

3   David Axe, 'Russia Has More Artillery Than Ukraine. But Russian Gunners have a Bad Habit of Shelling ... Nothing.' Forbes, December 18, 2022, https://www.forbes.com/sites/ davidaxe/ 2022/12/18/russia-has-more-artillery-than-ukraine-but-russian-gunners-have-a-badhabit-of-shelling – nothing/?sh=139e6cc2aed7

4   ISW Press, 'Russian Offensive Campaign Assessment, May 6,' Institute for the Study of War, May 6, 2022, at https://www.understandingwar.org/backgrounder/russian-offensive-campaignassessment-may-6

5   Vijainder K. Thakur, "Russia's 'Hard to Kill' Missile Succumbs to Ukrainian EW; Here's Why Iskander-M is No Longer The 'Sikandar'," *EurAsian Times*, February 16, 2025, https:// eurasiantimes.com/russias-hard-to-kill-missile-iskander-m-ukrainian-ew/.

6   Deepak Kumar,Impact of Artillery, Precision-Guided Munitions (PGMs) and Missiles,UKRAINE WAR Military Perspectives and Strategic Reflections, Manohar Parrikar Institute for Defence Studies and Analyses, New Delhi, 2024

7   Reuters, 'Ukraine Gets More U.S., German Rocket Launcher Systems - Minister, 'Reuters, August 1, 2022 at https://www.reuters.com/world/ukraine-gets-more-us-german-rocketlauncher-systems-minister-2022-08-01/

8   Cotovio, V., Sebastian, C., and Goodwin, A. Ukraine's AI-enabled drones are trying to disrupt Russia's energy industry. So far, it's working. CNN, April 2024.

9   Aleksandr Pinchuk, "Self-propelled howitzers strike without fail "(Ñàìîõîäíûå ãàóáèöû áüþò áåç ïðîìàõà), Krasnaia zvezda, No. 77, July 18, 2022

10   General Baluyevsky Spoke About the Revolution in Artillery." *RIA Novosti*, August 11, 2022, https://ria.ru/20220811/kvadrokoptery-1808761668.html

11   Emanuel Fabian IDF: 9,500 rockets fired at Israel since Oct. 7, including 3,000 in 1st hours of onslaught 9 November 2023 https://www.timesofisrael.com/liveblog_entry/idf-9500-rockets-fired-at-israel-since-oct-7-including-3000-in-1st-hours-of-onslaught/

12   Seth J. Frantzman,Israel-Hamas war: Meet the IDF artillery key to the Gaza war, The Jerusalem Post November 25, 2023, https://www.fdd.org/analysis/op_eds/2023/11/25/israel-hamas-war-meet-the-idf-artillery-key-to-the-gaza-war/

13   Agence France-Presse,Lebanon official media says Israeli fire wounds two amid ceasefire, Nov 28, 2024 https://www.al-monitor.com/originals/2024/11/lebanon-official-media-says-israeli-fire-wounds-two-amid-ceasefire

14   Charles Lister, Director, Syria and Counterterrorism and Extremism programs at the Middle East Institute,Why Assad's Regime Is Collapsing So Quickly, https://foreignpolicy.com/2024/ 12/05/syria-assad-regime-collapsing-quickly/

15   Scott R. Gourley, Army's Cannon, Propellant Updates Reflect Lessons Learned in Ukraine 1/ 24/2024 https://www.nationaldefensemagazine.org/articles/2024/1/24/armys-cannon-

propellant-updates-reflect-lessons-learned-in-ukraine

16   Samuel Bendett, Russian UGV developments influenced by Ukraine War, 19 June 2024 https:/ /euro-sd.com/2024/06/articles/38818/russian-ugv-developments-influenced-by-ukraine-war/

17   Ukraine: What are Himars missiles and are they changing the war? 30 August 2022 https:// www.bbc.com/news/world-62512681

18   Amoroso, D. and Tamburrini, G. Autonomous weapons systems and meaningful human control: Ethical and legal issues. Current Robotics Reports, 1(4):187–194, December 2020.

19   Danielle Kress and Maj. Henry Castillo for Army AL&T Magazine, February 14, 2025 https:/ /www.army.mil/article/283097/afatds_gets_an_upgrade

20   Simon Hilton,Nanotech at War, Feb 8, 2023, Polymer Nano Centrum https:// blog.polymernanocentrum.cz/nanotech-at-war/

21   Simon Hilton, Developing Nanotechnology in the Defence Industry, NANO CHEMI GROUP Blog, https://blog.nanochemigroup.cz/developing-nanotechnology-in-the-defence-industry/

22   U.S. Army DEVCOM Army Research Laboratory, Army, Argonne scientists explore nanoparticles for future weapon systems, February 25, 2021 https://www.army.mil/article/ 243587/army_argonne_scientists_explore_nanoparticles_for_future_weapon_systems

23   Brady, Matilda R., and Paul Goethals. "A comparative analysis of contemporary 155 mm artillery projectiles." Journal of Defense Analytics and Logistics (2019).

24   Inside GNSS, Russia Boasts 1-Meter Accuracy for New GLONASS-Guided, 200-Km Range Missile Squadrons,September 24, 2020, https://insidegnss.com/russia-unveils-new-glonass-guided-200-km-range-missile-squadrons/

25   Satellite Navigation Services, ISRO, https://www.isro.gov.in/SatelliteNavigationServices.html

26   A200 Rocket Weapon System, China Defence, https://www.militarydrones.org.cn/a200-rocket-launcherp00731p1.html.

27   What is the Wideband Global Satcom (WGS) Military Satellite Network?, 19 December 2024, https://maxpolyakov.com/what-is-the-wideband-global-satcom-wgs-military-satellite-network/ #:~:text=The%20Wideband%20Global%20Satcom%20(WGS)%20satellite%20network%2C%20laun ched%20in,of%20tactical% 20broadband%20communication%20needs.

28   Extended Range Cannon Artillery (ERCA), Director Operational Test and Evaluation (.mil) https://www.dote.osd.mil/Portals/97/pub/reports/FY2021/army/2021erca.pdf?ver=f1T3nU 9feKkHCzP0MyNwPg%3D%3D#:~:text=The%20ERCA%20system%20is%20an,to%20incre ase%20its%20lethal%20range.

29   Tracy Cozzens,Russia tests new GLONASS-guided missile, September 22, 2020, https:// www.gpsworld.com/russia-tests-new-glonass-guided-missile/

30   Byabenin, "Artillery Robots: From Mechatronics to Artificial Intelligence", Op. Cit.

31   Summary Of The Joint All Domain Command & Control (JADC2) STRATEGY, Department of Defense, March 2022, https://media.defense.gov/2022/Mar/17/2002958406/-1/-1/1/ SUMMARY-OF-THE-JOINT-ALL-DOMAIN-COMMAND-AND-CONTROL-STRATEGY.pdf

32   6x6 self-propelled howitzer 155mm CAESAR, Self-propelled howitzers, 13 October 2024, https://armyrecognition.com/military-products/army/artillery-vehicles-and-weapons/self-propelled-howitzers/6x6-self-propelled-howitzer-155mm-caesar#google_vignette

33   Serbia Trains and Deploys Latest-Generation of Nora B-52 Self-Propelled Howitzers, Defense News Army 2025, 28 Jan, 2025, https://armyrecognition.com/news/army-news/2025/serbia-trains-and-deploys-latest-generation-of-nora-b-52-self-propelled-howitzers

34   Jen Judson, Army races to widen the bottlenecks of artillery shell production, Oct 14, 2024, https://www.defensenews.com/land/2024/10/14/army-races-to-widen-the-bottlenecks-of-artillery-shell-production/

35   Army inaugurates Universal Artillery Projectile Lines facility, U.S. Army Public Affairs, May 29, 2024, https://www.army.mil/article/276727/army_inaugurates_universal_artillery_pro

jectile_lines_facility

36 European Commission, The Commission allocates €500 million to ramp up ammunition production, out of a total of €2 billion to strengthen EU's defence industry, Press release, Mar 15, 2024, Brussels, https://ec.europa.eu/commission/presscorner/detail/en/ip_24_1495

37 Kapil Kajal, South Korea to mass produce extended-range projectiles for K9 howitzers, 08 February 2024, https://www.janes.com/osint-insights/defence-news/defence/south-korea-to-mass-produce-extended-range-projectiles-for-k9-howitzers

38 Thorstein Korsvold, The range revolution, 6 February, 2019, https://www.nammo.com/story/the-range-revolution/

39 Breaking Defense, Outgunned, outranged: The Ukraine war shows where fires modernization is needed, https://info.breakingdefense.com/hubfs/GAMECHANGER_Artillery_Rheinmetall_Breaking_Defense.pdf

40 Gerry Doyle, Mariano Zafra, Adolfo Arranz and Jitesh Chowdhury, Israel's Iron Dome, April 18, 2024, https://www.reuters.com/graphics/ISRAEL-PALESTINIANS/IRAN-DEFENCE/mypmkljzopr/

41 Hilary Andersson,The agony of not knowing, as Mariupol mass burial sites grow, BBC Panorama,7 November 2022 https://www.bbc.com/news/world-europe-63536564

42 Rajesh Uppal, Russia's Electronic Warfare Dominance: A Comprehensive Overview, August 6, 2024 https://idstch.com/geopolitics/russias-electronic-warfare-dominance-a-comprehensive-overview/#:~:text=Russia's%20use%20of%20electronic%20warfare,strategically%20paramount%20for%20military%20forces.

43 Ukraine joins NATO counter-drone exercise for first time, NATO, 19 Sep. 2024, https://www.nato.int/cps/en/natohq/news_228911.htm#:~:text=From%2010%20to%2020%20September,enhanced%20deterrence%20and%20defence%20posture.

12

Soaring Smart: The Tech-Driven Future of Army Aviation

"If we lose the war in the air, we lose the war and we lose it quickly."
—Field Marshal Bernard Montgomery

Since its early 20th century beginnings military aviation has experienced numerous substantial transformations. Aviation's role evolved from simple reconnaissance and minor aerial skirmishes to extensive strategic and tactical operations during World War I and II. The employment of helicopters during the Korean War marked advent of a new era in military aviation, offering unparalleled flexibility in transporting troops, conducting medical evacuations and providing close air support.

The present day battlefields are characterised by a dynamic mix of manned and unmanned systems. However, traditional platforms, such as helicopters, retain their importance due to their versatility and human oversight, while UAVs and AI systems expand their operational roles to include both continuous monitoring and long distance targeted attacks. The global military aviation doctrines are being continuously shaped by the lessons from recent conflicts including the Iran Israel War, Russia Ukraine War, Israel Gaza War, US operations in Afghanistan against the Taliban, and Azerbaijan Armenia war. Today, the future of military aviation stands at an important crossroads, transformed by disruptive technologies such as artificial intelligence, unmanned aerial vehicles and autonomous systems.

Conventional doctrines face challenges as the battlespace gradually shifts from helicopter dominance to integrated systems comprising of drone swarms, AI targeting and stealth technology. The integration of multiple platforms with human machine collaboration stands to transform the character of military conflict in this modern era. This chapter examines how these emerging disruptive technologies affect military aviation by focusing on helicopters, UAVs, drones and AI enabled autonomous systems while delving into technological breakthroughs alongside their real world applications and evolving modern military aviation tactics. The discourse on fixed wing aircraft remains largely absent in this chapter, except in select instances, where it is pertinent to highlight specific trends or technological features, despite them being primarily the forte of Air Forces across the world.

HELICOPTERS: THE ENDURING VERSATILE WORKHORSE

Purely as a type of military vehicle, Helicopters are crucial to modern military operations, providing essential capabilities that fixed wing aircraft cannot. The progress in helicopter technology has benefited from the impetus in technological disruptive innovations that have considerably added to helicopter performance, survivability and operational capabilities. Smart avionics, EW suites, composite materials and stealth features are increasingly being incorporated into modern helicopters to enhance their probability of success in combat missions. While UAVs and autonomous systems capture the media space, helicopters continue to serve as the irreplaceable workhorses of military aviation. Their inherent versatility, coupled with the ability to operate in adverse conditions, makes them indispensable in certain critical operations. Versatility in Complex Environments In environments, where rapid mobility and flexible deployment are essential, Helicopters are the platform of choice. They can operate in confined urban spaces, mountainous terrains and rapidly evolving combat zones.

Vertical Take-off and Landing (VTOL): Helicopters can take off and land vertically, an essential capability in areas lacking proper helipads or airfields due to rugged terrain, enemy action or operational restrictions.

Hovering Capability: The ability of helicopters to hover and maintain a stable platform, enables helicopters to provide close air support, direct artillery fire, conduct recce, carry medevac operations and serve as a communications relay post.

Adaptability to Multiple Roles: Helicopters are designed with modular plug and play sections that can be configured as per requirement for various missions, troop transport, medical evacuation, disaster management, law and order, command and control, direct fire support or special heli borne operations.

Evolution of Helicopter Capabilities

Helicopters today, are equipped with digital sensors, advanced avionics and even few autonomous features. They can perform a multitude of roles:

Close Air Support: Attack helicopters such as the AH-64 Apache have been refined and configured to deliver precision strikes on enemy positions while manoeuvring in close proximity to friendly forces.

Troop Transport and Medevac: Helicopters such as the UH-60 Black Hawk and Cheetah/Chetak provide rapid mobility and medical evacuation capabilities in terrains such as the Saichen Glacier, where fixed wing aircraft cannot operate effectively.

Command and Control: Helicopters also serve as airborne command centres, enabling real-time communication between ground forces and higher headquarters. This role is critical in dynamic battlefields where rapid decision making is essential, such as mechanised operations in the deserts.

BATTLEFIELD IMPACT: HELICOPTERS IN REAL WORLD OPERATIONS

US Operations Against the Taliban

In the rugged and unpredictable terrain of Afghanistan, helicopters played a critical role by being the most sought after platform for rapid mobilisation, swift redeployment and logistic sustenance.

AH 64 Apache: The Apache is one of the best attack helicopters in service today, with advanced sensors and precision guided munitions. Firepower and survivability are paramount to the Apache's design concept. It has been used to accurately engage the enemy positions with extreme lethality. In addition to its striking capabilities, it possesses heavily armoured protective body and comprehensive EW countermeasures that enable successful operations within contested environments.[1] The accuracy provided by advanced sensors, targeting systems, as well as weapon integration aided greatly in neutralizing insurgent strongholds throughout Afghanistan's treacherous mountain ranges. While providing close air support (CAS), pilots were able to use on-board thermal imaging and radar systems to identify concealed enemy combatants within the rugged terrain. Once identified, guided munitions from the Apache, Hellfire missiles were used to neutralise the threats with pinpoint accuracy. It was instrumental in saving the lives of the crew as well as the ground forces it supported, as its heavily armoured design allowed it to survive in an environment where ground fire and IEDs were constant hazards. It delivered precision strikes against insurgent strongholds while manoeuvring in close quarters. Its ability to identify and engage targets rapidly was pivotal in neutralising enemy fire and protecting ground troops.[2]

UH-60 Black Hawk: The Black Hawk is revered for its flexibility, being utilized for troop transport, medical evacuation and logistics. As a workhorse, its capability in rugged terrains make it invaluable for operations that involve rapid insertion and extraction of troops, as well as sustaining extended operations in remote or urban environments. During US operations against the Taliban, Black Hawks were the workhorses of casualty evacuation and troop transport. In one critical operation, a Black Hawk was used to extract wounded soldiers from a hot zone under heavy enemy fire. Its ability to conduct vertical take-offs and landings in tight, rough terrain allowed it to access regions that fixed-wing aircraft could not. The advanced communications systems on board the helicopter made it possible to continue coordinating real-time actions with ground commanders to reposition troops rapidly and effectively despite the dynamic environment on the ground. It undertook vital medevac operations while under enemy fire, ceaselessly evacuating injured soldiers from hostile environments. The Black Hawk's versatility meant that logistical concerns never prevented did not hinder the rapid movement and reinforcement of troops.[3]

Sikorsky CH-53: Although primarily a heavy-lift helicopter, the CH-53 is also adapted for rapid deployment of special forces and critical equipment. Its high payload capacity and long-range capabilities enable it to perform a variety of support roles, from logistics to command-and-control functions.[4]

Integration with Long-Endurance UAVs: While UAVs like the MQ-9 Reaper supplemented surveillance efforts, helicopters remained essential for rapid response. Their ability to land, refuel and redeploy in real time provided a flexibility that unmanned systems, still under development, could not match.

Lessons Learned for Hybrid Operations: The challenges faced during these operations highlighted the need for better integration between manned and unmanned assets, a lesson that continues to shape US military doctrine today.

Azerbaijan–Armenia Conflict

While the success of Azerbaijani forces was largely driven by cost-effective UAVs like the Bayraktar TB2, helicopters played a supporting yet vital role.[5]

Logistical and Troop Support: Helicopters ensured rapid deployment and redeployment of forces and swift delivery of supplies, helping maintain the tempo of operations amid frequent advances and retreats. Their ability to operate in environments where UAVs might be vulnerable or less effective underscored the enduring value of manned rotary-wing platforms.

Combined Arms Warfare: Azerbaijan's Mi-24 Hind gunships provided close support during ground offensives, using 30mm cannons and rockets to suppress Armenian troops. Their presence further demoralised troops already reeling from drone strikes.[6]

Rapid Redeployment: Mi-17s airlifted Azerbaijani commandos to secure heights in Nagorno-Karabakh, exploiting gaps created by drone swarms.

Russian–Ukrainian Conflict

Helicopters have maintained a complementary role alongside a burgeoning drone fleet.

Russian Mi-28 and Ka-52: These platforms delivered direct fire support in contested zones. Their low-altitude flight capabilities allowed them to engage ground targets while attempting to avoid detection by enemy UAVs and ground-based air defence systems.

Integrated Operations: In the eastern regions of Ukraine, during intense firefights, Mi-28s were deployed to engage target enemy armour and strongholds. Their ability to quickly manoeuvre in response to emerging threats allowed them to deliver sustained and accurate fire while avoiding detection by adversary AD systems. Similarly, the Ka-52 "Alligator", with its tandem-seating arrangement, enabled the pilot and co-pilot (or weapon systems officer) to share situational awareness and coordinate attacks in real time, enhancing its effectiveness in the dynamic combat environment. As the UAVs carried out surveillance and precision strikes, helicopters continued to provide command & control linkages and logistical support. Their role in coordinating movement of ground forces and adjusting artillery fire based on real time intelligence proved critical during prolonged operations.[7]

Fire Support: Russia's Ka 52 Alligator and Mi 28NM provided close air support during the Siege of Severodonetsk, firing LMUR guided missiles from standoff ranges. However, early losses to Ukrainian MANPADS and TB2 drones forced a tactical shift toward indirect fire.[8]

Tactical Mobility: Ukraine's Mi 8 helicopters conducted audacious resupply missions to Mariupol and Snake Island, flying at ultra low altitudes to evade Russian radars. These missions underscored the helicopter's unmatched ability to insert/extract troops and deliver supplies in contested airspace.[9]

Israel–Gaza War

Despite the growing use of unmanned systems in this urban, densely populated conflict, helicopters continued to demonstrate their unique value.

Attack and Transport Helicopters: Helicopters carried out rapid insertion of special forces into hostile areas, where the threat envisaged demanded close human oversight and adaptability that unmanned systems could not have provided. Their ability to operate within dense urban areas, provide close air support and quickly extract personnel was indispensable to the success of fast paced operations by the IDF.

Recent conflicts find the AH 64 Apache to be one of the most agile and adaptable helicopters. The AH 64 Apache emerges as a versatile machine for urban warfare and anti-tank operations, featuring advanced targeting systems and lightweight yet potent weaponry. The Apache helicopter performed close air support missions during operations in dense urban environments like the Gaza strip. The helicopter's small footprint combined with its agility enabled it to function within skyscraper zones and narrow streets to execute attacks on enemy armoured units and fortified positions. The Apache's advanced avionics systems combined with its real time data link capabilities allowed seamless coordination with ground forces to deliver precision strikes with minimized collateral damage.

Urban Precision Strikes: The AH 64 Apaches used Spike NLOS missiles to eliminate Hamas operatives in Gaza's dense urban terrain. The ability to hover, adjust angles and modify missions during flight minimised civilian casualties in comparison to strikes by fixed wing aircrafts.[10]

Search and Rescue: The Sikorsky UH 60 Black Hawk, rechristened as Yanshuf in Israel has been the utility work horse of the IDF throughout the Gaza War, carrying out troop movement, casualty evacuation and resupply missions extensively. The crash of one of the helicopters in Sep 24 happened when it was on a casualty evacuation mission near the Philadelphi Corridor.[11]

Integration of Intelligence and Electronic Warfare: The optimum exploitation of UAVs and manned helicopters allowed Israeli forces to maintain comprehensive situational awareness. Their electronic warfare systems disrupted enemy communications and air defences, while real time data transfer enabled accurate and timely strikes.

BEYOND THE COCKPIT: NEXT GEN ARMY AVIATION

Emerging disruptive technologies have induced profound transformation in the landscape of army aviation. New advancements in aviation technology are improving helicopters, unmanned aerial vehicles and autonomous systems, opening up entirely new operational opportunities, along with significant doctrinal implications and tactical shifts. This section examines key technologies and their specific impact on helicopters, UAVs, and autonomous systems.

Helicopters: Adapting to Survive

Helicopters for long having been the backbone of tactical mobility and close air support, are undergoing radical upgrades to counter modern threats. Inspite of the rise of unmanned systems and autonomous technologies, helicopters continue to provide the human element which is most critical to decision making and adaptability, that ensures success in a dynamic and unpredictable battle environment.

Stealth and Speed: Stealth is becoming increasingly important for helicopters, particularly in roles such as deep strike special operations, where surprise is of paramount importance. Modern helicopters are being designed with radar absorbing materials, reduced infrared signatures and noise reduction technologies to minimise their detectability.

The development of the stealth variant of the Black Hawk helicopter, used by US in the raid that killed Osama bin Laden, is a prime example of stealth technology being integrated into helicopter design. The stealth helicopter has special rotor blades and other design elements that reduce its radar cross-section and acoustic signature, allowing it to penetrate heavily defended airspace without getting detected. The US Future Vertical Lift (FVL) program aims to replace aging platforms like the UH60 Black Hawk with next generation tilt rotors like the Bell V280 Valor, capable of 280 knots (twice the speed of conventional helicopters) and reduced radar cross sections.[12]

Russia has also modernized its helicopters including the Ka 52 Alligator and Mi 28NM by adding composite materials and Vitebsk electronic countermeasures. The electronic countermeasures help protect the helicopters by blocking enemy munitions such as missiles or drones.[13] Another important advancement is the compound helicopter that uses the standard rotor system in conjunction with extra propulsion (like pusher propellers/tilt rotors), allowing the helicopter to fly much faster and farther. The Sikorsky S-97 Raider and the Bell V280 Valor are examples of these advanced helicopters that use compound technology to exceed the limitations of their conventional siblings.[14]

AI and Sensor Fusion: AI and autonomous systems are becoming crucial to helicopter operations. AI driven flight control systems assist pilots by automating routine tasks, finding optimal flight paths and managing complex requirements during intense operations. Autonomous helicopters offer higher operational flexibility as they can also conduct missions without a pilot.

The Sikorsky MATRIX Technology is an autonomous flight control system that enables helicopters to operate with minimal pilot input, hereby enabling complex missions in challenging environments with reduced pilot fatigue. This technology is being integrated into both manned and unmanned helicopters, expanding the capabilities of rotary-wing platforms.[15] The AH64E Apache Guardian integrates AI to prioritize threats, while China's Z20 utility helicopter uses machine learning to analyse terrain data for safer low-altitude operations.[16]

Unmanned Helicopters: Unmanned helicopters, or rotary-wing UAVs, represent significant progress in helicopter technology. These platforms offer vertical take-off and landing like traditional helicopters but are operated remotely or autonomously, reducing risks to pilots. They are particularly useful for intelligence, surveillance & reconnaissance missions, cargo resupply in dangerous areas, and strike missions in high-threat environments. The Northrop Grumman MQ-8 Fire Scout, used by the US Navy until 2022, provided real-time reconnaissance and targeting support for naval and ground forces.[17]

EW: In Ukraine, Russia's Ka-52 helicopters initially suffered heavy losses to Turkish-made Bayraktar TB2 drones and Western-supplied MANPADS. However, retrofitted Ka-52Ms with L370Vitebsk ECM suites reduced losses by 60%, demonstrating the necessity of adaptive survivability tech.

Directed Energy: Experimental laser systems, like those tested on the Apache, are designed to destroy incoming missiles mid-flight, a critical defence in environments saturated with man-portable air defence systems (MANPADS).

Despite these advancements, challenges remain in helicopter technology. Developing and integrating new features such as stealth, compound propulsion systems, and AI-driven autonomy is costly and complex, requiring significant investment and testing. Additionally, helicopters remain vulnerable to modern air defence systems, including MANPADS and advanced radar-guided missiles, necessitating ongoing improvements in defensive measures.

Helicopters

Stealth features such as radar-absorbing materials and heat-suppressed exhaust systems reduce visibility to enemy defences, allowing these platforms to conduct covert insertions or reconnaissance missions in contested airspace. While fully stealthy helicopters remain rare – the cancelled Boeing-Sikorsky RAH-66 Comanche being a notable example – the design principles pioneered in such programs continue to influence modern helicopter development.[18] Current platforms like the Sikorsky S-97 Raider incorporate stealth elements to lower their signature, enabling rapid, discreet operations that evade traditional detection methods and enhance their utility in anti-access/area denial (A2/AD) scenarios.

UAVs

UAVs benefit immensely from stealth technology, which allows them to penetrate denied airspace for intelligence, surveillance, and reconnaissance (ISR) or strike missions, evading air defences with reduced signatures. The US RQ-170 Sentinel, a stealth reconnaissance UAV, has been deployed in high-threat environments such as Afghanistan and Iran, leveraging its low-observable design to gather critical intelligence without alerting adversaries.[19] This capability underscores how stealth UAVs are expanding the scope of covert operations, providing militaries with a tool to project power deep into enemy territory while minimizing risk.

Autonomous Systems

Autonomous systems push stealth technology further, combining low-observable designs with AI-optimized flight paths to avoid detection and execute deep Penetration missions. UK's BAE Systems Taranis, a stealthy autonomous combat demonstrator, capable of conducting reconnaissance or strike operations with minimal human input, exemplifies this potential. Its ability to navigate complex environments autonomously while maintaining a low profile exemplifies, how stealth and autonomy are converging to create platforms that can infiltrate and disrupt adversary defences with unparalleled skillfulness.[20]

Robotics

Robotics involves the development and deployment of machines designed to perform tasks autonomously or semi autonomously, often in environments too hazardous or challenging for human operators or soldiers. In military aviation, robotics is encouraging a shift towards greater automation and versatility, with significant implications for helicopters, UAVs and autonomous systems.

Helicopters

Robotics manifests in the form of robotic co-pilots and automated maintenance systems, both of which enhance operational efficiency. Robotic co-pilots, which the US Army is currently assessing, decrease workload of crew by automating navigation, communication, and system monitoring tasks and permitting human pilots to concentrate on tactical manoeuvres. Automated maintenance robots will assist in operational management as they will conduct inspections and maintenance on helicopters thereby reducing downtime and maintaining helicopter mission readiness. These advancements are poised to extend the operational life and versatility of helicopters, making them more adaptable to battlefield requirements.

UAVs

Since UAVs are basically robotic systems, advancements in robotics are amplifying their autonomy, payload capacity and mission flexibility. Thanks to sophisticated robotic controls, these platforms can now execute complex tasks, such as takeoff, landing and target engagement, with minimal human intervention. The MQ9 Reaper uses robotic automation for its takeoff and landing procedures, allowing its operators to focus on the mission and not the physical aspects of how to take-off and land.[21] Greater amounts of automation increases the duration a Reaper is able to conduct intelligence, surveillance and reconnaissance and strike missions and shows how robotics is changing the expanse of UAV operations.

Autonomous Systems

Autonomous systems represent the ultimate in military aviation, conducting a variety of combat, resupply and reconnaissance missions with little or no human intervention. The IAI Harop, an Israeli loitering munition, can autonomously engage targets, by loitering and remaining airborne until conditions are found to be optimal for target engagement. This capability illustrates how robotics and autonomy converge to perform high risk missions, without exposing the human operator while delivering precise, lethal effects, a clear precursor of the robotic revolution in warfare.

Swarm Technology

Swarm Technology is inspired by the natural phenomena like bird flocks or insect swarms, in which many autonomous agents can work together and reach a common goal through effective decentralized decision making. In military aviation, swarm technology is unlocking new tactical possibilities for helicopters, UAVs, and autonomous systems, amplifying their reach and impact.

Helicopters

Drone swarms launched or tethered to helicopters exploit the swarm technology that allows the deployment of micro UAV swarms to extend operational capabilities, such as reconnaissance or electronic warfare, beyond the helicopter's immediate range. These swarms can scout ahead, relay real-time intelligence, or jam enemy systems, providing a force-multiplying effect. The US is actively pursuing the Air-Launched Effects Projects, totalling $29.75 million which enables helicopter-launched UAV swarms, integrating them into operations to enhance battlefield awareness and support rapid decision-making, a development that promises to redefine the helicopter's role in combined arms operations.[22] India and China have already demonstrated swarm capabilities in the recent past and are refining their platforms.

UAVs

UAVs are ideally suited to swarm technology, which enables them to overwhelm enemy defences, conduct distributed attacks, or provide wide-area surveillance. By operating as a cohesive unit, UAV swarms can saturate air defences with sheer numbers, confuse targeting systems, or deliver coordinated strikes across multiple vectors. China's demonstrations of drone swarms as in the latest Jetank UAV, involving hundreds of UAVs acting in unison, highlight their potential for saturation attacks, offering a glimpse into how swarm tactics could reshape aerial combat by prioritizing quantity and coordination over individual platform capability.[23]

Autonomous Systems

Autonomous systems leverage swarm technology to self-organize and adapt, executing complex missions such as area denial or distributed sensing with remarkable efficiency. The US Defence Advanced Research Projects Agency (DARPA) is pioneering this approach through its OFFSET program, which develops swarm tactics for urban operations. These autonomous swarms can dynamically reconfigure to counter threats or achieve objectives, showcasing a future where collective autonomy could dominate contested environments, providing militaries with a scalable, resilient tool for modern warfare.[24]

Sensor Fusion

Sensor Fusion refers to the integration of data from multiple sensors, such as radar, electro-optical/infrared (EO/IR), and signals intelligence systems, to create a comprehensive, real-time operational picture. This technology enhances situational awareness and decision-making across helicopters, UAVs, and autonomous systems, enabling more precise and effective mission execution.

Helicopters

Advanced avionics systems fuse sensor data to improve target detection, tracking, and engagement, even in adverse conditions like darkness or inclement weather. This capability allows pilots to maintain a clear understanding of the battlefield, enhancing their ability to evade threats and strike with precision. The Apache AH-64E Guardian, used by the US Army, exemplifies sensor fusion in action, combining radar, EO/IR, and missile guidance data to enable all-weather operations, a critical advantage in dynamic combat scenarios.[25]

UAVs

UAVs rely on multi-sensor payloads to bolster their ISR capabilities, integrating visual, thermal, and signals data to provide a multi-dimensional view of the operational environment. This fused coherent data gives them the ability to detect targets, observe movement and support timely decision by commanders. The Turkish Bayraktar TB2, which has been used across various conflicts, including Ukraine, integrates EO/IR and synthetic aperture radar (SAR) systems that allow it to conduct reconnaissance and precision strikes very accurately, even in contested airspace.[26]

Autonomous Systems

Autonomous systems navigate and accomplish missions in challenging, unpredictable environments through sensor data fusion. This technology uses integrated data to formulate autonomous decisions. An example of this capability is demonstrated by the US XQ 67A Off-Board Sensing Station (OBSS) which provides a manned aircraft with fused sensor data to ensure that all members of the force are aware of the situation in the networked environment. By processing and disseminating this information autonomously, such systems amplify the effectiveness of broader military operations, illustrating how sensor fusion is a cornerstone of next generation aviation.[27]

Advanced Propulsion

Cutting-edge propulsion systems such as hybrid, electric and alternative fuel systems are enhancing the efficiency, range and stealth of military aviation platforms. These advances allow for better performance and reduced logistical burden. This shift is transforming the operational capabilities of helicopters, UAVs and other autonomous systems alike.

Helicopters

Hybrid and electric propulsion systems can provide an array of benefits from reduced noise signature and fuel consumption to increased range and endurance. The combined effects of these advancements will enable the helicopters to operate quietly, on missions well beyond current ranges and endurance limits. The Sikorsky S-97 Raider exemplifies this trend, employing a coaxial rotor system and pusher propeller that enhance speed and agility while optimizing fuel efficiency, achieving speeds over 200 knots, a leap forward in helicopter performance.[28]

UAVs

UAVs benefit from electric and solar propulsion, which extend endurance and reduce detectability, making them ideal for stealthy, long-duration missions. The Airbus Zephyr, a solar powered UAV, can remain aloft for months, conducting ISR missions over vast areas without the need for refuelling. This capability highlights how advanced propulsion is enabling UAVs to maintain persistent presence, a key requirement in modern surveillance and reconnaissance operations.[29]

Autonomous Systems

Leveraging efficient propulsion, Autonomous systems extend their operational range and loiter time, enhancing their ability to conduct prolonged missions with minimal external support. The MQ4C Triton, used by the US Navy, incorporates advanced propulsion that supports high altitude, long endurance ISR missions, often exceeding 24 hours. This endurance allows autonomous platforms to cover vast areas deep into enemy territory, providing continuous intelligence and strike capabilities in support of multi-domain operations.[30]

Quantum Technologies

Quantum Technologies employ the fundamentals of quantum mechanics to enable advancements in computing, sensing and communication that could transform military aviation. Though these technologies are still under development, the effects that quantum technologies could have on helicopters, UAVs, and autonomous systems are profound, especially as it relates to capabilities that could transform aviation operations entirely.

Helicopters

Quantum sensors provide the potential to navigate in GPS denied environments, using quantum inertial systems to accurately acquire position without the need for satellites, an important capability in contested environments where GPS jamming is expected. Similarly, quantum encryption provides unhackable links for communications, making the transmission of data secure from eavesdropping. It would be difficult to predict when we will see practical applications, but comprehensive research into quantum inertial navigation for suggests that we will have platforms operating with unprecedented independence and security in the foreseeable future.

UAVs

UAVs could also benefit from quantum computing, which could speed up mission planning and data analysis, leading to more timely decisions in rapidly changing environments. Quantum radar, which is still being developed, could allow UAVs to identify stealth platforms by taking advantage of quantum entanglement, effectively taking away the advantages of low observable platforms. Concepts for integrating quantum radar into UAVs are under exploration, suggesting a future where these platforms could neutralize one of the most significant technological edges in warfare.

Autonomous Systems

In the field of autonomous systems, quantum AI could greatly enhance the speed and complexity of the decision making process, processing large data sets to adapt to threats in real time. Quantum communication will provide instantaneous connections with secure, unhackable links between autonomous systems across the globe. Several nations are already researching these applications, but it will take years before they become a reality. But whenever that happens, quantum technology will propel autonomous systems to unprecedented levels of capability, and will position them as a central element in future military doctrine.

Hypersonic Helicopters and Aircraft

Hypersonic weapons as discussed earlier represent a new class of military platforms that operate at speeds greater than Mach 5, though Hypersonic glide vehicles, such as the Russian Avangard, are launched on ballistic missiles and then glide toward their target at hypersonic speeds, manoeuvring to avoid interception,[31] Though countries are actively pursuing hypersonic aircraft programs but hypersonic helicopters are still a distant reality.

The speed and manoeuvrability of hypersonic platfroms make them highly effective against time sensitive targets, such as enemy command centres, missile launchers, and critical infrastructure. Their ability to strike targets with precision and speed creates a significant challenge for existing air and missile defence systems, which may struggle to detect and intercept these fast-moving threats.

The development of hypersonic weapons and aircraft has significant implications for military aviation, particularly in the context of strategic deterrence and rapid response capabilities. Hypersonic weapons can be deployed from a variety of platforms, including aircraft, ships, and ground-based launchers, making them a versatile addition to military arsenals. May be in times to come the same could be launched from a helicopter too but helicopter due to its inherent stability aspects and comparative lower speed, may not be an ideal platform to launch a hypersonic projectile.

FLIGHT FORWARD: TRENDS DRIVING ARMY AVIATION

Looking to the future, the continued development of AI in military aviation is likely to drive significant advancements in autonomous systems. AI-powered drones and aircraft will become more capable, versatile, and reliable, allowing them to play an increasingly central role in military operations. The integration of AI with other emerging technologies, such as hypersonic weapons and stealth technology, will further enhance the capabilities of military aviation forces, providing them with a decisive edge in future conflicts.

The United States has been at the forefront of integrating emerging technologies into military aviation. Programs such as the Skyborg AI initiative, the development of hypersonic weapons, and the deployment of stealth UAVs demonstrate the US military's commitment to maintaining its technological edge in aerial warfare. The US has also made significant advancements in electronic warfare, with systems such as the EA-18G Growler providing critical EW capabilities to support air operations.[32]

China has rapidly modernized its military aviation forces, focusing on the development of advanced UAVs, stealth technology, and hypersonic weapons. The

People's Liberation Army Air Force (PLAAF) has deployed a variety of UAVs, such as the CH-4 and Wing Loong, for ISR and strike missions.[33] China is also investing heavily in hypersonic missile technology, with the DF-ZF hypersonic glide vehicle being a key component of its strategic deterrence capabilities.[34]

Russia has made significant strides in military aviation technology, particularly in the development of hypersonic weapons and electronic warfare systems. The Kinzhal hypersonic missile and the Avangard hypersonic glide vehicle are examples of Russia's focus on achieving rapid strike capabilities. Russia has also developed advanced EW systems, such as the Krasukha-4, which can jam and disrupt enemy radar and communication systems.

Several trends are likely to shape the development and deployment of aviation technologies over the next two decades.

Increased Autonomy and AI Integration: The integration of AI and autonomous systems into military aviation is expected to continue, with advancements in machine learning, data analytics, and robotics driving further improvements in the capabilities of UAVs, drones, and combat aircraft.

Enhanced Stealth and EW Capabilities: The development of new stealth materials, electronic warfare systems, and countermeasures will enhance the survivability of military aviation forces, allowing them to operate more effectively in contested environments.

Proliferation of Hypersonic Weapons: The deployment of hypersonic weapons by multiple nations will drive the development of new offensive and defensive capabilities, leading to a potential arms race in hypersonic technology.

Global Competition: The race to achieve technological superiority in military aviation will continue to be a key area of competition among leading military powers, with ongoing efforts to develop and deploy cutting-edge technologies that provide a decisive advantage in aerial warfare.

FROM ROTORS TO ROBOTS: EMERGING TECHNOLOGIES AND DOCTRINAL AND TACTICAL SHIFTS IN ARMY AVIATION

The integration of these emerging technologies into military aviation has not occurred in a vacuum; it has driven profound shifts in the doctrines and tactics that govern the use of helicopters, UAVs, and autonomous systems. These changes reflect the evolving nature of warfare, where adaptability, precision, and multi-domain integration are paramount. Below, we explore these shifts in detail, first addressing overarching doctrinal and tactical trends, then examining their specific manifestations in the latest aerial platforms.

Doctrinal Shifts

The incorporation of disruptive technologies has necessitated a rethinking of military doctrines, aligning them with the capabilities and challenges of modern aviation platforms. One of the most significant shifts is the integration of UAVs, which have transitioned from niche assets to central components of military operations. Doctrines now emphasize their role in providing persistent ISR, delivering precision strikes, and conducting electronic warfare, ensuring their utility across all phases of conflict – from peacetime surveillance to high-intensity engagements. This shift is evident in documents like the US Army's Roadmap for Unmanned Aircraft Systems 2010-2035 which outlines UAV employment in various environments.[35]

The rise of autonomous operations represents another doctrinal pivot, as militaries grapple with the implications of platforms that can operate with minimal human oversight. This has prompted the development of frameworks to address ethical and legal considerations, such as the U.S. Department of Defence's Directive 3000.09, which provides guidelines for autonomous weapons.[36] These doctrines balance the advantages of autonomy – such as reduced risk to personnel – with the need for accountability and human judgment, shaping how autonomous systems are integrated into broader strategies.

The concept of multi-domain operations (MDO) has also gained prominence, driven by technologies that enable seamless coordination across air, land, sea, space, and cyber domains. Doctrines now prioritize joint operational frameworks that leverage AI, sensor fusion, and advanced communication to synchronize effects across these domains, creating a cohesive battlespace where each element enhances the others. This holistic approach is critical in countering sophisticated adversaries who operate across multiple fronts.[37]

Integrated Manned-Unmanned Operations is one of the most significant doctrinal shifts that combines the strengths of manned and unmanned systems.

The "Loyal Wingman" Concept: Manned aircraft are increasingly taking on the role of mission commanders, delegating specific tasks to unmanned wingmen. Systems like the MQ-28 Ghost Bat[38] and HAL CATS Warrior[39] exemplify this trend, providing additional sensors, firepower, and redundancy. This integration allows a single pilot to command a network of autonomous systems, thereby extending operational reach and enhancing overall situational awareness.

Joint All-Domain Operations (JADO): Future warfare will span air, land, sea, space, and cyber domains. Doctrines are evolving to integrate assets from each domain into a cohesive fighting force. In this environment, the ability to rapidly re-task unmanned systems and integrate them with manned platforms is crucial for maintaining battlefield superiority.

Finally, the adoption of agile combat employment (ACE) reflects a doctrinal response to contested environments, emphasizing dispersed, rapid, and interoperable operations. Pioneered by the US Air Force, ACE leverages advanced propulsion, stealth, and autonomy to maintain operational tempo, allowing forces to adapt quickly to threats and sustain combat effectiveness.[40] Together, these shifts underscore a move toward flexibility and integration, ensuring that military aviation can meet the demands of an unpredictable and technologically advanced battlefield.[41]

Tactical Shifts

Parallel to these doctrinal changes, tactical adaptations have emerged to exploit the capabilities of new technologies, reshaping how helicopters, UAVs, and autonomous systems are employed. Swarm tactics, enabled by swarm technology, allow multiple platforms to overwhelm enemy defences, conduct distributed sensing or execute coordinated attacks. This approach saturates adversary response mechanisms by presenting a multitude of targets at a particular instant, as seen in recent drone swarm attacks and is becoming a cornerstone of modern aerial tactics. Enhanced survivability is another key tactical facet, driven by stealth, EW and cyber resilience.

Platforms now employ electronic deception, terrain masking and low observable design as well as manoeuvres to evade detection and withstand enemy countermeasures, enabling operations in high threat environments where traditional tactics might fail. Another hallmark of modern tactics is Precision engagement, with AI and sensor fusion enabling highly accurate targeting that minimizes collateral damage and maximizes mission success. The proliferation of precision guided munitions and real time targeting data, exemplified by platforms like the Bayraktar TB2, reflects this shift, allowing forces to strike with surgical precision even in complex urban settings.

Lastly, decentralized command and control leverages autonomy and advanced networks, allowing units to act independently based on real time intelligence, while maintaining alignment with overall operational objectives. This tactical flexibility, supported by systems like the MQ28 Ghost Bat, enhances responsiveness and resilience, enabling forces to adapt to fluid battlefield conditions without constant oversight from higher headquarters. These tactical shifts manifest uniquely across helicopters, UAVs and autonomous systems, as demonstrated by the latest platforms and their employment in contemporary operations. Below, we examine these changes in detail, highlighting specific examples that illustrate their practical application.

Helicopters

Under modern doctrines, helicopters have evolved into multi role platforms tasked with transport, attack, reconnaissance and EW, with special focus on rapid insertion and extraction from contested environments. Drawing on lessons from conflicts like Iraq and Afghanistan, the US Army's FM 3-04 (Army Aviation) codifies these priorities to position helicopters as agile, adaptable assets capable of overcoming sophisticated defences. This doctrinal evolution reflects the need to counter adversaries who use advanced air defences to deny access, requiring helicopters to operate in ways that survivability and maximize speed.[42]

As a preferred tactics, helicopters now conduct low level flying and terrain masking to evade radar and missile threats, leveraging UAVs for reconnaissance to extend their operational range. Smart cockpit systems, powered by AI, enhance pilot's decision making by processing real time data, while EW capabilities like those on the Ka 52 Alligator enhance survivability against electronic and missile threats. These tactics enable helicopters to penetrate hostile airspace, deliver forces or firepower and withdraw before adversaries can respond effectively.

The Kamov Ka52 Alligator which Russia is using in Ukraine employs its Vitebsk Electronic Warfare system including laser warning receivers and infrared jammers to avoid air defences, along with low level tactics to reduce exposure. Its ability to operate in high risk areas demonstrates how tactical innovations, underpinned by technology, allow helicopters to maintain relevance in modern warfare, even against an opponent with considerable ground and air capabilities.[43]

Another example that stands out is the Sikorsky S-97 Raider. A helicopter of the United States that incorporates stealth, AI-based avionics, and advanced propulsion to enable rapid, agile deployments. With a coaxial rotor system (the S-97 has no tail rotor) and a pusher propeller configuration, the Raider is capable of speeds over 200 knots, providing an elevated operational tempo for reconnaissance and light attack missions in contested airspace. The Raider's design aligns perfectly with the doctrinal demands of speed and flexibility, enabling it to insert special forces or conduct strikes in A2/AD zones.[44]

UAVs

'Intelligentization' i.e the Chinese concept of applying Artificial Intelligence's machine speed and processing power to military planning, operational command and decision support outlines the employment of UAVs in contested environments, highlighting their ability to maintain a continuous presence, deliver precision effects and counter enemy drones, a reflection of their growing centrality to military operations. This doctrinal focus ensures that UAVs are not just supplementary assets but primary tools for shaping the battlefield.[45]

Tactically, UAVs are being employed for layered reconnaissance, in tandem with high altitude platforms to provide a comprehensive operational picture, while combat UAVs leverage time sensitive targeting and coordination with manned assets for maximum impact. UAVs have thus become indispensable under contemporary doctrines, which emphasising their role in persistent ISR, rapid response strikes and counter-UAV strategies. The shift towards swarm tactics adds to the efficacy of UAVs and allows for overwhelming a defender, as well as conduct dispersed operations.

This approach enables UAVs to saturate the battle space, gathering intelligence over extended periods and striking with precision. The MQ9 Reaper, operated by the US, exemplifies these shifts, with upgrades that include autonomous navigation and targeting systems powered by AI. Capable of persistent ISR and precision strikes, the Reaper operates in contested airspace, aligning with doctrines that prioritize endurance and lethality. Its ability to loiter for hours and deliver guided munitions with pinpoint accuracy reflects the tactical emphasis on precision engagement, making it a versatile and effective asset in conflicts like Syria and Afghanistan.

Turkey's Bayraktar TB2, offers another compelling example, particularly from its operations in Russia Ukraine war. Combining sensor fusion with EO/IR and SAR systems, the TB2 conducts reconnaissance and precision strikes against armoured targets, demonstrating the tactical shift toward affordable, attritable platforms. Its low cost and high success rate, highlights how UAVs are redefining aerial warfare, offering cost effective solutions that rival more expensive manned systems.

Autonomous Systems

Doctrines for autonomous systems are still evolving, but they increasingly address the ethical and legal ramifications of lethal autonomy, as well as the integration of human oversight. The US. DoD's Directive 3000.09 provides a framework for their development and use, emphasizing accountability and rules of engagement while recognizing their potential to reduce risk to human personnel. This doctrinal groundwork ensures that autonomous systems are deployed responsibly, balancing innovation with strategic and moral considerations.[46]

Tactically, autonomous systems excel in mission planning, adapting to dynamic conditions without human input, and executing high-risk missions that minimize human exposure. They also support autonomous resupply and electronic warfare operations, leveraging their independence to operate in environments too dangerous or remote for manned platforms. These tactics capitalize on the unique strengths of autonomy, enabling rapid, decisive action in complex scenarios.

The Boeing MQ-28 Ghost Bat, developed collaboratively by Australia and the US, embodies these principles as a "loyal wingman" designed to enhance manned aircraft capabilities. Its autonomous systems allow it to conduct ISR, EW, or strike missions alongside human pilots, reducing their risk while amplifying overall effectiveness. This human-machine teaming aligns with emerging doctrines, showcasing how autonomy can extend the reach and resilience of air forces in contested environments.

The IAI Harop, an Israeli loitering munition, provides a stark tactical example, autonomously engaging radar targets to suppress enemy air defences (SEAD).[47] Capable of loitering until an optimal strike opportunity arises, the Harop demonstrates how autonomous systems can execute high-risk missions with precision, minimizing human involvement while delivering decisive effects. Its use in conflicts like Nagorno-Karabakh underscores the tactical advantages of autonomy, where independent operation enhances both survivability and lethality.[48]

Emphasis on Cost-Effectiveness and Attritability

Mass Deployment of Low-Cost UAVs: With prices for platforms like the TB2 and Shahed-136 being a fraction of the cost of a fighter jet, militaries can field large numbers of drones. This allows for saturation tactics, where overwhelming numbers force the enemy to expend resources on countermeasures.

Attritable Systems: The concept of "attritability" means that some systems are designed to be expendable. The loss of a low-cost UAV has far less impact on overall force readiness than the loss of a multi-billion-dollar manned fighter. This strategy reduces risk to high-value assets while maintaining continuous pressure on the adversary.

Budget Reallocation: As forces realize the benefits of integrating unmanned systems, defence budgets are increasingly directed toward research and development of these technologies, further accelerating their deployment.

Challenges to Flight to Future

Electronic Warfare and Countermeasures: As UAVs become more prevalent, adversaries are developing sophisticated jamming and counter-drone systems. Innovations such as fiber-optic guided drones and active electronic countermeasures will be critical in ensuring that unmanned systems remain effective in contested environments.

Ethical and Strategic Considerations: The increased autonomy of military systems raises profound ethical questions. Decisions about target selection, rules of engagement, and accountability in cases of collateral damage must be carefully managed to avoid unintended escalations or breaches of international law.

Training and Integration: Successfully integrating autonomous systems into traditional military structures requires extensive training and new operational concepts. Joint exercises, simulation environments, and updated command and control structures will be essential for realizing the full potential of these technologies.

Rapid Technological Obsolescence: The pace of technological advancement means that today's cutting-edge systems may become obsolete within a few years. Maintaining an agile acquisition process and continuous investment in R&D will be vital to staying ahead of adversaries.

THE INDIAN ARMY AVIATION CORPS

The Indian Army Aviation Corps (AAC), the youngest corps of the Indian Army has long been a critical component of the Indian Armed Forces, providing essential capabilities such as observation, reconnaissance, logistic support, casualty evacuation, and close air support.[49] Historically, the Corps which started with fixed wing aircrafts, now operates primarily with helicopters, relying on the Indian Air Force (IAF) for fixed-wing assets, attack helicopters, and advanced UAV operations.[50] However, the evolving nature of warfare, characterized by rapid technological advancements, increased operational demands, and the emergence of new domains such as cyber and space, has necessitated a re-evaluation of the role and structure of Indian Army Aviation.

As the Indian Army prepares for future conflicts, it faces crucial decisions regarding the future of its aviation capabilities. These decisions involve determining whether the AAC should focus exclusively on observation and logistic roles, expand to include fixed-wing assets and attack helicopters, or assume responsibility for all UAV and drone operations, which are currently shared with the Artillery and IAF.

The Traditional Role of Indian Army Aviation

Observation and Liaison Roles (OP/LP)

Since its inception, the AAC has primarily been tasked with observation and liaison duties, utilizing light helicopters to provide real-time intelligence and battlefield awareness to ground commanders. These roles are essential in supporting artillery spotting, coordinating troop movements, and ensuring effective communication across the battlefield. The AAC's success in these areas has been integral to the operational effectiveness of the Indian Army, particularly in mountainous and difficult terrain where ground-based observation is limited.[51]

Logistics and Casualty Evacuation

Another critical function of the AAC is logistical support, including the transportation of troops, equipment, and supplies to forward positions, as well as casualty evacuation (CASEVAC) from the battlefield. The Dhruv Advanced Light Helicopter (ALH) and Cheetah Helicopters have played a critical role in enabling the Indian Army to maintain operational continuity in isolated and inhospitable areas. They have played an hugely important role in providing the mobility and resilience to Indian forces in both peace and conflict alike.[52]

Incorporating Fixed Wing Assets and Attack Helicopters

One of the most important debate regarding the future of Indian Army Aviation is whether it should integrate fixed wing assets and attack helicopters into its operational structure. As of now, fixed wing assets in entirety and attack helicopters largely, are controlled by the Indian Air Force, which provides close air support and fixed wing reconnaissance for the Indian Army. However, there are strong arguments for bringing these capabilities under the AAC's direct control.[53]

Greater Operational Autonomy: By acquiring fixed wing assets and attack helicopters, AAC will be better enabled to exercise operational independence, thus allowing for the ability to respond quickly and in accordance with the operational needs of ground commanders. This will significantly reduce dependency on the IAF to deliver dedicated air support, thus allowing the Indian Army to effectively coordinate and employ these assets within the tactical battlefield.

Improved Close Air Support: Attack helicopters such as the AH64E Apache, which are largely operated by the IAF and few with the AAC, provide essential close air support to ground forces, particularly in anti tank and anti infantry roles and bringing them completely under AAC would lead to timely, cohesive and optimum utilisation of this resource. Bringing these assets under the control of the AAC enables closer integration with ground operations, improving the effectiveness of air-ground coordination and enhancing the overall combat power of the Indian Army.[54]

Comprehensive Battlefield Awareness: Fixed-wing assets, including surveillance and reconnaissance aircraft, offer extended range and endurance compared to helicopters, providing the AAC with the ability to monitor and dominate larger areas of the battlefield. Integrating these assets into the AAC would enhance its ability to gather and exploit intelligence, contributing to more informed decision-making and better battlefield outcomes.

Integrating UAVs and Drones into Army Aviation

A critical consideration is whether all UAVs and drones being currently operated by the Indian Army be integrated into the AAC. UAVs have become indispensable for almost all arms and services of the Army, offering capabilities such as ISR, target acquisition, direct attack and logistical sustenance. At present, these assets are spread among several branches of the military, but there are logical arguments that suggest to consolidate UAVs and drones into the AAC.[55]

Optimized Resource Allocation: A single corps managing all UAV assets could optimize the allocation of resources, ensuring that UAVs are deployed where they are most needed and reducing the risk of duplication or under utilization. This approach would also facilitate more effective training and maintenance programs, enhancing the overall readiness of UAV forces.

Unified Command and Control: Having all UAVs and drones within AAC would simplify command and control processes, decreasing the complexities of air space management and ensure a more integrated, combined approach to unmanned aerial operations. A single branch structure would provide for better integration and coordination of UAV missions to support ground operations.

Enhanced Integration with Helicopter Operations: UAVs can complement helicopter operations by conducting continuous reconnaissance, providing real time intelligence and adding to fire power in combat scenarios. Bringing UAVs into the AAC would allow for more efficient integration and coordination of both manned and unmanned platforms, enhancing overall operational capability of Army Aviation.

Strategic Implications and the Way Forward

The Indian Army Aviation Corps's future depends on making critical determinations about Army Aviation's role, capabilities and operational structure. Whether the AAC should maintain its tradition focus or change to include fixed wing assets, attack helicopters, or UAVs depends on multiple variables:

Status Quo

Maintaining the status quo, with continued focus on OP/LP and logistics roles while relying on the IAF for fixed-wing assets and attack helicopters, offers a path of least resistance. The AAC gets to continue to be the best at what it already does, without investing heavily on infrastructure or restructuring. However, it limits the AAC's ability to adapt to the future and take advantage of existing and emerging technologies.

Expanding Capabilities

Incorporating fixed wing assets, attack helicopters and UAVs into the AAC would provide potential for increased operational independence, seamless close air support and better battlefield awareness. But before the AAC can acquire this capability, it requires large investments in addition to rethinking on its current operational paradigms. If this capability is pursued, it will transform the AAC into a more capable and versatile force that is responsive to requirements modern warfare. These initiatives require large investments in infrastructure, training and maintenance too. The AAC would also need to develop new doctrines and standard operating procedures to incorporate these capabilities and integrate them into current operations. Finally, expanding the AAC's mandate could lead to inter service competition with the IAF, potentially complicating joint operations and resource allocation.

As the Indian Army prepares for the challenges of the 21st century, it must carefully weigh the benefits and risks of each approach, considering factors such as operational autonomy, resource allocation, and the need for flexibility in responding to emerging threats. Ultimately, the future of Indian Army Aviation will depend on its ability to adapt to new realities while continuing to provide critical support to ground forces across a wide range of missions.

The Indian Context

Army Aviation must become a seamless combination of manned and unmanned airframes with flexibility and persistent swift response in a broad range of operations and missions. In the future, modern rotorcrafts with advanced avionics, digital-cockpits, network-enabled data links will be the mainstay of aviation fleet, capable of providing real-time intelligence, situational awareness as well as support decision making by the pilots. Such enhancements will allow manned aircraft to operate more effectively in multi-domain environments, where rapid information exchange and precise targeting are paramount.

In tandem, the Army must give impetus to induction of autonomous and semi-autonomous unmanned air vehicles that can serve as forward observers, providing near round the clock surveillance data, carry out target acquisition and identification as well as conduct precise strikes while minimizing the exposure to own pilots. Integrating these platforms and systems into a common digital command and control network, preferably one that is interoperable with ground, naval, and space assets, will provide a cohesive operational picture that will act to ensure maximum effectiveness of the air power capability. More importantly, Army Aviation must increasingly adopt latest techniques based on data analytics and IoT sensors to ensure regular predictive maintenance, reduced down time and enhanced operational availability of critical assets. The operational tempo and effectiveness of Army Aviation will get redefined once the convergence of manned and unmanned aviation systems, underpinned by robust digital connectivity and advanced sensor technologies is achieved.

Army Aviation's primary functions remain mobility, reconnaissance & fire support and thus, its future significantly depends upon technologies which enhance the flexibility and survivability of aviation platforms as well as the crews operating them. In order for the existing helicopter fleet to adapt to modern threats, substantial and comprehensive upgrades to the helicopters are required. Retrofitting these helicopters with modern avionics, advanced navigation, targeting, communication and EW suites shall allow them to operate across variety of operational environments and enhance their all weather day night capabilities. Stealth upgrades, including radar absorbing materials and noise reduction technologies, reduce detectability, while enhanced defensive systems like missile warning receivers and countermeasures protect against aerial threats.

Retrofits enable a legacy or even a new but less capable platform to be transformed into a more lethal, survivable, helicopter configured to conduct any operation from transport to close air support. Unmanned Aerial Vehicles must become a cornerstone of the aviation fleet, expanding their deployment for reconnaissance, surveillance and precision strike missions. Combat drones starting from small tactical drones to larger armed drones, can fly into high risk environments, protected by enemy air defence networks and gather real time intelligence and deliver guided munitions without potentially risking a pilot. For the Indian Army, UAVs are particularly valuable in monitoring contested borders or supporting ground operations in rugged terrain, ensuring aviation remains a force multiplier across all theatres.

As we look to the future, vertical lift technologies have the potential to change the capabilities of Army Aviation. Next generation vertical lift helicopters, with better speed, range and payload potential compared to traditional helicopters, represent a huge leap in capability. These new platforms, potentially utilizing tilt rotor designs or hybrid designs, will be able to deliver troops or supplies to remote locations, cover large distances and perform responsive fire support with heavier weapon loads. An aircraft of this nature will allow the Army to quickly project power, across mountainous borders or coastal line, with new levels of flexibility. Investments in these capabilities will ensure that Army Aviation continues to be a dynamic and responsive force, readily able to meet the logistics and combat demands.

Integrating directed energy weapons with multiple aviation platforms is a disruptive solution for defeating aerial threats. High powered laser systems on helicopters or unmanned aerial systems can easily defeat incoming missiles, unmanned aerial systems, and even a low flying manned aircraft, at speeds and precision far greater than traditional air defense platforms. These weapons can either have a power source integral to them or can derive power from the power pack on the aerial platform housing them. Thus, the Army doesn't have to worry about continuous rearming of these platforms, which is limited only by the power capacity of onboard generators and not the payload capacity, providing near infinite defensive capacity against the growing threats of drone swarms, or precision guided munitions. Army Aviation will be able to enhance its survivability and protect both aircrews and ground forces by equipping its platforms with DEWs.

These technological advancements have precipitated doctrinal shifts toward greater integration, autonomy, and multi-domain operations, accompanied by tactical adaptations that emphasize precision, survivability, and decentralized command. As militaries worldwide adopt these platforms and strategies, they are redefining the nature of aerial warfare, achieving levels of operational superiority that were once unimaginable. Yet, this transformation is not without challenges, ethical dilemmas, cybersecurity risks, and the need for continuous adaptation loom large. As the pace of technological change accelerates, the ability to harness these advancements while addressing their implications will determine the future of military aviation, ensuring that it remains a decisive force in the conflicts of tomorrow.

Notes

1   Staff Sgt. Todd Pouliot 10th Combat Aviation Brigade, National Guard, Afghan aviators resupply remote bases with partnered close-air support, Nov. 6, 2013, https://www.nationalguard.mil/News/Overseas-Operations/Article/575241/afghan-aviators-resupply-remote-bases-with-partnered-close-air-support/#:~:text=of%20Afghan%20aviators.-,Capt.,is%20a%20very%20satisfying%20mission.%22

2   Kashish Tandon, US Uses Its Deadly Apache Gunship Helicopters To Disperse Crowd In Afghanistan, The Eurasian Times, August 16, 2021, https://www.eurasiantimes.com/watch-us-uses-its-deadly-apache-gunship-helicopters-to-disperse-crowd-in-afghanistan/

3   Abdul Rahmani and John McCain, "A New Platform and a New Plan for the Afghan Air Force", Small Wars Journal, January 17, 2018 https://smallwarsjournal.com/jrnl/art/a-new-platform-and-new-plan-for-the-afghan-air-force

4   Quigley, Matthew, W., Major, United States Marine Corps, Evolution of CH-53 Heavy Lift, Apr 25, 2017, https://apps.dtic.mil/sti/pdfs/AD1176588.pdf

5   Flight Lieutenant Chris Whelan, The 2020 Nagorno Karabakh War: Unmanned Combat Aerial Vehicles in Modern Warfare, Air and Space Power Review, Vol 25, No 2, https://www.raf.mod.uk/what-we-do/centre-for-air-and-space-power-studies/aspr/aspr-vol25-iss2-3pdf/#:~:text=Azerbaijani%20Unmanned%20Combat%20Aerial%20Vehicles,harbinger%20of%20things%20to%20come%3F

6   Tyler Rogoway, Azerbaijani Reporter Gets Way Too Close To Low Flying Hind Attack Helicopter, The War Zone, Jul 2, 2018, https://www.twz.com/21892/azerbaijani-reporter-gets-way-to-close-to-low-flying-hind-attack-helicopter

7   Gareth Jennings, Ukraine conflict: Russia's attack helicopter forces transition from 'easy target' to 'worst nightmare', Janes, 27 February 2025 https://www.janes.com/osint-insights/defence-news/defence/ukraine-conflict-russias-attack-helicopter-forces-transition-from-easy-target-to-worstnightmare#:~:text=At%20the%20outset%20of%20the,to%20maintenance%20or%20the%20like)

8   Ibid

9   John Spencer and Liam Collins, The Untold Story of the Ukrainian Helicopter Rescue Missions During the Mariupol Siege, Time, February 15, 2024 https://time.com/6694858/ukrainian-helicopter-missions-mariupol-siege/

10   Ashish Dangwal, Israeli AH-64 Apache Helos 'Rain Hell' On Hamas Militants With 30mm Chain Gun Fire & Hellfire Missiles, October 10, 2023 https://www.eurasiantimes.com/israeli-ah-64-apache-helos-rain-hell-on-hamas-militants-wit/

11   Clément Charpentreau, IDF helicopter crashes in Gaza during medical evacuation mission, Aerotime, September 11, 2024, https://www.aerotime.aero/articles/idf-helicopter-crash-gaza-rafah-medivac-mission

12   Mike Hirschberg, Achieving the Vision of Future Vertical Lift, May 22, 2024, https://www.forbes.com/sites/mikehirschberg/2024/05/22/achieving-the-vision-of-future-vertical-lift/

13   Ritu Sharma, With 18 Missiles 'Blocked', Russia's Vitebsk-25 Electronic Warfare System Saves The Day For Ka-52 Alligator, June 13, 2023, https://www.eurasiantimes.com/with-18-missiles-blocked-russias-vitebsk-25-ew-system/

14   Sydney J. Freedberg Jr, Bell V-280 Vs. Sikorsky-Boeing SB>1: Who Will Win Future Vertical Lift?, October 02, 2017 https://breakingdefence.com/2017/10/bell-v-280-vs-sikorsky-boeing-sb1-who-will-win-future-vertical-lift/

15   Sikorsky and DARPA Demonstrate Future Battlefield Logistics Missions with Autonomous Utility Helicopter, November 02, 2022, https://www.lockheedmartin.com/en-us/news/features/2022/sikorsky-and-darpa-demonstrate-future-battlefield-logistics-missions-with-autonomous-utility-helicopter.html

16   Defence News Aerospace 2024, China's new Z-20T assault helicopter challenges US MH-60L DAP for Special Operation Forces, 13 Nov, 2024, https://armyrecognition.com/news/aerospace-news/2024/chinas-new-z-20t-assault-helicopter-challenges-us-mh-60l-dap-for-special-operation-forces#:~:text=The%20Z%2D20T%20is%20equipped,such%20as%20the%20Z%2D20.

17   Richard R. Burgess, Senior Editor, Navy Is Sustaining 10 Operational MQ-8C Fire Scout UAVs; Rest in Storage, January 31, 2023 https://seapowermagazine.org/navy-is-sustaining-10-operational-mq-8c-fire-scout-uavs-rest-in-storage/

18   Peter Suciu, RAH-66: Why America's Comanche Stealth Helicopter Failed, July 1, 2024, https://nationalinterest.org/blog/buzz/rah-66-why-americas-comanche-stealth-helicopter-failed-207969

19   RQ-170 Sentinel Unmanned Aerial Vehicle, August 25 2020 https://www.airforce-technology.com/projects/rq-170-sentinel/?cf-view

20   Taranis, BAE Systems, https://www.baesystems.com/en/product/taranis

21   MQ-9 automated landing, takeoff, Aerospace Manufacturing and Design, November 29, 2018 https://www.aerospacemanufacturinganddesign.com/article/mq-9-automated-landing-takeoff/

22   Kris Osborn, Army Fast Tracking Helicopter-Launched Mini Drones to Seek out Targets, September 13, 2021, https://nationalinterest.org/blog/buzz/army-fast-tracking-helicopter-launched-mini-drones-seek-out-targets-193419

23   Ritu Sharma, Mother of All Drones – China Unveils Gigantic Drone Capable of Carrying Swarm of UAVs In Its Belly, November 19, 2024, https://www.eurasiantimes.com/mother-of-all-drones-china-unveil/

24   OFFSET Swarms Take Flight in Final Field Experiment, DARPA, Dec 9, 2021, https://www.darpa.mil/news/2021/offset-swarms-take-flight

25   John Keller, Boeing to build additional AH-64E attack helicopters with networked avionics, sensors, and communications, March 21, 2023, https://www.militaryaerospace.com/sensors/article/14291231/helicopters-avionics-sensors

26   TB2 Bayraktar Drone, Global Defence News, Army Recognition Group, 07 February 2025 https://armyrecognition.com/military-products/army/unmanned-systems/unmanned-aerial-vehicles/bayraktartb2#:~:text=The%20Bayraktar%20TB2%20is%20equipped%20with%20a%20suite%20of%20combat,tactical%20decisions%20on%20the%20battlefield.

27   Aleah Castrejon, AFRL releases video footage of XQ-67A first flight, Air Force Research Laboratory Public Affairs, June 26, 2024, https://www.afrl.af.mil/News/Article-Display/Article/3816823/afrl-releases-video-footage-of-xq-67a-first

28   S-97 RAIDER®, Expanding the Envelope for Future Vertical Lift, https://www.lockheedmartin.com/en-us/products/s-97-raider.html

29   Zephyr High Altitude Platform Station (HAPS), AIRBUS, https://www.airbus.com/en/products-services/defence/uas/zephyr

30   MQ-4C Triton BAMS UAS, Naval Technology, July 20 2023, https://www.naval-technology.com/projects/mq-4c-triton-bams-uas-us/?cf-view

31 Russia Expands Strategic Nuclear Arsenal with Avangard Hypersonic Missile Deployment, Defence News Army 2024, 24 Dec, 2024, https://armyrecognition.com/news/army-news/army-news-2024/russia-expands-strategic-nuclear-arsenal-with-avangard-hypersonic-missile-deployment

32 EA-18G Growler, Naval Air Systems Command, https://www.navair.navy.mil/product/EA-18G-Growler

33 Zaheena Rasheed, How China became the world's leading exporter of combat drones, Al Zajeera 24 Jan 2023, https://www.aljazeera.com/news/2023/1/24/how-china-became-the-worlds-leading-exporter-of-combatdrones#:~:text=For%20instance%2C%20the%20CH%2D4,the%20US%2Dbased%20think%20tank

34 DF-ZF Hypersonic Glide Vehicle, Missile Defense Advocacy Alliance, January 13, 2023, https://missiledefenseadvocacy.org/missile-threat-and-proliferation/todays-missile-threat/china/df-zf-hypersonic-glide-vehicle/

35 US Army's Roadmap for Unmanned Aircraft Systems 2010-2035, US Army UAS Centre For Excellence, TRADOC, https://irp.fas.org/program/collect/uas-army.pdf

36 DOD Directive 3000.09, Autonomy In Weapon Systems, Office of the Under Secretary of Defense for Policy, January 25, 2023, https://media.defense.gov/2023/Jan/25/2003149928/-1/-1/0/DOD-DIRECTIVE-3000.09-AUTONOMY-IN-WEAPON-SYSTEMS.PDF

37 Multi-Domain Operations in NATO – Explained, Allied Command Transformation, NATO's Strategic Warfare Development Command, October 5, 2023, https://www.act.nato.int/article/mdo-in-nato-explained/

38 Ghost Bat, Royal Australian Air Force, https://www.airforce.gov.au/our-work/projects-and-programs/ghost-bat

39 Hemanth C.S., HAL's latest Combat Air Teaming System completes crucial test ahead of upcoming Aero India, January 12, 2025, https://www.thehindu.com/news/national/hals-latest-combat-air-teaming-system-completes-crucial-test-ahead-of-upcoming-aero-india/article69090082.ece

40 Ryan Hefron, Program Manager, Tactical Technology Office, ACE: Air Combat Evolution, DARPA https://www.darpa.mil/research/programs/air-combat-evolution

41 ACE Program's AI Agents Transition from Simulation to Live Flight, DARPA completes first flight tests of air combat algorithms on specialized F-16 fighter jet, Feb 13, 2023, https://www.darpa.mil/news/2023/ace-program-transition

42 FM 3-04, ARMY AVIATION, Headquarters, Department of the Army, Washington, DC, 06 April 2020, https://armypubs.army.mil/epubs/DR_pubs/DR_a/pdf/web/ARN21797_FM_3-04_FINAL_WEB_wfix.pdf

43 Lyle Goldstein and Nathan Waechter, China Looks to Ukraine War for Guidance on Attack Helicopters, RAND, Feb 26, 2024, https://www.rand.org/pubs/commentary/2024/02/china-looks-to-ukraine-war-for-guidance-on-attack-helicopters.html

44 Joint Doctrine Note 1/20, Air Manoeuvre, Concepts and Doctrine Centre, UK Ministry of Defence, June 2020, https://assets.publishing.service.gov.uk/media/5ef4bf17d3bf7f713849f3d7/20200616-doctrine_uk_air_manoeuvre_jdn_1_20.pdf

45 The PLA: Close Combat in the Information Age and the "Blade of Victory", Mad, Scientist Laboratory, April 6, 2020, https://madsciblog.tradoc.army.mil/225-the-pla-close-combat-in-the-information-age-and-the-blade-of-victory/

46 DOD Directive 3000.09, Autonomy In Weapon Systems, Office of the Under Secretary of Defense for Policy, January 25, 2023, https://media.defense.gov/2023/Jan/25/2003149928/-1/-1/0/DOD-DIRECTIVE-3000.09-AUTONOMY-IN-WEAPON-SYSTEMS.PDF

47 Loitering Munitions: A "Must Have" In Combat, IAI, March 7, 2024, https://www.iai.co.il/news-media/features/loitering-munitions-must-have-combat#:~:text=A%20Legacy%20of%20Loitering%20Munitions,changer%20for%20its%20operating%20countries.

48 Flight Lieutenant Chris Whelan, The 2020 Nagorno Karabakh War: Unmanned Combat Aerial Vehicles in Modern Warfare, Air and Space Power Review, Vol 25, No 2, https://www.raf.mod.uk/what-we-do/centre-for-air-and-space-power-studies/aspr/aspr-vol25-iss2-3-pdf/

49 Sushant Kulkarni, Explained: What is Army Aviation Corps, the youngest Corps of the Indian Army?, The Indian Express, November 2, 2020 https://indianexpress.com/article/explained/explained-what-is-army-aviation-corps-the-youngest-corps-of-the-indian-army-6912908/

50 Rajiv Ghose, Army Aviation In India, *Air Power Journal*, Vol. 3, No. 3, MONSOON 2008, Centre for Air Power Studies, New Delhi, https://capsindia.org/wp-content/uploads/2022/10/Rajiv-Ghose-1.pdf

51 Army Aviation - Present Status and Modernisation, Lt General Naresh Chand (Retd), Aero India 2021 Special, SP Guide Publications. https://www.spslandforces.com/story/?id=748&h=Army-Aviation—Present-Status-and-Modernisation

52 Ibid

53 'Soldiers In The Sky' Vision for the next 25 years, Vayu Interview with Maj Gen PK Bharali, Additional Director General Army Aviation (ADG-AA), Vayu Aerospace and Defence Review, IV, 2011, The Society For Aerospace Studies, https://www.vayuaerospace.in/Issue/202005301456563072.pdf

54 Dinakar Peri, Army raises helicopter unit at Jodhpur to induct Apache attack choppers in May, March 16, 2024, https://www.thehindu.com/news/national/army-raises-helicopter-unit-at-jodhpur-to-induct-apache-attack-choppers-in-may/article67955092.ece

55 'Soldiers In The Sky' Vision for the next 25 years, Vayu Interview with Maj Gen PK Bharali, Additional Director General Army Aviation (ADG-AA), Vayu Aerospace and Defence Review, IV, 2011, The Society For Aerospace Studies, https://www.vayuaerospace.in/Issue/202005301456563072.pdf

13

Disruptive Technologies: Redefining the Future of Air Defense

"The sky is a vast and unforgiving ocean. To defend it, you must see all and kill what you see."

Since we are in the age of integrated battlefields with multi domain operations increasingly becoming the norm in war fighting, it almost becomes necessary to delve into the realms of air defence since any combat arm of the future will not be able to leverage its capabilities unless it is supported by aerial platforms or its adversary contested by such platforms. Air as medium shall be the first to be exploited for modern wars and last or never to be decoupled from war fighting efforts even in times of perfect peace as continuous ISR and Air Defence are as relevant during peace as they are inescapable in War. Emerging disruptive technologies are radically changing the character of air defence in the context of contemporary combat. These new technologies, which include enhanced propulsion, robotics, autonomous systems, artificial intelligence, hypersonics, quantum computing and electronic innovations, are drastically changing military capabilities and tactics and, consequently, the balance of power among states. Alongside these disruptive technologies, the advent of directed energy weapons, such as lasers and microwaves, is poised to transform the artillery landscape. These directed energy systems offer the potential for precision targeting, reduced collateral damage, and extended engagement ranges, challenging the dominance of traditional explosive-based munitions.

As countries integrate these transformative tools into their defence architectures, traditional air defence methodologies face unprecedented challenges, necessitating a thorough re-evaluation of operational doctrines, tactical frameworks, and technological deployments.

This analysis explores the game changing impact of these disruptive technologies on air defence by examining lessons from recent conflicts, discussing emerging weapon systems and innovations, evaluating advancements in air defence technologies, gauging their influence on military strategies, and projecting future trends and challenges. The deliberations span conflicts such as the Armenia-Azerbaijan war, the US-Taliban conflict, the Israel-Gaza war, the Russia-Ukraine war, and established rivalries between Israel-Iran and China-Taiwan, offering a comprehensive view of how technological disruption is reshaping air defence.

ECHOES OF COMBAT: TRANSFORMATIVE LESSONS FROM RECENT WARS

Recent conflicts have been a crucible for innovation and adaptation in air defence.

Armenia-Azerbaijan Conflict

In the Armenia–Azerbaijan conflict, mass and swift deployment of small, agile UAVs and loitering munitions forced operators to rethink traditional static defence postures. The conflict revealed that a dispersed, mobile sensor network integrated with real-time data fusion was essential for countering low-cost but highly lethal threats. The 2020 Nagorno-Karabakh war between Armenia and Azerbaijan showcased the disruptive potential of UAVs in modern air warfare. Azerbaijan effectively employed Turkish-made Bayraktar TB2 drones and Israeli-made loitering munitions to detect and destroy Armenian air defence systems, including surface-to-air missile (SAM) regiments positioned in mountainous terrain.[1] This multi drone strategy of employing reconnaissance drones for target identification and munition drones for destruction of the target enabled Azerbaijan to achieve suppression of enemy air defences (SEAD) with unprecedented precision at early stages of the conflict. Armenia's air defence architecture, was optimised for countering traditional manned aircraft and lacked a robust counter-strategy against small, agile drones. This conflict underscored the vulnerability of legacy air defence systems to low-cost, commercially available drone technologies and emphasised the urgent need for doctrines to incorporate countermeasures against large scale UAV attacks and loitering munitions.

In the 2020 Nagorno-Karabakh war, Azerbaijan leveraged drones like the Turkish Bayraktar TB2 and Israeli Harop loitering munitions to dismantle Armenia's air defenses. The Bayraktar TB2, an unmanned aerial vehicle (UAV), conducted reconnaissance and precision strikes, while the Harop, a "kamikaze drone," targeted radars and command posts.[2]

Armenia relied on aging Soviet systems like the S-300PS, with its 36D6 radar, and the 2K12 Kub, equipped with the P-40 radar, both optimized for larger, higher-flying aircraft. These systems struggled to detect and engage the small, low flying drones, which exploited the blind spots in radar coverage and overwhelmed the interceptor stockpiles.[3] The lesson was clear: modern air defence requires weapons like the 9K333 Verba MANPADS[4] or Pantsir-S1 hybrid gun-missile systems,[5] paired with radars such as the 96L6E,[6] capable of tracking smaller and stealthier targets in cluttered and dense environments.

US-Taliban Conflict

The US-Taliban war thus highlighted the importance of situational awareness and rapid reaction. While the emphasis was on counterinsurgency, the need to protect high-value assets from irregular aerial threats led to the deployment of mobile short-range air-defence units, which allow for quick relocation and target engagement. These asymmetric warfare lessons in the US-Taliban war have their own unique contributions to air defence. The Taliban never possessed a discerning air defence mechanism and relied on guerrilla warfare; they blended into civilian populations and exploited complex terrains. The US had virtually unchallenged air superiority, however; the challenge lay in the conduct of precision strikes, trying not to bring any collateral damage. Thus came the advanced targeting systems and intelligence, surveillance, and reconnaissance (ISR) means to minimize civilian fatalities. The conflict also underscored the integration of air defence with counterinsurgency operations that emphasize mitigating unconventional threats, as opposed to traditional aerial threats that require missile systems. Technologies such as artificial intelligence-driven analytics and autonomous ISR platforms promise greater situational awareness and foster ethical application of often-potent airpower.

The most prominently used air defense system by US forces in Afghanistan was the Counter Rocket, Artillery and Mortar System, commonly known as C-RAM. This system, based on the Navy's Phalanx Close-In Weapon System (CIWS), was specifically designed to counter indirect fire threats including rockets, artillery shells, and mortar rounds. The C-RAM system proved crucial during the final US withdrawal from Afghanistan in August 2021. US forces relied heavily on C-RAM to defend against rocket attacks at Kabul airport during the chaotic evacuation operations. The system operates as a "system of systems" that detects, warns, and intercepts incoming munitions using multiple layers of detection, with human operators certifying targets before the system fires hundreds of 20mm rounds from its six-barrel gun to detonate incoming threats in mid-air.[7]

The US Army deployed Avenger short-range air defense systems in Afghanistan as part of their mobile defense capabilities. The Avenger system, mounted on High Mobility Multipurpose Wheeled Vehicles (HMMWVs), features eight Stinger missiles and a .50 caliber machine gun, providing protection against low-flying aircraft, helicopters, cruise missiles, and unmanned aerial vehicles.[8]

Israel-Gaza Conflicts

During the Israel-Gaza confrontations, layered missile defence systems like Iron Dome and David's Sling countered saturation attacks by Hamas rockets and drones as well as missiles from far off Iran. Their Integrated Sensor and Shooter networks identified incoming targets and engaged multiple threats at the same time while underlining the role of counter-electronic warfare measures. The Israel-Gaza conflict has illustrated the transformative role of AI and autonomous systems in making air defence systems more agile, responsive and accurate. The Iron Dome missile defence system, which employs AI for analysis of incoming aerial threats and prioritises interceptions, helped safeguard civilian areas from Hamas rocket attacks. By tracking the trajectory and potential impact of every rocket, this system optimises resource allocation through selective interceptions, only of threats that posed highest and immediate threat. This integration of AI greatly enhances the speed and accuracy of decision-making which is most critical in massive and high speed engagements. Israel also used drone swarms for reconnaissance and for targeting strikes, effectively integrating autonomous systems into its air defence strategy. These trends illustrate how AI and autonomy can transform traditional air defence and scale and adapt it to confronting imminent threats.

The Iron Dome missile defence of Israel presents itself as an almost state-of-the-art air defence, an AI priority system to intercept projectiles targeted towards Israel. It boasts a success rate of 90%, prioritising threats through analytical investigation of trajectories, hence conserving resources.

Lesson: AI and autonomy increase the efficiency of air defence in high-intensity engagements.

Russia-Ukraine War

The Russia–Ukraine war has been characterised by the extensive use of drones, electronic jamming, and precision-guided munitions. The use of conventional defense systems was incidental while hybrid warfare became the main method of adversaries pursuing victory. The result was an accelerated shift towards networked and agile defence architectures. The subsequent Israel–Iran duels and the China–Taiwan tensions further cemented the view that future contestations would increasingly be dotted with high-speed, multi-domain warfare where even momentary lapses in air defence coverage could prove catastrophic. Air defence has been revolutionised by the current Russia-Ukraine war which has essentially become a real-world proving ground for disruptive technology. Both sides have largely relied on drones for reconnaissance, targeting, and post-strike damage assessment. Ukraine, thus, went on to acquire the capability to manufacture drones, including the Ukrjet UJ-22 and UJ-26, enabling them to carry out long-range missions into Russian territory,[9] while Russia has added the Iranian-made Shahed 136 drone known for flying stealthily and at low altitudes,[10] hence, evading radar detection. Electronic warfare has been a battle winning factor, with both nations using jamming and spoofing to disrupt drone navigation and communications. Ukraine's AI-powered drones enhance battlefield awareness, while Russia's countermeasures reflect the dynamic interplay of technology in modern air defence. The conflict underlines an urgent need for air defense systems to fuse strong EW capabilities for quick adaptation to evolving drone warfare strategies.

Russia deployed the S-400 Triumph system, reinforced with 48N6E3 missiles and supported by the 91N6E Big Bird radar, to create expansive anti-access/area denial (A2AD) zones, challenging Ukrainian aerial assets.[11] However, Ukraine, in turn, leaned on a mix of legacy systems like the S-300PT and newer Western-supplied weapons, such as the NASAMS, which uses the AN/TPQ-64 Sentinel radar and the Patriot PAC-3 with its AN/MPQ-65 radar.[12] These systems had effectively intercepted Russian cruise missiles and drones, but the ongoing conflict has made clear the superior importance of mobility and redundancy, mobile launchers, such as the NASAMS platform based on the HMMWV, and distributed radar networks have been indispensable under the constant barrages of artillery fire and electronic warfare attacks. As drones were extensively employed and countered with short-range systems such as the Strela-10, this further strengthens the case for the layered defensive architecture with agile, responsive weapons.

Israel-Iran Rivalry

The old rivalry between the two countries attained aggressive stance in the recent past in which, emerging technologies like AI, cyber warfare, ballistic missiles and advanced drones played significant roles. Iran has heavily invested in drone technology, developing UAVs for long-range strikes and surveillance. Israel, leveraging its advanced cyber capabilities, had been using offensive cyber operations to disrupt Iranian command and control systems,[13] and potentially targeted air defence networks to facilitate airstrikes on the Iranian nuclear facility as a retaliation to a massive drone and missile attack by Iran in support of the Hamas belligerence.[14] Conversely, Iran could deploy drones to overwhelm Israel's air defence systems, testing the Iron Dome's limits. This conflict highlights the potential for cyber-physical integration in air defence, where cyber-attacks degrade an adversary's detection and response capabilities, amplifying aerial threats.

Though de-escalation measures controlled further attacks from both sides but a future violent scenario between these countries will have Israel's David's Sling, with its Stunner missile and EL/M-2088 radar, alongside the Arrow-3 paired with the Super Green Pine radar,[15] face Iran's Fateh-313 ballistic missiles and emerging hypersonic threats. Iran might counter with its own Sayyad-4 missiles and Bavar-373 radar,[16] a domestic S-400 analogue.

China-Taiwan Tensions

Given the advanced technological prowess of both the countries and a scenario where diplomacy fails and China decides to forcefully realise it's One China dream, Hypersonic missiles and quantum technologies could evolve into game-changers, for both China and Taiwan in the near future. Exceeding Mach 5, with sea-gliding unpredictable trajectories, hypersonic glide vehicles developed by both countries stand as insurmountable challenges to traditional missile defense systems. Therefore, Taiwan given the geographical and military mass will have to rely on advanced detection and interception technologies, utilising possibly quantum radars for enhanced sensitivity and resolution.

Hence, as is very obvious a general arms race in the fields of hypersonic and quantum technologies, to continually develop highly responsive, accurate and effective air defence systems is on in this particular region of the South China Sea, with North Korea adding to the volatility. An armed conflict between the two China's will find Taiwan's Patriot PAC-3 and Sky Bow III, supported by CS/MPQ-90 radar,[17] against DF-17 hypersonic missiles from China tracked by PAVE PAWS radar,[18] emphasising the need for quality long-range, high-altitude interceptors and radars with enhanced detection against stealth and speed.

These conflicts collectively demonstrate that modern air defence must evolve to counter a broad spectrum of threats, ranging from low-cost UAV swarms to sophisticated hypersonic weapons, and must do so with enhanced agility, network integration, and electronic counter-counter measures.

GROUND-BASED TACTICAL AIR DEFENCE WEAPONS, SYSTEMS, AND RADARS

Ground-based tactical air defence weapons, systems, and radars form the backbone of military efforts to shield troops, installations, and civilian populations from an increasingly diverse array of aerial threats. These defences encompass a spectrum of equipment, including man-portable air-defence systems (MANPADS) like the FIM-92 Stinger,[19] short-range missile systems like the Pantsir-S1, and long-range missile platforms such as the Patriot and S-400, all supported by sophisticated radars like the AN/MPQ-53 and Giraffe AMB.[20] Designed to handle a staggering range of threats, everything from small, agile drones skimming the treetops to fast-moving jets and even ballistic missiles arcing down from the edge of space. To be capable of an effective layered air defence the modern day systems integrate cutting-edge sensors, command-and-control networks that allow for rapid decision-making, and a mix of kinetic weapons (missiles and guns) and emerging non-kinetic options (electronic jamming or directed energy) to neutralise threats in the most effective way possible.

The advancements from the archaic world wars in this field have been rapid and remarkable to say the least. During World War II, air defence meant cranking a 40mm Bofors gun, or a MMG platoon and hoping your aim was good enough to shoot down an enemy plane. Today, Air Defence Regiments have got networked missile batteries that share data in real-time, powered by phased-array radars that can track multiple targets simultaneously without missing a beat. The evolution from those early days to now reflects not just technological leaps but a fundamental shift in how militaries approach the challenge of defending the skies.

Modern ground-based air defence weapons have evolved from static, silo-based systems to highly mobile, networked platforms capable of rapid deployment across varied terrains. We've moved away from those old, static systems, the ones stuck in silos, waiting for something to happen. Now, we're dealing with highly mobile, networked platforms that can be deployed at a moment's notice, no matter the terrain. Take the Quick Reaction Surface-to-Air Missile (QRSAM)[21] and the Vertical Launched Short-Range Surface-to-Air Missile (VL-SRSAM),[22] for example. These systems are all about flexibility and multi-role capability. They're designed to take on a wide range of threats, everything from low-flying drones that sneak under the radar to high-speed jets screaming across the sky. And they don't operate in stand-alone mode; they're part of a layered defence strategy, working together to cover all the gaps.

What makes these systems really stand out is the latest tech behind them resulting in advanced guidance systems like active radar homing and data-linked command guidance. That means they can lock onto targets and stay on course, even when the enemy's throwing up decoys or jamming signals. Worth mentioning is the Akash missile family, our home-grown solution demonstrates commitment to self-reliance and rapid adaptation, ensuring that ground-based platforms can meet both conventional and asymmetric threats with precision and agility.[23]

In the realm of ground-based air defence, radar and sensor networks stand shoulder-to-shoulder with the weaponry in terms of pace of their evolution. Modern radar systems have graduated beyond the old two-dimensional detection methods absorbed the cutting-edge three-dimensional AESA (Active Electronically Scanned Array) technology.[24] This shift brings sharper resolution, a stronger shield against jamming attempts, and the ability to keep tabs on multiple targets all at once in a dynamic electronic environment. Furthermore, these radar systems have become smaller, stealthier and faster. They're built to be highly agile and mobile, allowing for rapid redeployment and increased survivability across varied terrains and weather conditions.

These systems rely on advanced digital signal processing and machine learning to cut through the clutter and zero in on threats as they unfold in real time. It's like giving the system a sharper instinct, capable of refined target discrimination and prediction capabilities in real time. And it doesn't end with radar alone. By collating data from radars, electro-optical sensors, and passive detection setups, create a full, seamless picture of the battlespace. This multi-sensor fusion synergises everything together, removing any gaps that a lone sensor might leave behind and keeping us one step ahead.

At the heart of today's ground-based air defence lies the digital command, control, communications, and intelligence (C3I) systems, pivotal to modern military operations. These systems connect missile launchers, radar installations, and mobile units into a unified network, allowing operators to quickly evaluate threats and assign interceptors precisely where they are needed most. This approach, centred on networked operations, harnesses real-time data streams, artificial intelligence, and robust encryption to guarantee that decisions are taken swiftly, with minimal latency. The rise of Joint All-Domain Command and Control (JADC2) underscores the push to weave ground-based systems into a wider, multi-domain operational tapestry, a clear indication of the military's drive to maintain superiority in complex environments.[25] In this setup, ground-based air defence is not a lone player; it is intricately interwoven with air, sea, and space-based sensors, forming a web of interconnected systems ready to tackle sophisticated adversaries operating across multiple domains.

New Innovations in Ground-Based Air Defense

Ground-based tactical air defense systems are being constantly upgraded, redefining their capabilities to better engage modern aerial threats. Using phased-array radar, such as AN/MPQ-53 with the Patriot system and 92N6E Grave Stone with the S-400, facilitates the electronic steering of beams for quickly providing a 360-degree view of the area. These radars are capable of simultaneously tracking many targets, as many as 100 in the Patriot's case, making them very capable, in a densely-packed multiple-threat scenario. Another AESA radar, Sweden's Giraffe 4A, integrates surveillance and fire control in one radar tracking both air and ground threats; obviating the use of two separate systems and boosting efficacy of short and medium-range systems like the RBS 70 NG missile launcher and improving the efficiency of layered air defence systems.[26]

Integrated air defence systems are another milestone in the integration of weapons, sensors, and command nodes into coherent networks. The US Army's Patriot batteries, integrating the PAC-3 missile, AN/MSQ-104 Engagement Control Station, and AN/MPQ-65 radar, employ Link 16 for data sharing and coordination with platforms such as THAAD's AN/TPY-2 radar.[27] Russia's Pantsir-S1 exemplifies hybrid integration, with the 57E6 missiles and twin 30mm autocannons employing dual-band radar for targeting both air and ground targets.[28] This provides for improved situational awareness and short-response time needed against fast-moving aerial threats like cruise missiles.

Directed Energy Weapons (DEWs) are posited to be game-changers as a cost-effective replacement of conventional munitions. The US Army's M-SHORAD system mounts a 50kW laser on a Stryker vehicle and uses the Multi-Mission Hemispheric Radar (MHR) to zap drones and missiles at short ranges.[29] Similarly, the UK's DragonFire laser is being developed to be integrated with existing radars like the Saab Giraffe, offering pinpoint accuracy with infinite shots assuming power is in constant supply.[30] Resultantly, by employing DEWs, swarms of drones may be thwarted in an economical way, wherein for instance, a $100,000 missile which destroys a $1,000 UAV will be replaced by a DEW costing a fraction with better strike probability and zero collateral.

Latest Air Defense Systems

Modern air defence systems have evolved significantly to counter an increasingly complex array of aerial threats, from drones and rockets to advanced missiles and stealth aircraft. The following sections covers some of the most prominent systems in use today, showcasing their capabilities and strategic importance.

Iron Dome (Israel)

Israel's Iron Dome is, on its part, the pinnacle of short-range air defense, designed to intercept incoming rockets, artillery shells and mortars with phenomenally high success rates. At the heart of it is an AI-driven EL/M-2084 radar that identifies and tracks incoming threats, feeding information into a complex battle management system that computes trajectories in real time. This system, which prioritizes threats, only engages incoming missiles expected to hit populated or critical areas, conserving its very expensive interceptors, each costing around $50,000. Hailing a success rate of over 90%, the Iron Dome has amply demonstrated its effectiveness in protecting Israeli cities from Hamas rocket onslaughts. Its integration of artificial intelligence for rapid, adaptive decision-making sets a benchmark for cost-effective defense in high-intensity, short-range threat scenarios.[31]

S-400 Triumf (Russia)

Russia's S-400 Triumf has established itself as a formidable long-range air defence system, designed to stop a wide range of aerial threats, including aircraft, drones, cruise missiles and ballistic missiles, at distances up to 400 kilometres. Its multi-layered architecture utilizes a variety of missiles, such as the long-range 40N6 for high-altitude targets and the agile 9M96 for lower-altitude threats, offering valuable operational flexibility. Russia has been extensively using the system like Syria and Ukraine, where the S-400 has been deployed to create denied access zones, deterring NATO. The system integrated with advanced electronic warfare systems, not only defends against physical threats but also disrupts enemy communications and navigation, cementing its role as a cornerstone of Russia's strategic defence architecture.[32]

NASAMS (Norway/US)

The National Advanced Surface-to-Air Missile System (NASAMS), a medium-range air defence solution optimised for countering drones, cruise missiles, and aircraft, is a joint venture between Norway and the United States. Utilising the Advanced Medium-Range Air-to-Air Missile (AMRAAM), it neutralises targets up to a range of 50 kilometres and excels in networked operations. Its modular design integrates seamlessly with multiple sensors and command structures, enabling rapid deployment and adaptability.[33] In Ukraine, NASAMS has safeguarded critical infrastructure against Russian missile barrages, demonstrating its effectiveness in real-world combat. By linking multiple launchers and sensors into a cohesive network, NASAMS represents the shift toward distributed, flexible air defence systems capable of responding to dynamic threats.

M-SHORAD (US)

The Mobile Short-Range Air Defence (M-SHORAD) system, is a mobile, multi-layered platform designed to counter the rising risk of low-altitude threats, such as drones and cruise missiles. M-SHORAD integrates traditional kinetic interceptors, Stinger missiles from Stryker vehicles with state-of-the-art directed energy weapons like high-energy lasers. These lasers present a low-cost means for countering drone swarms, assuring virtually unlimited "shots" in the battle since they work on a power supply compared to limited missile stocks. Their mobility also allows them to protect forces on the move. This combination of kinetic and non-kinetic defences places M-SHORAD as a forward-looking solution to increasing airborne threats.[34]

David's Sling (Israel)

David's Sling is Israel's medium-to-long-range missile defence system providing air defence capability between the short-range Iron Dome and long-range Arrow systems. It can effectively intercept targets such as ballistic missiles, cruise missiles and drones from 40 to 300 kilometres, using a two-stage interceptor armed with a multi-mode seeker for precision targeting suitable for engaging fast-moving, sophisticated threats. David's Sling offers active augmentations to Israel's multi-layered defence response by neutralising missile salvos launched at strategic targets during a conceivable Israel-Iran conflict. Its capability to integrate within a larger air defence network underscores its strategic importance.[35]

DEFENDING THE HORIZON: TRANSFORMATION IN AD ENVIRONMENT

The last decade has witnessed rapid technological advancements that have transformed both offensive and defensive capabilities in the aerial domain. Some of the innovations and systems leading this transformation are as under:

Unmanned Systems and UAVs

Unmanned aerial systems have been instrumental in transforming, weapons, systems and tactics of surveillance and targeting. Modern UAVs offer persistent intelligence, surveillance, and reconnaissance (ISR) capabilities while simultaneously serving as platforms for direct engagement. The fact that they become one of the most widely used combat platforms is the sole reason why counter-UAV systems are being developed at feverish face, which include kinetic interceptors, jammers, and directed-energy solutions. The adaptability of UAVs, in turn, forces aerial defence systems to constantly keep up by incorporating quick identification and tracking algorithms capable of differentiating between benign and hostile drones.

Loitering Munitions and AWS

Loitering munitions or "suicide drones" have emerged as an effective low-cost solution for precision strikes. These weapons are designed to hover over the target area and engage targets as opportunities arise. Their unpredictable flight paths and ability to evade traditional radar make them a particularly challenging threat, necessitating the development of high-speed, multi-target interceptors and AI enhanced situational awareness platforms.[36]

Electronic Warfare and Cyber Capabilities

The character of warfare has undergone profound changes and modern warfare being almost synonymous with the electronic spectrum. Electronic warfare systems jam, deceive, or disable enemy sensors and communication links, thus degrading their command and control capabilities. Alongside these systems, cyber capabilities are being developed to protect friendly networks and, where necessary, disrupt adversary systems. These integrated measures are an essential element of a resilient air defence posture.

Advanced Sensor Networks

From low probability of intercept (LPI) radars to distributed acoustic and optical sensors, the whole range of innovations in sensor technology is pushing up the capabilities of detection across the battlespace. The tentative possibility of quantum sensor integration offers unique precision and resilience to jamming, heralding a time when stealth and low observability shall be outclassed by revolutionary detection methods. Together, these emerging technologies are pushing the envelope of what is possible in air defence, creating both new capabilities and new challenges that require agile adaptation.

Artificial Intelligence

AI is revolutionizing air defence by accelerating decision-making and optimizing resource use. In real-time systems like the Iron Dome system, AI algorithms assess incoming projectiles' trajectories and identify only the threats that should be intercepted given their predicted impact. In that way, costly interceptors will not be wasted on lower-priority threats. AI also equips autonomous drones like the Bayraktar TB2 to identify and engage targets with minimal human input. In networked systems such as NASAMS, AI fuses information from multiple sensors into a shared battlefield image that enhances threat detection and coordinated response. While AI increases efficiency and responsiveness, it poses ethical concerns regarding autonomous lethal actions, underscoring the value of strong human oversight.

Quantum Technologies

Quantum technologies, built upon the foundation of quantum mechanics are poised to revolutionize air defence by fuelling advances in radar technology, communications, and computing. Quantum radar could identify stealth aircraft due to its ability to take advantage of quantum entanglement and effectively counter low observable technologies. During war, quantum cryptography provides unbreakable encryption thereby making critical communication secure. At the same time, quantum computing could drastically change cryptanalysis and electronic warfare as it has the ability to run huge amounts of data at incredible rates. Though still in early stages, these technologies promise to redefine detection and security, particularly against hypersonic and stealth threats, despite challenges related to cost and technical maturity.

Autonomous Air Defence Weapon Systems

Modern autonomous air defense systems increasingly incorporate artificial intelligence to enhance their decision-making capabilities. AI-powered systems can process vast amounts of data in real-time, distinguish between friend and foe, and prioritize threats based on their trajectory and speed. These systems can react within milliseconds, making them essential for countering hypersonic weapons that travel at more than five times the speed of sound.[37]

India's Akashteer system exemplifies this AI integration, providing fully automated detection, tracking, and engagement of enemy aircraft, drones, and missiles. The system integrates multiple radar systems and sensors into a single operational framework that enables automated, real-time engagement decisions.[38]

Hypersonic Threats: Making Traditional AD Obsolete

Most existing AD and BMD systems were designed for predictable, high-altitude ballistic threats, not for hypersonic weapons that blur the line between air defense and missile defense. Their inability to detect, track, and intercept hypersonic threats effectively makes them increasingly obsolete.[39,40] Hypersonic threats require integrated, multi-layered defense architectures combining space-based early warning sensors, advanced radars, AI-enabled tracking and decision-making, and new interceptor technologies capable of engaging manoeuvring targets at high speed.[41] The compressed timelines demand near-automatic, AI-supported decision-making to enable timely interception, while maintaining human oversight to manage the risks of rapid engagement decisions.[42] Defining characteristic of hypersonic projectiles which are challenging the AD architecture are as under:

Extreme Speed and Manoeuvrability: Hypersonic weapons travel at speeds above Mach 5 and can manoeuvre unpredictably during flight, unlike traditional ballistic missiles with predictable trajectories. This manoeuvrability complicates tracking and interception because defense systems cannot rely on fixed, calculable paths.

Low-Altitude Flight Paths: These weapons typically fly at altitudes between 20 and 60 km, higher than most aircraft but lower than the engagement envelopes of many ballistic missile defense (BMD) systems. This altitude range exploits a gap between air defense and BMD systems, making interception difficult.

Radar Horizon and Detection Limitations: Ground-based radars have limited line-of-sight due to Earth's curvature, restricting early detection of low-flying hypersonic weapons until they cross the radar horizon. This late detection compresses the available reaction time for defense systems, often to mere seconds.

Compressed Engagement Timelines: The high velocity drastically shortens the timeline for detection, decision-making, and interception. Traditional missile defense systems, designed for slower ballistic threats, struggle to respond within these compressed timeframes.

Inadequate Sensor and Tracking Technology: Hypersonic vehicles produce weaker infrared signatures than ballistic missiles, complicating space-based infrared detection. Their rapid manoeuvres also challenge continuous tracking across multiple sensors, requiring advanced data fusion and real-time processing beyond current capabilities.

Limited Coverage and High Costs: Systems like THAAD or Patriot can be adapted for hypersonic defense but only cover small areas. Defending large territories would require prohibitively large numbers of batteries, making comprehensive defense economically and logistically challenging.

Directed Energy Weapons: Using Concentrated Power

The entry of DE weaponry into AD systems would add a paradigm in land-based firepower, given that such weapons seek to change the face of warfare by virtue of their unique capabilities. Lasers and microwave systems, known as directed energy weapons (DEWs), represent state-of-the-art air defense technology that can neutralize threats with high precision and low collateral damage. The M-SHORAD system, for example, employs lasers to effectively counter drones and low-flying threats vis a vis conventional interceptors which are comparatively complex and costly, as there is no limitation on the "ammunition" to engage the threats, as long as power is available. The use of directed energy weapons (DEWs) is proven in sustained engagements, the most visible being the US Navy's High-Energy Laser Weapon System (HELWS), which has successfully neutralized UAVs.[43] Especially effective against drone swarms, DEWs address scenarios where kinetic systems might falter. However, limitations like power requirements and atmospheric interference must be overcome to unlock their full potential.

In contrast with conventional munitions based on kinetic energy or explosive force, DE weapons use highly focused beams of energy – lasers, microwaves, or particle beams – to neutralize targets at an unparalleled rate of speed and levels of precision. DE weaponry offers several key competitive advantages over ordinary artillery systems:

- **Speed-of-Light Engagement:** DE weapons, residing at speeds equal to or near light, have thus become a near-instantaneous means of putting targets ineffective. Due to high speeds of engagement, they are suitable for pinpointing moving targets, deterring hypersonic threats, and addressing saturation attacks.

- **Surgical Precision:** DE weapons can be configured to deliver almost no collateral damage. This precision makes them particularly well-suited for urban warfare, counter-terrorism operations, and other scenarios where minimizing civilian harm is paramount.

- **Scalable Effects:** The intensity and duration of DE beams can be adjusted to achieve a range of effects, from disabling electronic systems and detonating explosives to physically destroying targets. This flexibility allows for tailored responses across a spectrum of threats.

- **Deep Magazine Potential:** Unlike conventional artillery limited by ammunition storage and logistics, DE weapons can theoretically engage multiple targets consecutively, limited only by their power source. This "deep magazine" potential is particularly advantageous for sustained engagements and countering swarm attacks.[44]

A number of countries currently are developing and fielding DE weapon systems for artillery applications:

- **United States:** The U.S. Army is heavily investing in the development of laser weapons, especially mobile, high-energy systems for counter-rocket, artillery, and mortar missions. Examples of such systems are the High Energy Laser Mobile Demonstrator[45] and Tactical Ultrashort Pulsed Laser for Army Platforms.[46]
- **China:** Reports indicate that China is working towards high-power microwave weapons for anti-satellite and anti-missile defense-possibly even for artillery systems. Such weapons would disrupt or destroy enemy electronics and communications, crippling the ability of their forces to wage war effectively.[47]
- **Russia:** Russia claims that a mobile laser weapon system developed by the country, the Peresvet, is capable of engaging aerial targets and blinding satellites. While the particulars of this weapon remain hazy to date, the Peresvet reflects Russia's continuing interest in DE weapons for strategic applications.[48]

Future Developments and Challenges

Promising though they are, DEWs face numerous technological issues:

- **Power Generation and Storage:** The generation and storage of adequate energy for the functioning of high-energy lasers and microwave weapons continue to constitute a major challenge. Advances in battery technology, power generation, and miniaturization are needed if practical DE artillery systems are to be deployed.
- **Atmospheric Attenuation:** DE beams may experience attenuation due to certain atmospheric conditions (i.e., rain, fog, and dust) necessitating the application of adaptive optics, beam control techniques, and resilient targeting systems to combat these impairment mechanisms.
- **Thermal Management:** DE weapons produce significant heat, which demands advanced cooling systems to ensure overheating does not degrade the laser and the good thermal efficiency of the instruments. Efficient and compact thermal management must be devised to make DE weapons usable in mobile artillery platforms.

Development of Next-Generation Air Defence Systems

In response to the evolving threat landscape, defence research and development globally is focusing on next-generation air defence systems that seamlessly integrate the emerging technologies described above. Key innovations include:

Network-Centric Integrated Command and Control (C2)

Modern systems are designed with a distributed, networked architecture, with sensors, interceptors and command nodes integrated in real-time. This enables rapid threat detection, prioritisation, and engagement across the dynamic battlespace. In order to defend against the sophisticated electronic and cyber threats of modern era, systems such as the US Patriot Advanced Capability (PAC-3) and Terminal High Altitude Area Defense (THAAD) are being upgraded with more advanced data links and cyber-resilient architectures.

Autonomous Interceptor Platforms

The requirement to safeguard against hypersonic and agile threats has led to the development of autonomous interceptor systems that can engage multiple targets. The platforms use AI algorithms to process large numbers of sensor data and concurrently engage multiple threats, even in environments with significant decoy and electronic countermeasures. Thus, autonomous interceptors are an edge over human-operated systems which tend to be slow and prone to human fatigue during prolonged operations.

Mobile and Scalable Air Defence Units

Flexibility and rapid redeployment have become critical. Emerging systems are being designed with modular capabilities that allow rapid fielding and integration into various operational contexts. This includes mobile command centres and vehicle-mounted radars that provide localized, high-speed responses to low-altitude and rapidly emerging threats.

Integration of Directed Energy Modules

Future air defence platforms are expected to integrate directed energy modules as complementary layers to traditional kinetic interceptors. These modules, once fully operational, could provide a cost-effective method for intercepting swarms of drones or incoming artillery shells, supplementing missile-based systems and reducing overall engagement costs.

Enhanced Cyber and EW Resilience

Acknowledging the vulnerabilities that accompany networked environments, modern integrated air defence solutions are designed to withstand the wide spectrum of cyber and electronic warfare threats. These include multiple redundant means of communicating, ruggedized hardware, and responsive software that can react to jamming attempts and cyber intrusions.

These developments demonstrate a paradigm shift from isolated, platform-specific capabilities to an integrated, multi-domain defence network that can seamlessly address a range of emerging threats.

BATTLEFIELD BLUEPRINTS: EVOLVING DOCTRINE AND TACTICS

The infusion of disruptive technologies is reshaping air defence doctrines and tactics. Key impacts include:

Layered Defence

Constant research and development in ground-based air defence weapons, systems, and radars are shaping the military doctrine and the tactics of the future. Traditional layered defence – long-range S-400s, medium-range Patriots, and short-range Stingers – is rapidly being replaced with integrated networks where systems like IBCS connect disparate assets. This marks a shift to real time data sharing, permitting a Patriot battery to cue a NASAMS launcher based on an AN/TPQ-64 radar feeds. The weapon system is no longer isolated and dependent on a linear loop but is functioning in a web of intelligent information. Mobility is also key; systems like the truck-mounted Pantsir-S1 or the M-SHORAD's Stryker platform allow rapid redeployment, reducing vulnerability to enemy strikes.

AI and Automation are Streamlining Tactics

The Iron Dome's autonomous target selection, driven by the EL/M-2084 radar, exemplifies how machines can prioritize threats, freeing operators for strategic oversight. Similarly, the S-400's 55K6E command post uses automation to manage multiple engagements, cutting response times against saturation attacks. These advancements demand new training paradigms, as crews master complex interfaces while retaining decision-making authority in high-stakes scenarios, balancing technology with human judgment.

Decentralized and Agile Operations

Modern doctrines emphasize the need for decentralization to avoid single points of failure. Distributed sensor networks and mobile air defence units are now central to countering rapid, multi-vector attacks. Tactics now incorporate frequent repositioning and dynamic re-tasking of assets to respond to fluid battle conditions. The trend toward distributed lethality is exemplified by mobile platforms like M-SHORAD, which disperses air defence assets across smaller, agile units. This reduces vulnerability to precision strikes while maintaining shared situational awareness through networked systems. By decentralizing firepower, forces gain flexibility and resilience, adapting to dynamic threats with coordinated, distributed responses – a tactical shift that balances survivability and lethality.

Integrated Multi-Domain Operations

Air defence is no longer planned or executed in isolation. Modern warfighting involves coordinated operations across land, air, cyber, and space domains. The integration of AI-driven decision support tools allows commanders to process huge amounts of data and, consequently, strike back at threats nearly simultaneously across multiple domains, thus increasing overall operational effectiveness. Modern air defence systems, like Russia's S-400, require multilayer coordination not only across air, land, sea, and cyberspace domains, by integrating sensor and effectors for holistic coverage. This architecture ensures that threats are detected and countered along the various vectors in a coordinated operation, thus enhancing battlefield resilience. It is by linking interdependent systems to form a network that forces from multiple domains can counter these multi-dimensional attacks much more effectively, symbolizing a doctrinal movement towards an integrated cross-domain defence strategy.

Rapid Reaction and Autonomous Engagement

Emerging threats such as hypersonic missiles and swarm attacks by UAVs are causing a dramatic reduction in effective engagement timeframes. Accordingly, new doctrines emphasise rapid detection and autonomous engagement. Once fielded, autonomous systems, enabled with AI, will be equipped to execute engagement protocols autonomously without waiting for centralised command decisions, significantly reducing reaction time and increasing chances of success. AI is automating air defence and doing so in real-time. With such systems as the Iron Dome and NASAMS, an AI executes threat assessment for engagement with little to no human input, which reduces errors and speeds up the decision cycle. In high-intensity conflicts, making rapid decisions becomes crucial; this automation enhances operational efficiency and effectiveness on the battlefield, ushering in a tactical evolution towards increased reliance on smart, adaptive systems.

Electronic and Cyber Countermeasures

The persistent threat of electronic warfare and cyber-attacks has led to the integration of defensive EW measures into air defence tactics. This includes pre-emptive jamming, the use of decoys, and the development of cyber counter-offensive capabilities to ensure uninterrupted command and control during high-intensity conflicts.

Counter-Drone Tactics

The widespread use of drones has led to new tactics using a mix of directed energy weapons, electronic warfare and kinetic interceptors. This layered approach is designed for countering both single drones and drone swarms as their technology further evolves. The collective challenges of low cost, high numbers, and unpredictability of the drone technology can only be countered with doctrinal focus on flexibility and increased innovation.

These doctrinal shifts reflect a fundamental recognition that modern air defence must operate as an interconnected, agile system capable of responding to rapid technological change and multi-domain threats.

Modern counter-drone systems employ multiple autonomous detection and engagement methods. These systems integrate Radio Frequency Detection and Direction Finding (RFDD), video detection and imaging tracking, and radar systems to provide comprehensive autonomous protection. The systems can automatically detect, classify, track, and neutralize unmanned aerial threats through various means including jamming and kinetic engagement.

Autonomous sentry guns represent ground-based defensive systems capable of automatic target detection and engagement. The Samsung SGR-A1, deployed along the Korean DMZ, features integrated surveillance, tracking, firing, and voice recognition capabilities. These systems can operate autonomously with effective ranges up to 3.2 kilometers and provide continuous automated guard operations.[49]

While these systems operate autonomously, most retain some level of human oversight capability.[50] The concept of "meaningful human control" remains important in autonomous weapons deployment, with systems designed to allow human intervention when necessary. Many systems can operate in both fully autonomous and human-supervised modes depending on operational requirements.[51]

Modern autonomous air defense systems are designed for integration into broader command and control networks. These systems can share threat information across multiple platforms and coordinate responses to maximize defensive effectiveness while minimizing resource expenditure.[52]

CHALLENGES TO IMPREGNABLE AIR DEFENCE

While disruptive technologies offer a quantum leap in air defence capabilities, the path to their effective integration is laden with technical, operational, and ethical challenges.

Unmanned Systems

Despite their promise, there exist several barriers to the success of unmanned systems. The most prominent challenge is their vulnerability to electronic warfare measures as jamming, spoofing, and cyber-attacks can disrupt their operations. Additionally, the logistical burden of field maintenance and power requirements of large fleets, compounded with issues related to data security and interoperability across various platforms, need to be addressed. Regulatory and ethical considerations regarding autonomous engagement and accountability also remain unresolved, potentially hindering rapid field deployment.

Loitering Munitions and AWS

Loitering munitions are defined by their ability to loiter over a target area, allowing for precise timing of attacks. They are often small, unmanned aerial vehicles (UAVs) or drones equipped with a warhead, capable of waiting for the optimal moment to strike. For example, the Switchblade 300, a well-known loitering munition, can be launched from a backpack and is designed to engage targets beyond line of sight, indicating they can be either autonomous or non-autonomous.[53]

In contrast, AWS are defined by their autonomy in target selection and engagement, operating without human intervention. They encompass a broader range of systems, including drones, missiles, tanks, and cyber weapons, as long as they can make decisions independently. For instance, the Kargu-2 autonomous quadcopter, used in Libya, is an example of an AWS that can engage targets autonomously.[54]

Counter-countermeasures remains the primary challenge to the proliferation and advancement in loitering munitions. Already countries are developing systems and tactics to defeat loitering munitions, such as deploying decoys and employing advanced EW to disrupt their guidance systems. Also, the inherent "fire-and-forget" nature of AWS raises concerns about collateral damage and the need for stringent rules of engagement.[55] Ensuring high accuracy and reliable discrimination in a cluttered or urban environment remains a significant technical hurdle to wide scale employment of these munitions.

AI Enabled AD

Unquestionably, the biggest obstacle is still reliability – AI cannot afford to falter in contested environment where it's a matter of life and death. Adversarial attacks on vulnerabilities as well data poisoning or algorithmic bias could potentially cause distortion in threat assessments or even unintended engagements. Additional challenges range from how to deploy autonomous decision-making in lethal contexts while ensuring transparency and traceability of AI algorithms. Convincingly ensuring that these systems can be trusted to make life or death decisions without human intervention is a critical and ongoing concern.

DEWs

A number of technical constraints still need to be addressed. One of the biggest challenges is the production and storage of the required energy in a portable form factor. Moreover, DEWs have complex cooling systems for waste heat especially during sustained operations. The beam quality and effectiveness can be undermined by atmospheric interferences as well as adverse weather conditions. Finally, although DEWs are in theory cost per shot less than kinetic interceptors, so their inclusion alongside legacy systems and validation across a range of mission environments require extensive testing and iterative refinement.

Hypersonics

The biggest challenge in hypersonics is the sheer speed and manoeuvrability of hypersonic weapons, which compresses the window for detection, tracking, and engagement to mere seconds. Traditional radar systems may be unable to maintain track on such fast-moving targets throughout their flight path, necessitating the development of new sensor architectures and data processing algorithms. Moreover, the embedding of hypersonic interceptors in air defence networks will impose new and urgent demands for rapid communication and decision making on the existing command and control systems. More than budgetary and industrial challenges to large scale manufacturing, hypersonics also demands enormous research and development costs in the field.

Cyber Security

The principal challenge is that networked systems are inherently vulnerable to cyber threats. The more our air defence gets networked and depends on digital communications, the greater the necessity to secure it against disruptive actions. Similarly, the very technologies we develop to defend own networks, can be leveraged by adversaries as they attempt to disrupt or infiltrate our EW domain. Establishing connectivity versus operational security and data availability will be a tug-of-war which incessantly requires investment in in both hardware resilience and advanced encryption methodologies.

Quantum Sensing

Quantum sensor technologies, while promising, are still in relatively early stages of development and face significant technical and cost-related barriers. Issues such as environmental sensitivity, noise management, and the need for cryogenic cooling in some systems could impede widespread deployment. Interoperability between quantum and conventional sensor networks is another challenge, as is the development of algorithms capable of processing the enormous amounts of data these advanced sensors will generate. Ensuring that these systems can be reliably integrated into existing command and control frameworks is essential, but remains a daunting engineering task.[56]

Non-State Actor Access

Advanced technologies reaching non-state actors, such as Hamas and Hezbollah, complicate air defense by introducing asymmetric threats. Adaptable countermeasures are essential to counter these groups' growing capabilities, ensuring defenses remain effective against unconventional adversaries.

Cybersecurity

As air defence systems operate in a highly networked, they are more vulnerable to cyber-attacks on their critical components inducing malfunction or completely shutting them down. The security protocols required by these systems need to be highly resilient and adaptive in order keep them functional and reliable amidst an interconnected battlespace. Ensuring the cybersecurity of integrated systems, while maintaining rapid data flows, poses a key challenge to designing cyber defences that can evolve in tandem with emerging threats.

Training and Adaptation

Operators of modern air defence systems require specialised protracted training in order to make them competent to handle complex, technology-driven systems under pressure. To stay abreast with the rapid technological changes, continuous training, including simulations, is required for personnel in order to maintain operational readiness in an evolving air defence landscape.

Cost and Proliferation

There exists huge cost disparity between advanced systems like the Iron Dome and inexpensive threats like drones, underscoring the need for developing affordable weapons and systems. Directed energy weapons offer a scalable solution, however, their development costs still remain high. Addressing this financial challenge is crucial to maintaining effective air defence against proliferating low-cost threats, specially by nations who are economically weak.

Interoperability and Global Collaboration

As multi-domain threats become more complex, interoperability among allied air defence systems will be crucial. Collaborative efforts in technology sharing, joint training and integrated command structures will help create a unified defence against common adversaries.

Ethical and Strategic Considerations

The deployment of autonomous systems and directed energy weapons also raises ethical questions. Establishing clear rules of engagement, ensuring accountability and developing strategic frameworks that balance rapid technological innovation with responsible use will be critical for future operations.

Each of these challenges highlights the need for continuous innovation, strategic foresight, and international cooperation. While emerging disruptive technologies offer significant enhancements in air defence capabilities, they also introduce new vulnerabilities that must be managed through advanced planning, robust training, and adaptive doctrines.

The Indian Context

The future of air defence for the Indian Army hinges on developing a comprehensive, multi-layered air defence framework capable of intercepting a diverse array of aerial threats from low-flying drones to hypersonic missiles. The modern air defence grid capable of thwarting today's and future's aerial threat would involve transforming the existing ground-based systems to allow them to function in a digitally networked environment with AI enabled integrated advanced radar arrays, missile interceptors and electronic systems. Modern radar architectures, such as active electronically scanned arrays (AESAs), should be deployed widely to ensure high-resolution target detection and tracking, even in cluttered or contested electromagnetic environments. These radars must be connected to the command centres and missile batteries through encrypted communication channels facilitating rapid data fusion and enabling real-time engagement decisions powered by artificial intelligence enabled decision support systems. The integration of directed energy weapons such as high-energy lasers capable of neutralizing incoming projectiles almost instantaneously, into the air defence envelope will further augment traditional kinetic interceptors, providing a versatile and cost-effective solution to counter saturation attacks akin to those conducted by Iran over Israel, which had a mix of autonomous drones and ballistic missiles. Lastly, mobility and rapid deployment remain key to an effective air defence architecture and air defence systems should be designed for quick redeployment to plug any gaps in the defensive grid, ensuring continuous coverage over critical assets and operational theatres. By creating a layered, digitally interconnected air defence system, the Indian Army can significantly enhance its ability to detect, engage and neutralize sophisticated aerial threats.

Army Air Defence must keep pace with constantly evolving aerial threats from autonomous drones to hypersonic missiles to effectively protect both the airspace and ground forces. Integrated Air Defence Systems are the building blocks of this defensive framework. They combine radars, missile launchers, and gun systems into a single, interconnected architecture. This integration allows for real time data sharing and well-coordinated fire plans between dispersed air defence units in a theatre of operations, enabling them to engage multiple aerial threats simultaneously ranging from high altitude bombers, stealth aircraft, or low flying drones. For the ground troops, operating in areas vulnerable to air strikes or missile barrages, such systems provide comprehensive coverage, seamlessly linking short-range point defences with long-range interceptors to create a layered shield that maximizes protection and minimizes gaps in coverage.

With hypersonic weapons on the rise, hypersonic missile defence systems have become an inescapable priority. The fact that hypersonic threats can destroy their target up to three to five times faster than a traditional ballistic threat with far greater manoeuvrability, further complicates their interception by traditional methods. In today's battlefield, AI enabled countermeasures are the primary and most effective means necessary to defend against hypersonic threats minimizing human error and delay. AI algorithms are essential to predict trajectories of large number of projectiles in a saturation attack and to immediately launch interceptors within very thin timeframes. These systems can analyze incoming data speed, altitude and flight patterns to calculate optimal engagement points and launch the most suitable kinetic or energy-based countermeasures to destroy the threat before it reaches its target. India's regional military adversaries may soon acquire and perfect hypersonic capabilities, thus investing in these technologies and systems ensures that Army Air Defence can safeguard critical infrastructure and troop concentrations, maintaining deterrence against cutting-edge weaponry.

In the modern era of multi domain operations, advancement in cyber and electronic warfare are important capabilities to disrupt an adversary's air operations and defend one's own airspace. The Army can deploy advanced jammers to vitiate enemy's radar, navigation and communication networks and considerably degrade adversary's ability to coordinate air strikes or suppress our air defences. Cyber operations can target an adversary's command-and-control systems and infuse delays, confusion or misinformation, disrupting the adversary's aerial campaign. This approach increases the survivability of own air defence units in conjunction with creating opportunities for friendly forces to exploit the adversary's disarrayed systems. For the Indian Army, it is imperative to gain competence in these areas to negate technologically advanced opponents and ensure dominance in the electromagnetic spectrum.

Key to survival in a dynamic battlefield is flexibility. Therefore, agile and mobile systems must remain at heart of our Air Defence strategy. Mobile units, whether wheeled or tracked platforms, can engage and redeploy swiftly to protect advancing forces, defend vulnerable sites, or respond to rapidly changing threats, such as surprise air raids or missile launches. These systems made potent with compact radars, highly responsive missile and energy weapons provide the much needed flexibility during intense operations across varied terrain from deserts to forested valleys. In India, where terrain and threat profiles vary widely, mobility ensures that air defence remains responsive and effective, providing commanders with the flexibility to prioritize protection where it is most needed, thereby enhancing overall battlefield resilience.

Keeping Indian Skies Safe

Operation Sindoor, conducted by India, offers several important lessons on air defense drawn from the conflict with Pakistan, particularly in countering aerial threats including drones, UAVs, and missile strikes.

Effectiveness of a Layered and Integrated Air Defense Architecture: India's success in thwarting Pakistan's retaliatory drone and UCAV (Unmanned Combat Aerial Vehicle) attacks was largely due to a comprehensive, multilayered air defense system.[57] This included a mix of legacy systems like Pechora and OSA-AK, modern indigenous systems such as the Akash surface-to-air missile, and low-level air defense guns (LLAD guns).[58]

The Integrated Air Command and Control System (IACCS) played a pivotal role by providing real-time threat identification, assessment, and coordinated interception across multiple domains. This net-centric approach ensured seamless communication and operational synergy between the Army, Air Force, and Navy air defense units.[59]

Blending Legacy and Indigenous Systems for Optimal Coverage: The operation demonstrated that a blend of battle-proven legacy systems and advanced indigenous technologies can create an effective air defense shield. The Akash missile system, with its electronic counter-countermeasures (ECCM) and ability to engage multiple targets simultaneously, was particularly effective in protecting vulnerable points.[60] Legacy systems like Pechora and OSA-AK complemented newer platforms, ensuring coverage across different altitudes and threat profiles.[61]

Countering Emerging Threats: Drones and Loitering Munitions: Pakistan's use of drones, loitering munitions, long-range rockets, and commercial quadcopters highlighted the evolving nature of aerial threats. India's air defense successfully neutralized these, showcasing the importance of counter-unmanned aerial systems (C-UAS) integrated within the broader air defense network.[62] The use of electronic warfare assets alongside kinetic air defense weapons was critical in detecting and defeating these small, agile threats.[63]

Importance of Real-Time Command and Control: The operation underscored the necessity of centralized command with decentralized execution, enabling rapid decision-making and flexible responses to fast-evolving aerial threats. The IACCS facilitated this by integrating sensors, shooters, and command nodes into a unified operational picture, allowing for timely and precise engagement decisions.[64]

Technological Edge and Electronic Warfare: India's ability to bypass and jam Pakistan's Chinese-supplied air defense systems (such as the HQ-9) demonstrated the critical role of electronic warfare and cyber capabilities in modern air defense.[65] The operation exposed vulnerabilities in Pakistan's air defenses, particularly the

HQ-9 system, which performed poorly under electronic attack, emphasizing the need for continuous technological upgrades and countermeasures.[66]

Operational Synergy Across Services: Operation Sindoor was a showcase of jointness and synergy among the Indian Army, Air Force, and Navy, with coordinated air defense efforts protecting critical infrastructure and enabling offensive air operations.[67] This multi-domain integration is essential for modern warfare, where air defense must be synchronized with offensive airpower and ground operations.[68]

The concept of JADC2, developed by the US Department of Defense, aims to connect sensors and communications from all military services into a single network, enabling unified command and control across land, sea, air, space, and cyber domains. The Akashteer system, integrated with the IACCS, had delivered 107 units by November 2024, with plans for 455 units by April 2027, and achieved a 100% kill rate in Operation Sindoor.[69] This success of Akashteer in Operation Sindoor makes a compelling case for pursuing this integration through all services and arms to create the capability to achieve desired results in the multidomain operations.

New disruptive technologies are not only revolutionising the air defence domain by throwing up unprecedented opportunities but at the same time also posing significant challenges. Lessons from recent conflicts around the globe, highlight the need for developing air defence systems capable of defeating threats ranging from cyber-attacks to hypersonic missiles. Innovations in AI, autonomy, electronic warfare and quantum tech are driving the next wave of capabilities, necessitating a doctrinal and tactical rethink. As technology evolves, air defence must also remain agile and keep pace, absorbing technological advancements and at the same time effectively address ethical, operational and cybersecurity issues. The future hinges on anticipating and adapting to these disruptions and ensuring force protection in a complex and contested environment. The dynamic interplay between these technologies and their countermeasures will shape the future battlefield, making continuous innovation and adaptive doctrinal evolution imperative for maintaining fool proof air defence in the face of rapidly evolving threats.

Notes

1   Shaan Shaikh and Wes Rumbaugh, The Air and Missile War in Nagorno-Karabakh: Lessons for the Future of Strike and Defense, December 8, 2020, https://www.csis.org/analysis/air-and-missile-war-nagorno-karabakh-lessons-future-strike-and-defense

2   Flight Lieutenant Chris Whelan, The 2020 Nagorno Karabakh War: Unmanned Combat Aerial Vehicles in Modern Warfare, *Air and Space Power Review*, Vol 25, No 2, https://www.raf.mod.uk/what-we-do/centre-for-air-and-space-power-studies/aspr/aspr-vol25-iss2-3-pdf/

3   Michael Kofman, A Look at the Military Lessons of the Nagorno-Karabakh Conflict, December 14, 2020 https://www.russiamatters.org/analysis/look-military-lessons-nagorno-karabakh-conflict

4   9K333 Verba (SA-29) Russian Man-Portable Infrared AD System, OE Data Integration Network

https://odin.tradoc.army.mil/WEG/Asset/9K333_Verba_(SA-29)_Russian_Man-Portable_Infrared_Homing_Surface-to-Air_Missile_(MANPAD)

5 Missile Defense Project, "Pantsir S-1," Missile Threat, Center for Strategic and International Studies, May 4, 2017, last modified July 6, 2021, https://missilethreat.csis.org/defsys/pantsir-s-1/.

6 Christian Wolff, 96L6E "Cheese Board", https://www.radartutorial.eu/19.kartei/02.surv/karte044.en.html

7 How C-RAM rocket defence system helped US forces to exit Kabul safely, Sambad English Bureau, 31 Aug 2021, https://sambadenglish.com/how-c-ram-rocket-defence-system-helped-us-forces-to-exit-kabul-safely/

8 Avenger Air Defense System, U.S. - Air Defense, Intercept, July 2, 2020 https://missiledefenseadvocacy.org/defense-systems/avenger-air-defense-system/

9 Long-Range Ukrainian Strike Drones, Which are Hitting Targets Deep Inside Russia, Defense Express, February 5, 2024 https://en.defenceua.com/weapon_and_tech/long_range_ukrainian_strike_drones_capable_of_striking_deep_inside_russia-9409.html

10 Oded Yaron,Gold for Drones: Massive Leak Reveals the Iranian Shahed Project in Russia, HAERTZ, 21 Feb 2024, https://www.haaretz.com/israel-news/security-aviation/2024-02-21/ty-article-magazine/gold-for-drones-massive-leak-reveals-the-iranian-shahed-project-in-russia/0000018d-bb85-dd5e-a59d-ffb729890000

11 S-400 Triumph Air Defence Missile System, Army Technology, February 3 2020, https://www.army-technology.com/projects/s-400-triumph-air-defence-missile-system/

12 Giorgio Di Mizio & Michael Gjerstad, Ukraine's ground-based air defence: evolution, resilience and pressure, The International Institute for Strategic Studies, MILITARY BALANCE BLOG, 24th February 2025 https://www.iiss.org/online-analysis/military-balance/2025/02/ukraines-ground-based-air-defence-evolution-resilience-and-pressure/

13 Report: Mossad carried out covert sabotage operations against Iranian air defenses, long-range missiles, 13 June 2025, https://www.timesofisrael.com/liveblog_entry/report-mossad-carried-out-covert-sabotage-operations-against-iranian-air-defenses-long-range-missiles/

14 How Israel's Operation Rising Lion Dismantled Iran from Within: A Case Study in the Art of Deception,Hudson Institute, Jun 13, 2025, https://www.hudson.org/defense-strategy/how-israels-operation-rising-lion-dismantled-iran-within-case-study-art-deception

15 What are Israel's Iron Dome, David's Sling, Arrow and Thaad missile defences?, BBC, 16 October 2024 https://www.bbc.com/news/world-middle-east-20385306

16 LT Gen (DR) VK Saxena, PVSM, AVSM, VSM (Retd), How Come Iran's Air Defences Succumbed on 26 Oct 2024? A Brief Analysis, Centre For Joint Warfare Studies, November 26, 2024 https://cenjows.in/how-come-irans-air-defences-succumbed-on-26-oct-2024-a-brief-analysis/

17 Tianran Xu, Taiwan's Air and Missile Defence. Part 2: Patriot PAC-2 and PAC-3, 9 October 2024 https://platform.opennuclear.org/thoughtroom/quick-takes/taiwans-air-and-missile-defence-part-2-patriot-pac-2-and-pac-3

18 Charles Engelbrecht, China and Russia's Hypersonic Missiles, Domino Theory, Feb 2, 2023 https://dominotheory.com/china-and-russias-hypersonic-missiles/

19 Marshall Brain & Patrick J. Kiger, How Stinger Missiles Work, Feb 27, 2024 https://science.howstuffworks.com/stinger.htm

20 US to Protect Air Bases Against Ballistic Threats with Saab's Advanced Giraffe 4A Radar Systems, Global Defense News, 11 Dec, 2024 https://armyrecognition.com/focus-analysis-conflicts/army/defence-security-industry-technology/us-to-protect-air-bases-against-ballistic-threats-with-saabs-advanced-giraffe-4a-radar-systems

21 Joe Saballa, India's Quick Reaction Surface-to-Air Missile Completes Trials, The Defense Post, September 13, 2022, https://thedefensepost.com/2022/09/13/india-surface-missile-trials/

22   Hemant Kumar Rout, Back-to-back successful tests of VL-SRSAM validate upgraded missile components, The New Indian Express, 13 Sep 2024, https://www.newindianexpress.com/nation/2024/Sep/13/back-to-back-successful-tests-of-vl-srsam-validate-upgraded-missile-components

23   Pawan Atri, Lessons from Ukraine Conflict Helped India Develop Akash Missile, 02.04.2024 https://sputniknews.in/20240402/lessons-from-ukraine-conflict-helped-india-develop-akash-missile-7013258.html

24   Design and Test Challenges of the Modern Electronically Scanned Array, National Instruments, https://www.ni.com/content/dam/web/pdfs/design-and-test-challenges-of-the-modern-esa.pdf?srsltid=AfmBOorhcUfmcuCgIeo109Qc2R6_sLLRvavF6KQNAUZRjRye PK80WGBD

25   Summary of the Joint All-Domain Command & Control (JADC2) Strategy, Department of Defense, March 2022 https://media.defense.gov/2022/Mar/17/2002958406/-1/-1/1/SUMMARY-OF-THE-JOINT-ALL-DOMAIN-COMMAND-AND-CONTROL-STRATEGY.pdf

26   Matejcek, Miroslav & Sostronek, Mikulas. (2023). The Ground Based Air Defence Solutions. Science & Military. 18. 21-28. 10.52651/sam.a.2023.1.21-28.

27   Missile Defense Project, "Patriot," Missile Threat, Center for Strategic and International Studies, June 14, 2018, last modified August 23, 2023, https://missilethreat.csis.org/system/patriot/.

28   Missile Defense Project, "Pantsir S-1," Missile Threat, Center for Strategic and International Studies, May 4, 2017, last modified July 6, 2021, https://missilethreat.csis.org/defsys/pantsir-s-1/.

29   Ashish Dangwal, Laser Weapons On Stryker AFV: U.S. Army Deploys DE M-SHORAD To Middle East To Counter UAV Threats, November 16, 2024 https://www.eurasiantimes.com/laser-weapons-on-stryker-afv-u-s-arm/

30   Advanced future military laser achieves UK first, Defence Science and Technology Laboratory and Ministry of Defence, 19 January 2024 https://www.gov.uk/government/news/advanced-future-military-laser-achieves-uk-first

31   Ray, Michael. "Iron Dome". Encyclopedia Britannica, 5 Mar. 2025, https://www.britannica.com/topic/Iron-Dome. Accessed 11 March 2025.

32   Missile Defense Project, "S-400 Triumf," Missile Threat, Center for Strategic and International Studies, May 4, 2017, last modified July 6, 2021, https://missilethreat.csis.org/defsys/s-400-triumf/.

33   NASAMS Air Defence System, Integrated air and missile defence, Kongsberg Defence & Aerospace https://www.kongsberg.com/kda/what-we-do/defence-and-security/integrated-air-and-missile-defence/nasams-air-defence-system/

34   Saab Mobile Short Range Air Defence System (MSHORAD), Sweden January 26 2024, https://www.army-technology.com/projects/saab-mobile-short-range-air-defence-system-mshorad-sweden/

35   "David's Sling - Medium to Long Range Defense." Rafael Advanced Defense Systems. Accessed March 11, 2025. https://www.rafael.co.il/system/medium-long-range-defense-davids-sling/.

36   insideFPV, In Depth Exploration of Kamikaze Drones, Nov 23, 2024 https://insidefpv.com/blogs/blogs/what-are-suicide-drone-in-depth-exploration-of-kamikaze-drones?srsltid=Afm BOopA1fXRmEvr5IhKXmgfiv-Y5Ih-lUypG9Ef3ajR6VOCCl7j3OX1

37   Vivek Kaul, Role of AI and Robots in Air Defense System, State Times, May 23, 2025 https://statetimes.in/role-of-ai-and-robots-in-air-defense-system/

38   Akashteer: The Unseen Force Behind India's New War Capability, Ministry of Information & Broadcasting, 16 MAY 2025, https://www.pib.gov.in/PressReleasePage.aspx?PRID=2129132

39   Countering the hypersonic threat, Doug Richardson, 20 June 2024, https://euro-sd.com/2024/06/articles/38866/countering-the-hypersonic-threat/

40   Anand Ramachandran, Countering Hypersonic Threats: Integrating Advanced Artificial Intelligence and FPGA Technologies for Next-Generation Missile Defense Systems, 07 May

25, https://www.linkedin.com/pulse/countering-hypersonic-threats-integrating-advanced-anand-ramachandran-ma9pe/

41 Hypersonic Defence, Chapter 4: Engagement of hypersonic targets, 09 Oct 2024, https://www.thalesgroup.com/en/worldwide/defence-and-security/magazine/hypersonic-defence-chapter-4-engagement-hypersonic-targets

42 Anand Ramachandran, Countering Hypersonic Threats: Integrating Advanced Artificial Intelligence and FPGA Technologies for Next-Generation Missile Defense Systems, 07 May 25, https://www.linkedin.com/pulse/countering-hypersonic-threats-integrating-advanced-anand-ramachandran-ma9pe/

43 Ferreira, David & Marcell, Frederick. (2009). Navy High Energy Laser Weapon System. Naval Engineers Journal. 105. 105-109. 10.1111/j.1559-3584.1993.tb02303.x.

44 Directed energy weapons, National Research Council of Canada and Defence Research and Development Canada, https://science.gc.ca/site/science/en/safeguarding-your-research/guidelines-and-tools-implement-research-security/emerging-technology-trend-cards/directed-energy-weapons

45 U.S. Army Space and Missile Defense Command/ Army Forces Strategic Command, https://www.smdc.army.mil/Portals/38/Documents/Publications/Fact_Sheets/Archived_Fact_Sheets/HELMD.pdf

46 David Hambling, Senior Contributor, New U.S. Army Laser Machine Gun Fires 'Bullets' Of Light, Mar 11, 2021 https://www.forbes.com/sites/davidhambling/2021/03/11/us-army-develops-laser-machinegun-firing-light-bullets/

47 Cindy Hurst, China Unveils New High-Power Microwave Weapon Systems, February 15, 2025 https://fmso.tradoc.army.mil/2025/china-unveils-new-high-power-microwave-weapon-systems/

48 Peresvet Russian Mobile Laser System, ODIN - OE Data Integration Network (.mil), https://odin.tradoc.army.mil/WEG/Asset/Peresvet_Russian_Mobile_Laser_System

49 Samsung Techwin SGR-A1 Sentry Guard Robot, https://www.globalsecurity.org/military/world/rok/sgr-a1.htm#google_vignette

50 A Hazard to Human Rights. Autonomous Weapons Systems and Digital Decision-Making, April 28 2025, https://www.hrw.org/report/2025/04/28/hazard-human-rights/autonomous-weapons-systems-and-digital-decision-making

51 Amitai Etzioni, PhD, Oren Etzioni, PhD, Pros and Cons of Autonomous Weapons Systems, May-June 2017, https://www.armyupress.army.mil/Journals/Military-Review/English-Edition-Archives/May-June-2017/Pros-and-Cons-of-Autonomous-Weapons-Systems/

52 "National Advanced Surface-to-Air Missile System (NASAMS)". Missile Defense Advocacy Alliance. September 2022. https://missiledefenseadvocacy.org/defense-systems/national-advanced-surface-to-air-missile-system-nasams/

53 Lethal Autonomous Weapon Systems (LAWS), United Nations Office for Disarmament Affairs, https://disarmament.unoda.org/the-convention-on-certain-conventional-weapons/background-on-laws-in-the-ccw/

54 Kelsey Atherton, Loitering munitions preview the autonomous future of warfare, August 4, 2021, https://www.brookings.edu/articles/loitering-munitions-preview-the-autonomous-future-of-warfare/

55 Andy Wilson "Fire and Forget: Dissecting the Thin Line Between Self-Homing Missiles and Guided Munitions." LinkedIn. Accessed March 11, 2025. https://www.linkedin.com/pulse/fire-and-forget-dissecting-thin-line-between-self-homing-missiles-0irtc/.

56 Texeira, Kalea. "Unveiling the Future of Warfare: Quantum Sensing and Precision." LinkedIn. Accessed March 11, 2025. https://www.linkedin.com/pulse/unveiling-future-warfare-quantum-sensing-precision-kalea-texeira-rc8qe/.

57 Operation SINDOOR: Forging One Force, The Synergy of India's Armed Forces, PIB Delhi, 18 MAY 2025, https://www.pib.gov.in/PressReleasePage.aspx?PRID=2129453

58 Operation SINDOOR: The Rise of Aatmanirbhar Innovation in National Security India's growing technological self-reliance, 14 May 2025, https://static.pib.gov.in/WriteReadData/specificdocs/documents/2025/may/doc2025514554901.pdf

59 Operation SINDOOR: Forging One Force, The Synergy of India's Armed Forces, PIB Delhi, 18 MAY 2025, https://www.pib.gov.in/PressReleasePage.aspx?PRID=2129453

60 Ibid

61 Operation SINDOOR: The Rise of Aatmanirbhar Innovation in National Security India's growing technological self-reliance, 14 May 2025, https://static.pib.gov.in/WriteReadData/specificdocs/documents/2025/may/doc2025514554901.pdf

62 Operation SINDOOR: Forging One Force, The Synergy of India's Armed Forces, PIB Delhi, 18 MAY 2025, https://www.pib.gov.in/PressReleasePage.aspx?PRID=2129453

63 Operation SINDOOR: The Rise of Aatmanirbhar Innovation in National Security India's growing technological self-reliance, 14 May 2025, https://static.pib.gov.in/WriteReadData/specificdocs/documents/2025/may/doc2025514554901.pdf

64 Ajay Ahlawat, Lessons for airpower from Operation Sindoor – unified command to tech advancements, 25 May, 2025 https://www.perplexity.ai/search/what-air-defence-systems-did-u-JbtpEY3KTx668Yd5aGZ0hw

65 IDRW, Operation Sindoors Fallout: Pakistan's Chinese HQ9 air defense system proves vulnerable, casting shadow on Beijing's defense deals https://idrw.org/operation-sindoors-fallout-pakistans-chinese-hq-9-air-defense-system-proves-vulnerable-casting-shadow-on-beijings-defense-deals/#google_vignette

66 Ibid

67 Operation SINDOOR: Forging One Force, The Synergy of India's Armed Forces, PIB Delhi, 18 MAY 2025, https://www.pib.gov.in/PressReleasePage.aspx?PRID=2129453

68 Ajay Ahlawat, Lessons for airpower from Operation Sindoor – unified command to tech advancements, 25 May, 2025 https://www.perplexity.ai/search/what-air-defence-systems-did-u-JbtpEY3KTx668Yd5aGZ0hw

69 "Army begins induction of Akashteer system to sharpen air defence posture". Hindustan Times. 4 April 2024. Retrieved 28 April 2024. https://www.hindustantimes.com/india-news/army-begins-induction-of-akashteer-system-to-sharpen-air-defence-posture-101712238995413.html

14

Space: The Ultimate High Ground
for the Army

"Morality doesn't exist in space, only in the spaces between people."

—Alex Latimer

Space warfare refers to the employment of space-based platforms and technologies for military operations. This domain has evolved considerably from its original focus on reconnaissance and communications throughout the Cold War, to encompass an area of contention, where nations battle for supremacy through developing space-based niche technologies. Notably armies, the land component of warfighting is dependent on space for communications, navigation, and intelligence, which are imperative to successful ground operations. New and emerging disruptive technologies, like artificial intelligence, quantum computing and hypersonic technologies, are radically transforming where and how missions will be conducted (especially in space), creating new capabilities, as well as new challenges. Emerging technologies will have implications not just for the employment of space-based weapons, but for the broader idea of weaponising space, and developing counter-space capabilities, which may have direct implications for the land forces. The Army has, historically, been perceived operating in the land-warrior domain, while space has been viewed and reserved as a specialised, separate area leveraged primarily by air and space component forces. However, In the Multi-Domain Operations (MDO) era we live, where all elements of warfare become more intertwined, the Army's need to integrate with space-based capabilities has become an accepted necessity. MDO integrates land, sea, air, cyber and space into a viable operational continuum of all domains and thrusts space-based technologies into a highly relevant, and necessary component for land troops and systems.[1]

Satellites now support crucial functions such as communication, navigation, intelligence and reconnaissance, assisting armies in coordinating complex operations, conducting precise strikes and maintaining round the clock situational awareness of their theaters in complicated contested environments. As a consequence, the Army's effectiveness on the ground is closely linked to the Army's capacity to leverage and protect space based assets. Today's Army units rely on expensive and extensive communication, geolocation and intelligence satellite networks for real-time intelligence, precise navigation and secure communication, allowing them to conduct rapid, coordinated actions in a complex, dynamic battlefield. These technologies not only enhance the Army's situational awareness and decision-making processes but also serve as critical force multipliers that bridge the gap between traditional land combat and the expansive, data-driven world of space. It is therefore, important to understand how the evolution of space-based capabilities, though not historically within the Army's purview, are introducing new possibilities into land warfare and redefining the strategic and operational landscape of ground operations.

The endlessly escalating adoption of new technologies increases this reliance even more. Developments in the field of artificial intelligence, quantum computing, hypersonic weapons and advanced cyber capabilities are transforming space operations. They not only improve the Army's capabilities, providing new levels of precision and connectivity, but also introduce new vulnerabilities, such as anti-satellite weapons and cyber-attacks that could interfere with the ability of land forces to operate effectively. For an Army that may not actually control space assets, the ability to understand the implications of these technological changes is necessary to gain an edge over current day adversaries. This chapter will aim to understand the complex implications of these changes, using insights from recent wars, development of innovative weapons and systems, and their implications on military doctrine and tactics. Furthermore, it delves into counter-space operations, the use of space as a weapon, and the future trends and challenges that lie ahead, emphasizing the army's perspective in leveraging and adapting to these changes. Understanding the importance of space within the domain of warfare cannot be underestimated, given its integral nature to modern day land operations. This chapter underscores why military strategists, particularly those focused on land warfare, must prioritize space in their planning and why this exploration is essential to understanding the future of conflict.

SPACE WARFARE 101: WHAT THE ARMY NEEDS TO KNOW

In modern warfare, the Army can hardly ignore space and its associated technologies to effectively conduct land operations. Space, while not a traditional Army function, has since become an element of Multi-Domain Operations (MDO) and has modified the way land forces operate. Specifically, satellites provide vital support to missions, supporting communications, intelligence collection and navigation, enabling effective and coordinated operations despite varied and difficult operating environments.[2] This section examines the Army's dependence on space through three key areas: satellite-based communication, Intelligence, Surveillance, and Reconnaissance (ISR), and navigation and positioning.

Satellite-Based Communication

The bedrock of Army's operational construct, Satellite Communications (SATCOM) provides secure and reliable connections to disparate battlefields that may not be effectively served by terrestrial systems.

Importance to Tactical and Strategic Operations: At the tactical level, SATCOM enables real-time coordination among ground units, supports command and control (C2), and integrates assets like unmanned aerial vehicles (UAVs), artillery, and air support into a unified effort. Strategically, it links theater commanders with national authorities, facilitating swift decision-making and synchronized operations across multiple theaters.

Challenges in Contested Environments: The threat of adversary's capability to disrupt or jam the satellite signals is a serious challenge to ground operations. To address these threats, the Army is focusing on next-generation resilient SATCOM systems, including anti-jamming capability, redundant satellite networks as well as terrestrial back-up for communication in a contested environment.

Intelligence, Surveillance, and Reconnaissance (ISR)

Space-based ISR assets deliver critical intelligence, offering the Army unmatched situational awareness to monitor adversaries and assess operational environments effectively.

- **Space-Based ISR Capabilities:** Satellites incorporating new advanced sensors (electro-optical, infrared and synthetic aperture radar (SAR)) provide persistent wide area surveillance. Such systems follow enemy movement, track supply lines and detect environmental changes that may affect land operations.
- **Impact on Operational Planning:** The data from space-based ISR and situational awareness empowers Army to plan, execute and modify their missions. The commander gets a real time battlefield picture and can pre-empt threats, allocate resources in optimal way and refine targeting decisions. Operational success is greatly improved by such a capability, which magnifies situational understanding.

Navigation and Positioning

The Global Positioning System (GPS), a space-based navigation network, is vital to the Army's land operations, providing precise location and timing data for a range of military functions.

- **GPS in the Modern Warfare:** GPS is nowadays pivotal in troop and vehicle navigation; artillery targeting and the coordination of air and ground forces. Precision-guided munitions (PGM)s with GPS promise accurate delivery of lethal power with significantly less collateral damage, introducing paradigm shifts in modern land warfare.
- **Threats and weaknesses:** Receipt of signals from GPS systems means that such signals are vulnerable to enemy jamming and spoofing capabilities. Militaries are working to address these weaknesses by leveraging alternative solutions such inertial navigation systems (INS) and regional satellite networks that would provide operational resilience in a GPS denied environment.

DISRUPTIVE TECHNOLOGY IN SPACE WARFARE

Space Operations are not the business of Army but as space-based technologies are increasingly being integrated within Multi-Domain Operations (MDO), it demands a comprehensive understanding of emerging disruptive technologies in space. These advancements have the potential to reshape land warfare by enhancing the Army's capabilities and introducing new vulnerabilities that adversaries might exploit. This section explores four critical areas – anti-satellite weapons, satellite jamming and cyber warfare, space-based laser weapons and directed energy weapons, and artificial intelligence and autonomous systems – and their implications for Army operations on the ground.

Anti-Satellite Weapons (ASAT)

Anti-satellite (ASAT) weapons are designed to neutralize or destroy orbiting enemy satellites which adversary requires for its communication, ISR and navigation functions. Such weapons include the Kinetic Weapons, which are destructive in nature and result in dangerous fragments of satellites and Non-Kinetic Weapons, such as those hitting satellites functionality either via jamming or cyberattacks.[3] The US unveiled its ability to destroy space satellites with a missile launched from an F-15 fighter jet as part of the first kinetic ASAT in 1985.[4] Since then, other countries like Russia, China and India have expanded their ASAT programs demonstrating an increasing threat to space assets. If an enemy were to kill or shut-down satellites, backup systems to support the land forces and substitute battle-ready satellites could become vital contingency plans ensuring effective responses to a highly contested space.

Satellite Jamming and Cyber Warfare

Satellite jamming and Cyber Warfare are major challenges to the Army's secure and reliable communication networks required for land operations. Jamming disrupts satellite signals, while cyberattacks target control systems to compromise or disable space assets.[5] The Army is hardening its systems and networks through countermeasures such as frequency hopping and spread spectrum techniques that increase resilience of the signal against jamming, in addition to the adoption of enhanced cybersecurity best practices such as encryption and AI-aided intrusion detection. As adversaries continue to improve upon their own disruptive capabilities, the Army must increase strategic investments towards defensive technologies protecting operational continuity when space-based support is contested, safeguarding integrity of its communication and command structures.

Space Based Laser Weapons and Directed Energy Weapons

Directed energy weapons (DEW) and space based laser weapons are frontier technologies with the potential to influence land warfare, even if not directly managed by the Army. Laser weapons deliver concentrated light beams to disable or destroy targets, such as satellites supporting Army operations or enemy missile threats, while DEWs encompass broader energy forms like microwaves and particle beams to disrupt electronics or neutralize threats.[6] These technologies are being developed by major powers like the United States, China, Russia and India, which may add to the potency of their missile defence systems and also provide capability to hit with precision from space. However, challenges such as power generation, cooling, and targeting accuracy in space, combined with international concerns about space weaponisation, complicate their deployment. Nevertheless, the Army needs to be prepared for a battlefield where these technologies are capable of disrupting its space-dependent operations, or offer new strategic advantages.

Artificial Intelligence and Autonomous Systems in Space

Introduction of artificial intelligence (AI) as well as autonomous systems in the space operations will result in substantial opportunities and threats for Army. AI boosts satellite management and speeds up information analysis for intelligence and improves space situational awareness thereby adding to the effectiveness of the land operations.[7] Autonomous systems are capable of doing things like satellite servicing or debris removing, which would be great because there are critical space assets that we can't afford to lose and their maintenance through human intervention is extremely dangerous, time consuming and costly.[8] But these technologies also create new risks – like the ability of an enemy to exploit your digital footprint and ethical challenges around autonomous military decision-making. Major military powers are leveraging AI to optimize satellite constellations and ISR missions, but the Army must balance these benefits with robust security measures and ethical guidelines to ensure reliability and accountability in an increasingly autonomous space domain.

THE PAST OF THE FUTURE: SPACE IN RECENT CONFLICTS

Recent military engagements highlight space's essential function in contemporary land combat as armed forces depend more on space-based resources to achieve operational victory. Operation Desert Storm (1990-1991) represented a crucial turning point as it introduced space-based technologies into large-scale military operations. The Global Positioning System (GPS) emerged as a critical navigation and targeting tool that allowed coalition land forces to manage troop movements across expansive desert areas by providing precise positional data and operational timing.[9] During the 2003 Operation Iraqi Freedom, the Tomahawk cruise missile represented GPS-guided munitions that enabled precise strikes on Iraqi targets while reducing collateral damage and enhancing operational effectiveness.[10] The integration of space assets into tactical planning transformed land warfare by enabling armies to achieve enhanced coordination while minimizing friendly fire incidents and executing more effective enemy strikes.

The 2020 Armenia-Azerbaijan conflict over Nagorno-Karabakh demonstrated how unmanned aerial vehicles (UAVs) relied on space-based systems to perform essential roles that affected ground operations. Azerbaijan specifically utilized Turkish-manufactured Bayraktar TB2 drones that depend on satellite links for both operations and data transmission. These drones performed reconnaissance missions while acquiring targets and executing precision strikes which altered battlefield dynamics with their attacks on Armenian armour and artillery units.[11] Effective UAV operations depended on stable satellite connections which showed how space resources directly enable terrestrial military activities. The necessity for military forces to integrate UAVs with space-based systems emerges as a crucial factor to develop reconnaissance and strike abilities while boosting situational awareness and facilitating faster battlefield responses. Both sides employed commercial satellite imagery to track troop movements and evaluate damage which demonstrates how space-derived intelligence has become accessible to non-state actors and smaller nations because companies like Maxar deliver detailed imagery that military forces can use for their operations.

The US-Taliban war (2001-2021) witnessed extensive employment of space assets as sources of intelligence, surveillance, and reconnaissance (ISR); satellites enabled drone strikes, while coordinating troop movements in remote areas. In Afghanistan's unforgiving landscape, using traditional communications was problematic, making satellite connections crucial to keeping command and control.[12] Satellites gave real-time imagery and communication links, of vital importance to operations, and the GPS enabled platforms allowed precision strikes to be planned and executed to minimize collateral damage. For land forces, this meant enhanced situational awareness, improved coordination, and the ability to operate effectively in isolated regions, transforming army operations by relying on space for logistical support and tactical decision-making.

Space-based data for tracking helped Missile Defence systems such as Iron Dome, to successfully improve interception rates protecting civilian infrastructure as well as vital military establishments.[13] Satellites gave early warning and trajectory information allowing the system to prioritize threats and conserve interceptors demonstrating integration of space assets with urban warfare which demands high level of precision and lethality.[14] For armies, it's about shielding themselves from aerial threats, enabling ground troops to engage in urban warfare without the constant worry of rocket strikes, while ISR inputs from space get channeled into command centers for real-time decisions and analysis.

US Space Command has also noted that the Russia-Ukraine war (2022-present) demonstrated the strategic value of space for land forces. Ukraine was able to use commercial imagery from Maxar for intelligence gathering, while Starlink provided internet service after terrestrial networks were degraded by a series of Russian attacks.[15] In contrast, Russia depended on electronic warfare to jam their GPS signals, impacting navigation and communication for Ukrainian forces, which disrupted drone missions and troop movements. Starlink's low Earth orbit (LEO) satellites deployed by SpaceX provided resilient communication indicating commercial space's military utility for armies, with over 50,000 terminals supplied to Ukraine since the start of war.[16] This conflict illustrated the dual-use nature of space assets, with both military and civilian applications, and the vulnerability of space systems to adversarial actions, forcing armies to adapt tactics, such as using alternative navigation methods when GPS is jammed.

Simmering conflicts, like Israel-Iran 2024 or China-Taiwan 2023-24, suggest space could be pivotal for land warfare, given current tensions. In an Israel-Iran scenario, Israel's advanced space program, including reconnaissance satellites, could be crucial for monitoring Iranian military activities, providing intelligence to ground forces for preemptive strikes or defense. Iran's efforts to develop its own space capabilities indicate a potential for space-based competition, with both sides potentially targeting each other's satellites, impacting army operations. Similarly, in a China-Taiwan conflict, China's extensive satellite network, including the Beidou navigation system, would likely support invasion plans, providing real-time data for troop movements and logistics, while Taiwan might rely on allied space assets for defense, with space denial strategies affecting land operations.

These conflicts illustrate the increasing reliance on space assets for land warfare, from navigation and communication to intelligence gathering and protection from aerial threats. As space becomes more integral to army operations, nations are investing in technologies to protect their own assets while developing capabilities to deny adversaries the use of space, directly impacting ground forces' effectiveness.

Orbital Firepower: The Next Generation of Space Systems

Emerging technologies have introduced new tools in space warfare, as it enhances military potentials and adds another level of ambiguity into strategic calculations for armies. A deluge of commercial satellite imagery has opened the floodgates to high resolution data aiding in Earth observation, which has profound consequences for military intelligence and the land operations. Imagery resolution from companies like Maxar Technologies is detailed enough to detect minutest of troop movements, observe military bases and evaluate battle damage. Open Source Intelligence (OSINT) analysts used images from Maxar to track Russian military buildups on Ukraine's border in the lead-up to Russia-Ukraine war, most notably providing advanced warning of potential invasions. To armies, near real-time data provides decisive insight about what the enemy is doing, where they are going and their infrastructure and drastically influencing how armies plan and execute battles with satellite imagery being used for tactical level decisions.

Mega Constellations

Low Earth orbit (LEO) mega constellations are networks of hundreds or thousands of small satellites providing global services. Modern militaries are keenly interested in these constellations for secure communications, surveillance, and resilience. Major programs include the US SpaceX Starlink/Starshield system, and China's Qianfan, Guowang and private Yinhe networks. Governments are also exploring government-owned constellations (analogous to Luxembourg's GovSat or the UK's Skynet 6) to support defense users. SpaceX's Starlink constellation, consisting of thousands of low Earth orbit (LEO) satellites, has introduced a new paradigm in satellite communications, directly benefiting land forces. Unlike traditional geostationary satellites, LEO satellites offer lower latency and higher bandwidth, making them ideal for real-time applications in conflict zones. Throughout the course of the Russia-Ukraine war, Starlink was a lifeline of internet access for Ukrainian forces and civilians that had terrestrial networks disrupted by Russian attacks ensuring communication for command and control, coordinating humanitarian efforts, and countering Russian propaganda. The insertion of Starlink terminals in Ukraine underscored the strategic value that commercial space assets offered to armies with over 22,000 terminals operational by Dec 2022, enabling ground troops to maintain operational continuity in contested areas.[17] This trend also suggests a future where commercial entities will play an increasingly larger role in supporting army operations, raising questions about security and regulation, with land forces needing to integrate these assets into their communication strategies.

Starlink (SpaceX) A commercial LEO broadband network. As of mid-2025 it has over 7,600 active satellites (each 300-700 kg). SpaceX launches these nearly entirely with its Falcon 9 rockets. Starlink satellites use phased-array antennas in Ku/Ka (and emerging V-band) frequencies with inter-satellite laser links, enabling high-speed broadband globally. The constellation delivers multi-Gbps per user throughput and low latency (about 20–40 ms), supporting video, data and even mobile phone links. SpaceX's Redmond facility handles manufacturing and network control. More than 12,000 satellites are planned in total (with proposals up to 34,000).

Starshield (SpaceX/USG) A militarized variant of Starlink for US and allied defense. Starshield satellites are similar hardware but hardened and equipped for missions like target tracking, signals intelligence, missile warning, and secure communications. As of early 2025, at least 118 Starshield satellites have been launched (for example, SpaceX delivered 22 on an NRO mission in Jan 2025). The U.S. Defense Department plans to field 100+ Starshield satellites in its own constellation by 2029. This effort is driven by the Pentagon's shift to "proliferated LEO" (pico- and nano-satellites in LEO), which offer greater resilience and coverage than a few large GEO satellites.

Both Starlink and Starshield sats are launched on SpaceX Falcon 9 rockets (and booked on Falcon Heavy). SpaceX builds and controls the network. For military use, Starshield satellites and terminals are owned/controlled by the U.S. government. The U.S. Space Force and allied commands operate ground hubs, and the Defense Information Systems Agency contracts with SpaceX to provide commercial services under strict security conditions. Amazon's Kuiper project (3,200 planned sats) is another U.S. constellation in development and may eventually provide an alternative LEO service, but it is not yet on orbit.

Qianfan ("Thousand Sails") – A Chinese state-led LEO internet constellation run by Shanghai Spacecom (Spacessail). It has launched 72 satellites to date (January 2025) on Long March rockets. Qianfan plans a 14,000+ satellite network (targeting 600 in orbit by end-2025, 14,000 eventual). These satellites (flat-panel design) likely use Ku/Ka bands and 5G-spectrum links for high-throughput broadband and military datalinks.

Guowang ("National Network") – A new Chinese super-constellation under China SatNet (a government-owned company). In Dec 2024 China launched the first 10 Guowang spacecraft on a Long March 5B. China has filed with the ITU for 12,992 satellites total (split into two constellations: 6,080 and 6,912 sats). China SatNet expects to deploy all 13,000 Guowang satellites by 2034 (with 6,500 by 2032). Chinese media stress this will be a "broadband mega constellation" akin to Starlink.

Hongyun/Hongyan (legacy) – China's earlier LEO projects (by CASIC and CASC) aimed for hundreds of satellites. For example, CASC's Hongyan program envisioned 60 sats in Phase 1 and hundreds in total. These plans were revised or consolidated into Guowang/SatNet projects, but they show the scale of China's ambitions.

Yinhe (GalaxySpace) – A commercial Chinese constellation by GalaxySpace (private). It uses a Q/V-band network integrating 5G technologies. GalaxySpace has launched 8 prototype Yinhe satellites since 2020 and aims for 1,000 satellites eventually (making it another "mega-constellation"). Yinhe sats are 100–200 kg with flat antennas. They demonstrate capabilities like Q/V and Ka-band links (tested high-bandwidth video calls). China uses its Long March family for these constellations. Qianfan sats (Group 6) have flown on the solid-fuel Long March 6A. Guowang's first batch flew on Long March 5B. New variants like Long March 6C/6D (smaller, liquid/solid hybrids) and private rockets (e.g. Kuaizhou, Ceres by Galactic Energy) will help raise large numbers of sats. China is rapidly expanding launch capacity (new sites in Hainan, Jiangxi, sea launches planned) to meet these goals.

As of mid-2025, the global race to establish large low-Earth orbit (LEO) satellite constellations for broadband and surveillance capabilities is well underway. SpaceX's Starlink leads the field with approximately 7,600 satellites launched, aiming for a target of over 12,000 in its first phase. Its military-oriented counterpart, Starshield, has seen the deployment of around 118 satellites, with over 100 more planned to support national security objectives.

China has accelerated its efforts through multiple programs. The Qianfan constellation has already launched 72 satellites, out of a planned 14,000, reflecting an ambition to rival Starlink in both scale and utility. The Guowang system, also part of China's strategic push into space, currently has 10 satellites in orbit, with a projected goal of around 13,000. Meanwhile, China's Yinhe constellation has deployed 8 satellites and plans to scale up to 1,000 in the near future.

All major LEO satellite constellations being developed for military or dual-use purposes are designed with advanced multi-gigabit-per-second data capabilities. SpaceX's Starlink satellites each support tens of Gbps using Ku and Ka band beams, as well as inter-satellite laser links, enabling high-speed data relay across the globe and delivering around 100 Mbps to individual user terminals, even in remote or denied environments. Chinese systems, including Qianfan and Guowang, utilize similar Ku and Ka band technologies and are increasingly experimenting with emerging 5G-compatible frequencies. Notably, Galaxy Space's Yinhe constellation employs Q and V bands, allowing ultra-high-frequency links that could support low-latency battlefield communications and real-time intelligence. In contrast, Indian military satellites such as Cartosat and RISAT are focused on high-resolution imagery and reconnaissance, transmitting a few hundred Mbps back to ground stations, sufficient for imagery but not yet optimized for high-throughput tactical networking or persistent broadband coverage across operational theatres.

India: Emerging Efforts

India does not have its own LEO broadband megaconstellation. It relies on traditional GEO sats (GSAT series) for comms and MEO sats (NavIC/IRNSS) for navigation. India's military has used limited foreign LEO services (e.g. leasing Inmarsat or commercial satcom) for niche roles. In 2023–2024 the government granted a license to SpaceX's Starlink for civilian Internet, under strict security rules (e.g. allowing lawful interception).

Recognizing vulnerabilities, India has accelerated development of LEO constellations for reconnaissance (not communications). Under the Space-Based Surveillance (SBS-III) program, India plans 52 new satellites (optical, radar, SIGINT) for "near real-time" monitoring. ISRO and private industry will build 21 satellites initially, with 31 more to follow. The Defence Space Agency will command these, aiming for rapid launch capability. These sats (similar to Cartosat/RISAT) provide sub-meter imagery and signals intelligence, enhancing border and maritime awareness. India's workhorse is the PSLV rocket (medium lift), which can place a few hundred kg to LEO. PSLV has launched record clusters (e.g. 105 small sats in 2017). Heavier GSLV rockets serve GEO payloads. ISRO also launched OneWeb (648-sat internet) tests as a rideshare, but India's own constellation efforts remain small compared to US/China.

India is opening space to private constellations. For example, in late 2024 thirty Indian firms expressed interest in building Earth-imaging constellations. Companies like Pixxel (hyperspectral imaging) and Bellatrix/Aurora (Ka-band broadband) are planning small LEO networks. However, none approach "mega" scale. Future launch goals include supporting these startups: PSLV and private rockets (AgniKul, Skyroot, etc.) may deploy dozens of microsats per year, but India has no announced plan for a 10^4-sat network.

India, while not yet active in the LEO broadband satellite segment, is focusing on strategic space capabilities. It has zero LEO broadband satellites deployed so far, but has outlined an ambitious plan to deploy a 52-satellite constellation spanning low Earth orbit (LEO), medium Earth orbit (MEO), and geostationary orbit (GEO) to strengthen its intelligence, surveillance, reconnaissance (ISR), and secure communication capabilities. As of June 2025, none of these new satellites have been launched, but the planned system under the Strategic Broadband Satellite (SBS-III) project worth approximately Rs 27,000 crore, is designed to provide layered, persistent coverage for the armed forces. Currently, India relies on a limited number of military-specific satellites such as the GSAT-7 series, which supports naval and tri-service communication, and the RISAT series, which provides radar-based surveillance. While these assets serve key operational needs, the envisioned 52-satellite constellation would represent a major leap in India's space-based military capabilities, aligning with global trends toward proliferated satellite networks for real-time, resilient battlefield awareness and secure data exchange.

India's launch vehicles, such as the PSLV (1.75 tons to LEO) and GSLV Mk III (8 tons), are significantly outmatched by SpaceX's Falcon 9 (22.8 tons), Falcon Heavy (64 tons), and the planned Starship (150 tons). China's Long March 5 (25 tons) and future Long March 9 (140 tons) further widen the gap, limiting India's capacity to deploy large-scale constellations. India's SBS-II program includes six surveillance satellites, while the planned SBS-III aims for 52, none launched by June 2025. In contrast, Starlink's 8,949 satellites and China's Guowang, with 29 launched and 13,000 planned, demonstrate a vast difference in scale, impacting India's military surveillance and communication capabilities.

Geopolitical and Military Implications

Strategic Deterrence and Resilience: Large LEO constellations make space-based C2 difficult to deny. A single ASAT weapon can't easily knock out an entire network of thousands of satellites. As one U.S. official noted, "the proliferated nature" of Starshield-style LEO means it is far more resilient than traditional few large GEO satellites. In deterrence terms, adversaries must consider that U.S. forces could still communicate and observe even if some satellites are attacked. China's military fears the opposite: RAND analysts report that PLA writings say Starlink-like constellations "undermine the PLA's preferred method of operations". Beijing has therefore made "Project SatNet" (Guowang) a national priority to avoid such a disadvantage. India, with its smaller constellation, remains more vulnerable to communication loss and hence is strengthening other links (undersea fiber, high-frequency comms) to back up space.

Command and Control: Ubiquitous LEO comms allow global real-time links for military units. For example, Space Force tests show Starlink enabling high-bandwidth video and data links to ships and aircraft in remote areas. In conflicts (e.g. Ukraine), portable Starlink terminals gave troops and civilians Internet where none existed. China envisions its LEO nets supporting PLA forces anywhere (on land, sea, and even along the Belt & Road), and may integrate them with 5G networks for soldier-worn devices. India's forces currently rely on older GEO/terrestrial comms; after Starlink's approval, India's services like Reliance Jio and Airtel plan to sell LEO-based broadband, which the military could co-opt under emergency policies. India is also eyeing a future quantum communication satellite (for secure C2), but this is still 2–3 years off.

Space-Based Surveillance: LEO constellations are not just for communications. China is also investing in large imaging/sensing constellations: for example, CASIC plans a 300-satellite "very low orbit" (300 km) network by 2030 for ultra-fast imaging and communications. India's SBS-III (52 sats) will give its military near-real-time imagery of enemy movements. The U.S. already uses many commercial and military LEO imagers (Planet, Maxar, NRO's new cubesats) to support ISR. Denying an adversary such coverage (by jamming or destroying sats) is a new front in space warfare.

Anti-Satellite Weapons and Countermeasures: The proliferation of LEO constellations changes the ASAT equation. The U.S. destroyed an old satellite in LEO in 2008 (Operation Burnt Frost) without disrupting other satellites. China's 2007 ASAT test and recent laser/explosive tests show it can target LEO. But with hundreds of small satellites, China's and India's networks would require many ASAT shots to cripple – making the act of attacking them more escalatory. For defense planners, this means LEO systems are partly self-healing: a few satellites are lost (e.g. to space debris or ASAT) and more can be launched quickly. The U.S. Congress has even funded studies for space-based missile defense on Starshield satellites, showing confidence in using LEO nodes as weapon platforms.

Hybrid and Cyber Warfare: In modern hybrid warfare, satcom is a double-edged sword. Adversaries (including non-state proxies) could disrupt or spoof LEO networks. India and the US worry about jamming and cyber-attacks on satellite links. To counter this, military LEO systems use encryption (Starshield) and are developing anti-jam antennas. Conversely, satellites can enable asymmetric tactics: e.g. tying remote insurgent forces into a command network, or broadcasting propaganda if terminals fall into the wrong hands. Many LEO sats also carry sensors, raising counterintelligence issues: for example, the US has launched small LEO sensors to track other satellites. China could similarly use its satnet for signals intelligence.

The future launch goals of leading space powers reveal the rapidly widening gap in military-use LEO constellations. SpaceX plans to continue launching Starlink satellites at a high cadence, with dozens of missions annually and a long-term vision of deploying up to 42,000 satellites. Complementing this, the US Department of Defense, through its Space Development Agency, is aggressively pursuing a Proliferated LEO or PLEO architecture, aiming to deploy hundreds of satellites by 2028. This includes a mix of Starshield units and other specialized military constellations for missile tracking, encrypted communication, and global ISR. China, in response, plans to launch about 200 LEO satellites each year, targeting a total of nearly 13,000 satellites across its Qianfan and Guowang networks by 2034, which would match or exceed Starlink in scale and potentially offer overlapping civil-military capabilities. India's launch capacity remains modest by comparison, limited to a few LEO missions annually. Even with emerging private-sector participation, it will take several years for India to deploy even a few hundred small satellites, constraining its ability to match the scale, resilience, and real time global reach of American and Chinese mega constellations.

ASAT Systems

As discussed earlier, Anti-satellite (ASAT) weapons is another key innovation, where various states acquire the ability to destroy or derange enemy satellites which pose a direct threat to army operations. Chinese Test of an ASAT in 2007, destroyed a defunct weather satellite in space and generated large amount of space debris resulting in international backlash from other nations.[18] In fact, Russia, US and India have all conducted ASAT tests, with Mission Shakti by India in 2019 demonstrating our capability to hit a low orbit satellite.[19] They would be able to take out communication, navigation and ISR satellites with these game changing weapons which takes away an adversary's ability to conduct warfare effectively, be it on land with GPS navigation and/or targeting. In the case of armies, without satellites the primary means of communication, navigation and intelligence could be degraded, severely impacting operations, necessitating alternative methods like inertial navigation or terrestrial backups.

The militarization of space has emerged as one of the most significant security challenges of the 21st century, with anti-satellite (ASAT) weapons and fractional orbital bombardment systems (FOBS) representing critical components of modern space warfare capabilities. Anti-satellite weapons are specialized technologies designed to disable, destroy, or interfere with satellites in orbit for strategic or defensive purposes. These systems have evolved from Cold War-era concepts into sophisticated weapons platforms that pose significant threats to the satellite-dependent infrastructure upon which modern military operations and civilian communications rely.

The development of ASAT capabilities represents a fundamental shift in military strategy, as nations recognize that space-based assets have become critical vulnerabilities in modern warfare. Satellites provide essential services including surveillance, communication, navigation, and early warning systems, making them high-value targets for adversaries seeking to degrade an opponent's military effectiveness. The proliferation of ASAT technologies among major powers has created new dimensions of strategic competition and raised concerns about the long-term sustainability of space operations due to the debris created by destructive testing.

Kinetic Energy ASATs

Kinetic energy anti-satellite weapons represent the most direct approach to satellite destruction, employing physical impact to destroy target satellites. These systems utilize missiles that collide with satellites to destroy them, with the kinetic energy from high-speed collision being sufficient to completely obliterate the target. The impact generates orbital debris, which can pose long-term hazards to other space assets and create cascading risks for the entire orbital environment.

Direct-ascent ASAT weapons attempt to strike satellites using trajectories that intersect the target satellite without placing the interceptor into orbit. Ballistic missiles and missile defense interceptors can be modified to act as direct-ascent ASAT weapons provided they have sufficient energy to reach the target satellite's orbit. These systems require sophisticated guidance and tracking capabilities to successfully intercept satellites traveling at orbital velocities of approximately 17,000 miles per hour.

Directed Energy Weapons

Directed energy weapons represent a non-kinetic approach to satellite attack, using highly focused energy without solid projectiles to damage targets. These systems include lasers, microwaves, particle beams, and other forms of concentrated electromagnetic energy. Directed energy weapons (Lasers) are being developed for satellite targeting, with debris-free alternatives to kinetic ASATs.[20] Directed-energy systems, including high-energy lasers with sensor blinding or satellite component degradation capability are being developed by the US, Russia and China for applications in counter-space operations. The primary advantage of directed energy weapons is their ability to operate at the speed of light with minimal atmospheric effects when used in space. Although these weapons are at the early stages of development, they offer the potential to alter the calculus of space warfare due to their ability to provide focused, scalable answers to threats, even with the challenges of power distribution and weather interference. Hardening satellites and developing other communication channels are among the defensive measures that armies need to focus on in order to protect space support systems against such attacks.[21]

Laser ASAT Systems

Ground-based laser anti-satellite systems represent a significant category of directed energy weapons, with several countries developing high-power laser capabilities for space applications. The United States developed the MIRACL (Mid-Infrared Advanced Chemical Laser) system, which became operational in 1980 and could produce over a megawatt of output for up to 70 seconds. MIRACL was successfully tested against the MSTI-3 satellite in October 1997, demonstrating the ability to blind satellite sensors at ranges of over 400 kilometers.

The Soviet Union developed the Terra-3 laser system at the Sary-Shagan test site in Kazakhstan, which was designed to create laser installations capable of hitting ballistic missile warheads and potentially satellites. The Terra-3 program developed laser ranging capabilities and laser damage systems, representing early Soviet efforts to weaponise directed energy technology. Russia continues to develop laser ASAT capabilities with the Peresvet laser weapon system, which became operational in 2019 and is deployed to protect mobile ICBM launchers.

Electromagnetic Pulse (EMP) Weapons

Electromagnetic pulse weapons constitute a particularly concerning category of ASAT systems, designed to maximize the electromagnetic pulse component of nuclear or non-nuclear explosions to disrupt critical infrastructure. These weapons create massive energy waves when detonated, potentially crippling vast swaths of commercial and government satellites. EMP weapons consist of three different pulse components known as E1, E2, and E3, each causing different types of damage and allowing subsequent components to cause greater damage. Russia has been reported to be developing nuclear space weapons that would destroy satellites by creating massive electromagnetic energy waves when detonated, potentially crippling satellite networks that support global communications, navigation, and financial systems. Chinese researchers have also claimed development of high-power microwave (HPM) weapons capable of producing electromagnetic pulses with intensity similar to nuclear explosions, using advanced phased-array transmission technology to precisely focus energy.

Electronic Warfare and Jamming Systems

Satellite jamming represents a form of electronic anti-satellite attack that interferes with communications traveling to and from satellites by emitting noise of the same radio frequency within the field of view of satellite antennas. Unlike kinetic or directed energy weapons, jamming does not physically damage satellites but creates entirely reversible attacks that can be terminated by turning off the jamming signal. There are two main types of satellite jamming: uplink jamming, which interferes with signals going from ground stations to satellites, and downlink jamming, which disrupts transmissions sent from satellites to ground-based receivers. Uplink jamming is considered more difficult because greater transmitter power is required to reach satellite transponders, but it can be more impactful due to its ability to degrade satellite signals for all users.

Co-Orbital ASAT Systems

Co-orbital ASAT attacks involve another satellite in orbit being used to attack a target satellite. The attacking satellite is first placed into orbit, then later manoeuvres into an intercepting orbit using sophisticated on-board guidance systems to successfully steer into the path of another satellite. Co-orbital ASATs can remain dormant in orbit for days or even years before being activated, making them particularly difficult to detect and counter. A co-orbital attack can utilize various methods, including simple space mines with small explosives that follow the orbital path of the targeted satellite and detonate when within range. Another co-orbital attack strategy employs kinetic-kill vehicles (KKV), which are objects designed to collide with target satellites. The Soviet Union developed one of the most sophisticated co-orbital ASAT systems, with interceptors designed to approach satellites within one or two orbits (1.5 to 3 hours) before detonating explosive charges that damage targets with shrapnel.

Fractional Orbital Bombardment Systems (FOBS)

The Fractional Orbital Bombardment System represents a unique weapons delivery concept that fundamentally differs from traditional intercontinental ballistic missiles and anti-satellite weapons. FOBS involves launching nuclear warheads into partial Earth orbit, where the trajectory launches a warhead and its delivery vehicle into an orbit of approximately 100 miles (160 kilometers) but breaks from that orbit before completing a full revolution of Earth. This system had no range limit and could target any point on Earth, with flight paths that did not reveal target locations until payload impact. The primary advantage of FOBS lies in its ability to approach targets from unexpected directions, particularly from the south or north poles, rather than along traditional ballistic trajectories. This capability allows FOBS to bypass existing missile defense systems and early warning networks, making it extremely difficult to predict timing and target selection until the final approach phase. The orbital mechanics of FOBS provide unlimited striking range and the capability to strike from any direction while concealing target locations until payload deployment.

Distinction Between FOBS and ASAT Systems

The primary distinction between Fractional Orbital Bombardment Systems and Anti-Satellite weapons lies in their intended targets and operational objectives. FOBS are designed specifically to deliver nuclear warheads against terrestrial targets using partial orbital trajectories, while ASAT weapons are designed to incapacitate or destroy satellites for strategic or tactical purposes. FOBS utilize orbital mechanics to approach Earth-based targets from unexpected vectors, whereas ASATs target satellites operating in various orbital regimes.

FOBS systems complete partial orbits before deorbiting to strike terrestrial targets, while ASAT systems either intercept satellites directly from ground-based launch platforms or operate as co-orbital systems that manoeuvre to engage satellite targets. The payload and warhead designs also differ significantly, with FOBS carrying nuclear warheads optimized for ground targets, while ASATs may employ kinetic kill vehicles, explosive fragmentation warheads, or directed energy systems optimized for satellite destruction.

Legal and Treaty Implications

The legal status of FOBS and ASAT systems under international law presents different challenges and interpretations. The Outer Space Treaty of 1967 prohibits the placement of nuclear weapons in orbit around Earth, creating legal ambiguity regarding FOBS since these systems complete only partial orbits. The Soviet Union attempted to sidestep treaty conflicts by arguing that FOBS did not make complete orbits and therefore did not violate the Outer Space Treaty. ASAT weapons face different legal constraints, with no specific treaty prohibiting their development or testing, though destructive ASAT tests have been criticized for creating long-lived debris that threatens other space assets. The United States became the first country to adopt a voluntary moratorium on destructive direct-ascent ASAT missile testing in April 2022, citing concerns about space debris and the long-term sustainability of space operations.

Contemporary Threats and Future Implications

The proliferation of anti-satellite weapons among major powers has created new strategic challenges and risks for space-based infrastructure. The potential for cascading space debris from ASAT testing could cause Earth to suffer from Kessler syndrome, where collisions create more debris that causes additional collisions in a self-sustaining cascade. This scenario could render certain orbital regions unusable for extended periods, effectively denying space access to all nations.

The development of nuclear-powered EMP weapons in space represents a particularly concerning escalation, as such weapons could create electromagnetic pulses capable of disabling vast numbers of satellites simultaneously. These weapons would cross dangerous thresholds in nuclear weapons deployment and could cause extreme disruptions to modern life in unpredictable ways. The reversible nature of electronic warfare attacks against satellites also creates new possibilities for temporary disruption without permanent destruction.

The evolution of FOBS technology, particularly China's development of hypersonic glide vehicles using fractional orbital trajectories, represents a significant challenge to existing missile defense systems. These systems combine the unpredictable approach vectors of traditional FOBS with modern hypersonic manoeuvrability, creating weapons that are extremely difficult to intercept or counter. The implications for strategic stability and deterrence relationships remain unclear as these technologies continue to develop.

An additional emerging threat is that of cyber warfare against space systems, which could well affect Land Forces without any physical destruction of space based assets. The suspected Viasat attack during outset of war between Russia and Ukraine severed Ukrainian communications showing weaknesses in satellite control systems.[22] Cyber-attacks can be used to compromise data, take control of a satellites or knock out ground stations, hence a useful tool to de-orbit space-assets cheaply. For the armies, protecting these ground-based systems that manage space assets is imperative, with implications on operational security and the need for robust cybersecurity measures, such as secure communication protocols and AI-driven threat detection systems.

These innovations highlight the dynamic nature of space warfare, with commercial, kinetic, cyber, and energy-based technologies reshaping military strategies for land forces. The integration of these systems into army operations requires new approaches to command and control, with implications for training, planning, and operational resilience, ensuring armies can leverage space assets effectively while mitigating risks.

ARMING THE HEAVENS: THE DEVELOPMENT OF SPACE-BASED WEAPONS

Space-based weapons are an extremely contentious issue regulated by international treaties such as the Outer Space Treaty of 1967 which bans the deployment of weapons of mass destruction in outer space, creating ambiguity for other types of armaments.[23] This legal vacuum has provoked the potential development of numerous space based weapons ranging from kinetic anti-satellite (ASAT) weapons, which destroy satellites by smashing into them, and non-kinetic such as lasers or electromagnetic pulses (EMPs) that disable satellite functions without creating debris. Development of these weapons means that space is transforming into an operationally contested domain, and the more an army depends on space based assets, greater the risk of having to operate in degraded/ denied space support environments.

China, Russia, US and now India has conducted launches of kinetic weapons like ASAT missiles in order to show their capabilities. The international opprobrium of China's 2007 ASAT test which created over 3,000 pieces of orbiting debris not only has environmental impact but also strategic implications for land forces depending on satellite navigation and communication.[24] Russia's Nudol system and the US's Ground-Based Interceptor program targeted at satellites in low Earth orbit both could disrupt army operations by disrupting GPS and ISR functions.[25] Other non-kinetic options such as high-powered microwaves and space based laser for missile defense are being considered offering debris less options that could protect or even endanger the land forces.

Speculations about the military utility of the United States X-37B spaceplane (an unmanned orbital test vehicle) with potential applications for land warfare are rampant. Since its first launch in 2010, it has flown several long durations classified missions (reconnaissance, satellite servicing or even offensive capability), with US Space Force highlighting testing of its reusable spacecraft technology. Possible benefits to armies would involve more effective intelligence or the danger to its own space assets requiring defensive considerations.[26] Russia and China's co-orbital ASAT tests, where satellites manoeuvre near others, further complicate the landscape, with potential to inspect, interfere with, or destroy adversary assets, impacting land forces' reliance on space for operational continuity.[27]

Shijian-21: Orbital Manoeuvring and Satellite Interaction Capabilities

China's Shijian-21 satellite, launched in October 2021 from Xichang Space Launch Center aboard a Long March 3B rocket into geosynchronous transfer orbit, has demonstrated concerning capabilities for satellite manipulation and orbital operations. Officially described by China's Xinhua News Agency as an On-Orbit Service, Assembly, and Manufacturing (OSAM) satellite designed to test and verify space debris mitigation technologies, the Shijian-21 has exhibited behaviors that suggest significant dual-use potential for counterspace operations.

The satellite gained international attention when it demonstrated the ability to dock with and physically manipulate other satellites in orbit. In January 2022, ExoAnalytic Solutions reported that Shijian-21 went "missing" from its orbital slot to dock with the defunct Beidou G2 navigation satellite, utilizing the inability of optical satellites to track space objects during daylight hours to conduct covert operations. Following the docking manoeuvre, Shijian-21 moved the Beidou G2 satellite to an orbit 3,000 kilometers higher, releasing it into graveyard orbit before returning to geosynchronous orbit.

The Shijian-21 mission has raised significant concerns among Western intelligence analysts due to its demonstrated capability to manipulate satellites without prior notification or transparency regarding its operations. The U.S. Office of the Secretary of Defense specifically mentioned Shijian-21 in its 2022 China Military Power Report, noting that China has launched multiple satellites to conduct scientific experiments on space maintenance technologies while conducting research on space debris cleanup.

United States Project Silent Barker

Project Silent Barker represents a critical United States space surveillance initiative developed jointly by the U.S. Space Force and the National Reconnaissance Office to address growing threats from Chinese and Russian space vehicles in geosynchronous orbit. The classified satellite constellation is specifically designed to track objects and potentially nefarious activities approximately 22,000 miles above Earth, providing unprecedented space domain awareness capabilities that complement existing ground-based radar and low-Earth orbit surveillance systems.

The Silent Barker program emerged as a direct response to concerning behaviors exhibited by adversary satellites, including incidents where Russian satellites trailed U.S. spy satellites in what officials described as "unusual and disturbing" behavior. The constellation serves as a replacement for the aging Space-Based Space Surveillance System satellites, which are scheduled to reach the end of their operational lifespan in 2028, necessitating the development of more advanced space surveillance capabilities.

The first Silent Barker satellites were successfully launched on September 10, 2023, aboard a United Launch Alliance Atlas V rocket from Cape Canaveral Space Force Station in Florida, marking the beginning of operational testing for this critical space surveillance constellation. The launch utilized one of the last Russian-made RD-180 engines in the US inventory, highlighting the transition period in American space launch capabilities. The complete Silent Barker constellation is planned to be fully operational by 2026, providing comprehensive coverage of geosynchronous orbit activities.

Space-based weapons are big ethical and legal debates, with the UN increasingly failing to build a consensus over the long debated treaty to prevent an arms race in space (PAROS).[28] Among armies, these weapons mean a need to reinvent the doctrinal wheel, developing resilient space architectures and training for GPS-degraded or denied environments to ensure effectiveness in contested space domains.

GUARDIANS OF THE GALAXY: MASTERING COUNTER-SPACE DEFENSE

Counter-space operations purposefully are designed to deny adversaries utilization of space capabilities through a variety of kinetic and non-kinetic means, each with benefits and downsides for land forces. Some kinetic anti-satellite (ASAT) weapons, such as those recently tested by China in 2007 and by India in 2019, used missile systems to destroy satellites. Co-orbital ASATs, or satellites that manoeuvre near other satellites in the same orbit, are a less destructive development, and lend themselves to disabling others through various means such as deploying nets, robotic arms, or a directed energy weapon with high-powered microwaves.[29] Russia has conducted tests that suggest close approaches to US satellites, which offers limited capacity to disable space based systems impacting land forces that regularly rely on satellite capabilities for communication, intelligence and targeting. Directed energy (DE) weapons could blind or degrade satellite sensors or damage satellite component capabilities without destroying them. Russia's Peresvet laser is an example of a DE system that is fielded with counter-space roles; it could impact army operations by disrupting satellite support.[30]

In various conflicts, including the Russia-Ukraine conflict, electronic warfare techniques have been used, including jamming GPS signals, to disrupt Ukrainian military navigation and communications, particularly impacting drone operations and ground troop movements. There have been reports from Ukrainian officials noting the impact of Russian electronic warfare capabilities. For armies, this means adapting tactics, such as using alternative navigation methods or enhancing electronic countermeasures. Cyber-attacks on the ground control of satellites can degrade data or take control of satellites, a tactic reportedly employed during the alleged Viasat attack at the beginning of the war in Ukraine, disrupting communications and demonstrating vulnerabilities, requiring land forces to protect ground-based systems with robust cybersecurity measures.

Each counter-space attack method has implications and risks to strategic stability. Kinetic attacks create space debris that damages and collaterally destroys other spacecraft instead of its intended target. Co-orbital systems that allow for space weaponisation risk blurring distinctions between civilian and military purpose, and non-kinetic options provide flexible and controllable responses. Yet all of these implications and risks have bearing on military operations, highlighting the need for international norms that deter military escalation, with the UN discussing measures to limit ASAT tests and enhance safety through space transparency.

STELLAR STRATEGIES: ADAPTING DOCTRINE AND TACTICS TO SPACE

Emerging technologies that are increasingly being employed in space warfare are revolutionizing army doctrines and tactics, and require getting accustomed to a contested and congested space domain. Resilient architectures in space, using smaller and cheaper satellites to mitigate vulnerability to ASAT attacks, are on the rise, prime example being the success of the Starlink Constellation. For land forces, this provides access to a more resilient communication and navigation system, as distributed systems are harder to destroy than a single larger satellite, which improves survivability. Nations are spending vast amounts on creating the capability of rapidly launching satellites to make up for satellites lost to enemy actions with companies such as Rocket Lab offering rapid launch services.[31]

Space situational awareness (SSA) is essential for detecting threats in orbit, including debris or hostile actions aimed at disrupting and/or damaging space-based capability which directly affects the land forces operations. To illustrate, the US Space Surveillance Network (SSN) currently monitors and tracks some 27,000 plus objects in orbit. This generates tracking and orbital collision avoidance data aiding army's operational planning and success. For example, the army can by integrating SSA into command structures can obtain levels of situational awareness that will allow the real-time adjustments to operations, rerouting of convoys and adjusting fire support based on satellite status.[32]

The integration of space with the air, land, sea, and cyber domains creates a more comprehensive conception of space, as space is now an integral part of army strategy. This holistic approach requires seamless coordination and information sharing across different branches, with joint exercises incorporating space scenarios. For instance, the US Indo-Pacific Command's exercises include the use of space-based ISR and communications, highlighting the need for integrated command structures, with land forces relying on space for situational awareness and coordination.[33] Historically, land forces have utilized the value of space-based technology to provide surveillance, whether for the purpose of situational awareness, or for effecting advanced missions with command and control. From a doctrinal standpoint, armies have been developing counter-space strategies that involve two approaches: defensive strategies and counter-space capabilities. Defending space assets involves their 'hardening' against attacks, while counter-space capabilities deter attacks and provide the capacity for a military to respond to space-based threats to own training and operations by an adversary.

The intent to pursue pre-emptive offensive counter-space capability is already underway. Nations are working on the development of capabilities that are aligned with the denial of space to adversaries. Capabilities rooted in the electronic spectrum to include hacking and data manipulation, all fall within this domain and will certainly influence land forces. This move represents a doctrinal change that will include rapidly responding to losses, as well as a reduction in reliance on space based capabilities with some redundancy or alternative mechanisms. For example, soldiers will have to train to operate in GPS denied environments, demanding higher level of coordinated human behavior and instead relying on basic navigation with maps and compasses. The implementation of AI and machine learning into space operations is helping our decision making, with autonomous satellites even able to 'sense' threats, and reconfiguring networks, supporting land forces with real-time data and communication.

Cosmic Conflict: Opportunities and Barriers

Several trends are influencing the future of space warfare for land forces, with consequences for security, sustainability, and governance. The commercialization of space, driven by companies like SpaceX, Blue Origin, Rocket Lab and Virgin Galactic, is increasing the number of space missions and satellites, many of which could become targets or collateral damage in a conflict, thereby affecting the operations by militaries.[34] An apt example of this is SpaceX's Starlink constellation with a planned six thousand satellites by 2025, a number which they have already surpassed due to overwhelming demand, and the security of commercial assets and their military potential being a growing concern with land forces having to integrate these into future communication capabilities.[35]

The integration of advanced disruptive technologies into space operations marks a new age of capabilities and risk, with implications for military strategy, global security and the long-term sustainability of space as shared universe. These technologies, which include anti-satellite (ASAT) weapons, directed energy capabilities, artificial intelligence and autonomous vehicles, will radically change how countries can project power and maintain dominance in space. However, we need to reconcile this with the risks that these technologies pose, ranging from increasing tensions between space-faring countries to threatening the orbital environment itself.

This section explores the dual nature of these technological advancements, offering a detailed assessment of their capacity to enhance space operations while grappling with the complexities and uncertainties they engender. The deployment of emerging technologies in space operations brings a wealth of strategic and tactical advantages, fundamentally altering the capabilities available to military forces.

Enhanced Space-Based Capabilities: The emergence of anti-satellite weapons (ASAT), satellite jamming technologies, and space-based lasers means that militaries have highly effective tools to both protect their own satellites, and undermine those of their adversaries. These capabilities are especially relevant to what is now an increasingly likely contested space environment in which the ability for militaries to communicate or navigate or gather intelligence can likely turn the tide of terrestrial battles. ASAT weapons can, for example, destroy an adversary's satellite, and cut off its ability to transmit critical information or support functioning of ground based systems. In addition, space-based laser systems provide precision and speed to threat engagement, where threats like enemy satellites or ballistic missiles can be controlled with a high degree of effectiveness. Together, these technologies bolster a nation's capacity to assert space superiority, an increasingly vital component of modern warfare, ensuring that military forces can operate with confidence in an ever more contested domain.

Increased Autonomy and Efficiency: The integration of AI and autonomy into space operations marks a quantum leap, making missions faster and reducing dependence on human intervention. AI systems can analyze huge amounts of sensor data, identify faults and provide relevant, real-time information to speed up the decision-making process. Autonomous spacecraft used for satellite servicing, debris removal, and/or tactical manoeuvres can operate without continuously depending on ground control, which is particularly helpful due to the time delays caused by long distances in orbit. The incorporation of autonomy increases efficiency and capability with resilience because it can respond to unforeseen complications and modify plans based on changing mission requirements. Decreasing reliance on human guidance, facilitates more flexible and sustainable space operations, enhancing a nation's ability to maintain a persistent presence in orbit.

Enhanced Space Situation Awareness (SSA): AI and autonomous systems significantly improve space situation awareness through a more detailed and comprehensive view of the orbital environment. This provides valuable offensive and defensive capabilities by being able to continuously observe your space assets, monitor areas for space debris and detect any potential adversarial actions such as jamming or natural hazards. AI algorithms are excellent at being able to synthesize and analyze multiple sources of data – radar, optical sensors and satellite telemetry – to identify potential collision conditions or even the faintest piece of evidence of adversarial behavior being exhibited. This level of awareness provides military commanders the utmost confidence for managing assets in space, enabling them to take preemptive actions against vulnerabilities or assess an emerging threat posture in the event of a dangerous situation. Improved SSA is not only beneficial for defense; it also keeps safety in mind for civilian space endeavours, which is important as the number of new satellites and spacecrafts increase. Improved SSA will prevent catastrophic events from occurring with so many new assets in orbit.

Speed-of-Light Engagement: Directed energy weapons (DEWs), such as high-powered microwaves and lasers, introduce a revolutionary element to operations in space, with the possibility of instantaneous engagement capability that surpasses the speed of traditional kinetic systems. In the vastness of space, dangerous threats can appear in seconds; therefore, having the ability to engage at the speed of light creates a game changing advantage. DEWs can disable enemy satellites, intercept missile launches, or destroy an enemy platform, while generating little or no debris, unlike the destructive characteristics of kinetic engagement. The blend of speed, precision and low debris associated with DEWs makes them enticing for protection of a friendly platform or to deliver offensive combat power. The emergence of DEWs will fundamentally change space warfare, shifting the emphasis from brute force to rapid, targeted interventions that preserve the operational integrity of the orbital environment.

At the same time, it introduces a host of disadvantages that could destabilize the delicate balance of power in space and beyond, necessitating careful consideration by policymakers and military planners alike.

Weaponisation and Expansion of Space: The proliferation aided by rapid commercialization of offensive space capabilities such as ASAT weapons and DEWs could lead to a worrisome escalatory cycle that commits space to military competition rather than collaboration. As states field systems with greater capabilities to protect their interests, the risk of escalation increases as states will need to defend system upgrades from adversaries. An escalatory cycle of capabilities could lead to the weaponisation of space, with capabilities able to execute attack or counter before achieving escalation thresholds.[36] All the while, erosion of decades of diplomatic work to maintain space as a 'peaceful' domain may further support other conflicts. Meanwhile, the lack of relevant international treaties to regulate offensive space capabilities creates further complexity and a greater threat; misunderstandings or tests of offensive capabilities (such as exploding a satellite) could easily lead to conflicts well beyond nation-states. Further, such incidents would risk an armed conflict that would further erode trust between spacefaring states and jeopardize global security.

Space Debris and Its Impacts on the Environment: Deployment of kinetic ASATs will break satellites into fragments and aggravate a top-priority orbital debris issue that has got the space-faring community on high alert. Every successful kinetic ASAT generates thousands if not more high velocity fragments, many of which linger in orbit for years, threatening operational satellites, crewed missions, and even the International Space Station.[37] As fresh debris pile up, all of this eventually raises the chances of a cascading event known as the Kessler syndrome when collisions spawn more collisions and make critical space orbits unusable.[38] For military forces dependent on space-based systems, this environmental degradation could cripple their operational capacity, while civilian sectors, such as telecommunications and weather forecasting, face similar disruptions. The long-term health and sustainability of space will be eroded, pushing nations to measure the relative military utility of weapons such as these against the consequent destruction of global environment.

Cyber and AI Vulnerabilities: As militaries of the world rely increasingly on autonomous systems and AI enabled space-based platforms that underpin land warfare, they tend to face an expanding array of cyber and algorithmic threats that continue to eat away at operational effectiveness of their weapons and systems. The adoption of AI enhances capabilities such as real-time satellite imagery analysis, predictive threat modeling, and automated command-and-control (C2) functions, offering land forces unparalleled situational awareness and responsiveness. However, this reliance exposes critical systems to a range of cyberattacks, including hacking attempts that could hijack satellite controls, data breaches that compromise mission-sensitive intelligence, and malware that disrupts communication networks essential for coordinating ground manoeuvres. For instance, a cyber-attack into space-based navigation could feed artillery units with spoofed position data causing inaccurate targeting or causing a degree of operational chaos. In addition, subtle issues within AI systems are algorithmic mistakes that stem from bad programming, data sets which are biased or poorly tested machine learning models and could result in false threat evaluations, wrong target identification or resource allocation errors. They would probably lead to catastrophic results during high intensity contests, such as friendly fire incidents or collapse of logistics support chains that further magnify the chaos on the battlefield.

A comprehensive multi-pronged cybersecurity defence coupled with an uncompromised commitment to human oversight is required to inoculate against these threats. Advanced encryption must secure data transmissions between space assets and ground forces, while continuous monitoring systems – augmented by AI-driven threat detection identify and neutralise threats before they cause any serious damage. However, given the complexity and self-evolving character of contemporary cyber threats – risks cannot be mitigated entirely with technology alone. Human operators, equipped with the authority and training to intervene in autonomous processes, serve as a critical failsafe, particularly in scenarios where AI-driven decisions carry life-and-death consequences, such as targeting or resource allocation. Establishing protocols for seamless human-machine collaboration, where personnel can override faulty algorithms or validate AI outputs, ensures that technological efficiency does not come at the expense of operational reliability. For the Army, unaddressed cyber and AI vulnerabilities in space systems could sever the lifeline of intelligence and communication that land forces depend upon, rendering them vulnerable to adversaries' adept at exploiting these weaknesses. Strengthening cybersecurity and maintaining human oversight are thus not optional enhancements but essential imperatives for preserving the Army's ability to operate effectively in a space-dependent battlespace.

Multi-Domain Integration: The future of warfare hinges on the Army's ability to achieve seamless integration across space, cyber, and conventional domains, a paradigm shift that has the potential to revolutionize land operations while posing significant challenges to operational cohesion and strategic success. Multi-Domain Operations (MDO) envisage a common battlefield where space-based sensors, cyber resilient operators and terrestrial forces would inter-operate to produce synchronised effects in time and space, allowing for quick reaction to threats and effective exploitation of fleeting opportunities. This integration is vital for land forces that rely on space assets for navigation, communication, and intelligence, while simultaneously depending on cyber resilience to protect these systems from disruption. However, the road to such convergence is encumbered with numerous hurdles. Huge technical challenge of creating interoperable systems from satellite networks to ground based command systems, as disparate platforms must share data seamlessly while having differing architectures, protocols and operational speeds. The technical challenge in applying interoperability to embedded data is simple to articulate but hard to execute. Any disruption of this interoperability, such as delay in relay of live satellite imagery to a manoeuvring infantry unit would be a significant operational problem, that exposes forces or prevents them from taking advantage of enemy weaknesses.

Beyond technology, multi-domain integration necessitates the creation of joint operational doctrines that harmonize the distinct cultures, priorities and capabilities of the Army, Air Force, Navy and Space arm. Historically, land forces have operated with a terrestrial focus, however, MDO demands a higher level of awareness, a way where space-based assets combined with cyber operations shape the battlefield. The change will need cross-community interoperability (in this case, of response tools, such as platforms for fusing domain-centric data) and more systemic training of individual soldiers exposed to multi-domain environments. Realistic simulations and joint exercises, where space assets are contested, cyber networks are breached, and land forces must adapt, are indispensable for building proficiency and trust across domains. The strategic cost of not being able to integrate this way is horrendous: adversaries who are able to integrate their capabilities across domains could outmanoeuvre a fragmented force, their kill chain, degrade intelligence surveillance and reconnaissance (ISR) networks, or place entire units out of the fight. Conversely, successful integration transforms space and cyber into force multipliers, amplifying the Army's ability to project power and maintain dominance. For land operations, this means leveraging satellite-enabled precision and cyber-secured networks to outpace opponents – an advantage that hinges on overcoming the technical, doctrinal, and cultural barriers to multi-domain unity. In an era of increasingly sophisticated threats, this integration is not merely a goal but a prerequisite for sustaining the Army's strategic advantage.

Complexity and Cost: The pursuit of advanced space technologies demands extraordinary resources, from the financial investments required for research and development to the sophisticated infrastructure needed for deployment and maintenance. AI-driven systems, directed energy weapons and resilient satellite networks entail complex engineering challenges, necessitating cutting-edge facilities, skilled expertise and sustained funding. For many nations, these costs may prove prohibitive, deepening the divide between established space powers and emerging players thereby giving undue advantage and leverage to select few. This may also force the emerging players to go in for quick fixes which may be unreliable and dangerous at times. Moreover, the complexity of unreliable systems introduces operational risks – software glitches, hardware failures or integration issues could compromise missions or lead to costly setbacks. This resource-intensive nature underscores the strategic trade-offs involved, as nations must balance the allure of technological superiority against the practical realities of affordability and reliability.

Ethical and Legal Implications: With space including the use of AI and autonomous systems in operations, especially in offensive roles, there are very tough ethical and legal questions arise. The ambiguity in international laws of armed conflict ensues as autonomous platforms with target selection and engagement capabilities blur the lines of accountability. Should an AI-driven satellite mistakenly attack a neutral asset or escalate a minor incident into a major confrontation, attributing responsibility becomes murky, challenging traditional notions of command and control. Furthermore, the susceptibility of these systems to hacking and spoofing makes matters worse since it opens the possibility for adversaries that could aim to sow chaos. We urgently need robust ethical and legal standards to guide the development and use autonomous technologies in space in order to prevent the unchecked development of from eroding moral credibility and inviting unintended consequences. The future of space warfare requires international cooperation to manage these challenges, with initiatives like the Artemis Accords aiming to establish norms for responsible space behaviour, though geopolitical tensions, such as US-China rivalry, complicate efforts, with calls for transparency and confidence-building measures to prevent escalation. The space will remain a critical enabler and a possible vulnerability for military forces, hence investments into survivable systems, counter-space capabilities development and training for space domain operations of ambiguous nature must be continued in order for our forces to remain effective in contested domains.

Galactic Leap: Emerging Trends in Space Warfare

With the ever increasing relevance of space domain in military strategy, a few emerging trends are set to guide the path of space technology development and operations over the next two decades in many aspects of national security. Not only will these trends reshape the future of space operations, but they will also fundamentally change the way the Army will conduct land warfare. In an age where critical space-based capabilities form the foundation for a host of functions ranging from communication, navigation and intelligence; the Army must adapt and refine its doctrines, capabilities, partnerships accordingly. This section explores four pivotal trends – the proliferation of space-based weapons, advancements in artificial intelligence (AI) and autonomy, an intensified focus on space resilience, and the imperative for international cooperation and governance – and analyzes their far-reaching implications for the future of military operations on land.

Proliferation of Space-Based Weapons

The ongoing advancement and deployment of space-based weapons, including anti-satellite (ASAT) systems, directed energy weapons (DEWs), and other sophisticated technologies, herald a significant escalation in the weaponisation of space. This is the inevitable result of an international consensus that "space control can mean the difference between victory and defeat in terrestrial wars" and thus countries are increasing their space arsenal with weapon systems capable of denying or destroying an adversary's space capabilities. For the Army, this development introduces profound challenges to its operational framework. Space-based systems are integral to land warfare, providing real-time intelligence, precision navigation, and secure communications – capabilities that could be jeopardized by ASAT weapons capable of physically disabling satellites or DEWs that impair their functionality with focused energy bursts. The expected disruption of these assets could complicate the synchronization of military operations, execution of precision strikes and situational awareness especially in high intensity conflict. Additionally, the likelihood of conflict transiting from space to Earth exists as weapons in orbit may be primary targets by adversaries to degrade land-based systems. To address this, the Army must prioritize the development of countermeasures, such as alternative communication channels and ground-based navigation systems, while integrating space-domain awareness into its strategic planning. The growing prevalence of space-based weapons underscores the urgency of viewing space not merely as a support domain but as a contested battlefield with direct implications for land warfare success.

Advancements in AI and Autonomy

The integration of artificial intelligence (AI) and autonomous systems into space operations represents a transformative leap forward, promising to enhance the sophistication and efficiency of military missions with minimal human oversight. In due course, the space will be operated by AI-driven platforms taking centre-stage – analyzing petabytes of satellite data; optimizing mission planning through simulations and responding to threats in real-time. AI can enhance space situational awareness by rapidly processing imagery and signals intelligence, enabling commanders to anticipate adversary actions and allocate resources more effectively. Autonomous satellites, capable of independent navigation, self-repair, or defensive manoeuvres, could maintain operational continuity even under duress, ensuring uninterrupted support for ground forces. Yet, these benefits are tempered by inherent risks. The increased reliance on AI introduces vulnerabilities to cyberattacks that could corrupt data or hijack autonomous systems, potentially turning them against friendly forces. Furthermore, the delegation of decision-making to machines raises ethical questions, particularly in scenarios involving lethal force or strategic escalation. The Army must therefore invest in robust cybersecurity protocols and establish clear guidelines for human oversight, ensuring that AI and autonomy amplify rather than undermine its operational effectiveness. By harnessing these technologies judiciously, the Army can position itself to thrive in a future where speed, precision, and adaptability define success on the battlefield.

Focus on Space Resilience

Space is quickly becoming contested and congested domain with resilient space-based assets being one of the highest priorities to support military operations. Given the Army's reliance on satellites for everything from GPS-guided munitions to tactical battlefield communications, these operations will remain vulnerable to jamming, hacking and kinetic ASATs. In response, future efforts will center on fortifying space infrastructure through advanced technologies and contingency planning. Anti-jamming measures, such as adaptive frequency modulation, will safeguard communication links against electronic interference, while satellite hardening – through shielding or redundant subsystems – will enhance durability against physical and electromagnetic assaults. The development of rapid satellite replacement capabilities, including small, deployable constellations, will further ensure that the Army can restore functionality swiftly following an attack. Beyond technical solutions, resilience demands a broader strategic shift. The Army must diversify its reliance on space by cultivating terrestrial alternatives, such as inertial navigation systems or high-altitude platforms, to maintain operational tempo when orbital assets are compromised. In order to accomplish this, we need to work with industry partners to drive innovation and with our allies to leverage their capabilities. By prioritizing space resilience, the Army can mitigate the risks of operating in a degraded environment, preserving its ability to project power and achieve dominance in land warfare under increasingly adversarial conditions.

International Cooperation and Governance

The accelerating pace of technological disruption in space clearly underscores that international cooperation and governance is needed to ensure the space domain remains stable and secure, which underpins our future. Without cohesive norms and agreements, the unchecked proliferation of space weapons and the accumulation of orbital debris – exacerbated by kinetic ASAT tests – could precipitate miscalculations, escalate tensions, or spark an arms race with catastrophic consequences. For the military, whose operational success hinges on reliable access to space-based capabilities, supporting diplomatic efforts to regulate this domain is a strategic imperative. International frameworks that limit the deployment of offensive systems, establish protocols for responsible behavior, and mitigate the debris threat will reduce the likelihood of conflict and preserve the orbital environment for military use. Such efforts could include treaties banning certain ASAT capabilities or multilateral initiatives to enhance space traffic management, ensuring that key orbits remain viable for satellite operations. The Army must advocate for these measures through its broader defense establishment, participating in forums that shape global space policy and fostering partnerships with allies to align priorities. By contributing to a rules-based order, the Army can help secure space as a domain that enhances rather than endangers land warfare, reinforcing its role as a stabilizing force in an increasingly complex geopolitical landscape.

The Indian Context

There is a growing recognition that space has become an arena of contestation, and the Indian Armed Forces must develop sufficient leverage in space in order to support its missions. This requires investments in space-based surveillance and early warning capabilities, and secure communications. Space-based capabilities must deliver a persistent, real-time understanding of the battlespace to the Army. Through the launching of indigenous satellites with various sensors, the Army will able to continuously monitor hostile activities so it can get an understanding of enemy threats as they manifest. In order to further enable the Army's understanding through the use of space-based sensors, space-based assets must be integrated into the larger joint all-domain command and control (JADC2) network so that the different combat arms can share data and service expertise can further add to collective analysis creating common multi domain situational and operational awareness. As adversaries increasingly develop credible ASAT capabilities, the armed forces should consider developing counter-space capabilities, such as hardening of satellites, rapid replacement, and electronic countermeasures, to guarantee its forces continuous support from its space based assets. The development of counter-space capabilities is not only critical for our strategic autonomy in space but also reinforces the overall resilience of the Army's operational networks. By cultivating a robust space-based capability, the Indian Army can extend its reach, secure its communications, and enhance its ability to operate in a multi-domain environment where space plays an integral role in intelligence, surveillance, and reconnaissance.

Space has become a decisive theatre in modern warfare, and the Indian Army must be prepared to exploit it for strategic advantage. Space-based reconnaissance is a top priority that must be addressed immediately. Develop and increase the satellite constellations required to provide real-time intelligence, surveillance, and reconnaissance (ISR) capabilities. These satellites must be outfitted with high-resolution imaging, thermal sensing, radar and signals intelligence systems and have the ability to image vast swathes of country such as India's northern borders or maritime zones. In addition to ISR, these satellites provide secure communication and navigation capability, enabling precise targeting and coordination across all combat arms. The Army is using and could exploit a range to engage and acquire threats with extended targeting reach. If the Army is able to conceptualise and implement an effective space strategy, then it can have persistent, all-weather situational awareness of enemy activities across the operational area; ensuring that commanders can take informed decisions based on intelligent data from multiple space based sensors.

Shielding these assets and countering adversary space capabilities require the development of anti-satellite capabilities. India demonstrated anti-satellite (ASAT) technology successfully in 2019. It was a strong start, but further investments are required in order to have a viable offensive and defensive capability. For defending own space based assets, the primary means are hardening the satellites against electronic attacks, or kinetic interceptors whereas deploying ASAT weapons, such as rockets, DEW's and co-orbital systems to destroy the enemy satellite forms the offensive part of counter space capabilities. This dual role will protect India's space infrastructure that is fundamental to communication, navigation, and intelligence, while additionally deterring adversaries who are relying on space to advance their own operations.

As Earth's orbit becomes increasingly congested, it will require advanced space situational awareness to maintain operational continuity of space assets. Advanced tracking systems must be employed to track space debris, hostile satellites, and natural hazards to take preemptive action to avoid collisions that could damage or disable critical infrastructure. A situational awareness capability will ensure Indian satellites remain functional in an increasingly congested orbital environment that involves competing demands. This situational awareness will not only preserve Indian investments in space, but also support broader military planning, such as timing satellite-dependent operations to avoid disruptions, thereby sustaining the Army's reliance on space in contested scenarios. Not directly falling into the Army's mandated role, but by virtue of being a key beneficiary of space based assets, the Indian Army will have to constantly remain in the loop on the functioning, efficiency, health, integrity and security of our space-based assets.

Finally, quantum communication is the next major innovation in secure battlefield command and communications, using quantum mechanics to ensure unhackable ultra secure communication over vast distances across the globe. As opposed to traditional methodologies which rely on expensive encryption technologies, quantum communication creates a secure signal by using entangled particles to communicate information that is impossible to hack and if the adversary attempts to intercept the communication, that breaks the entangled signal and alters it, which sends a detectable alert to the sender. In 2016, China launched a satellite named Micius, which successfully demonstrated quantum communication between cities that were thousands of kilometers apart.[39] India is also working on its own projects in this area and once the technology stabilizes, the Indian Army, will be communicate sensitive orders, intelligence and targeting data even under intense electronic warfare without any threat of interception or jamming. While still in its early stages, investing in quantum research positions the Army to lead in next-generation communication, providing a strategic edge in coordinating dispersed forces across vast and hostile theatres.

With the modernisation of warfare, the importance of space technology in land operations has become increasingly critical. Space is an enabler for the Army, as it provides the communication, intelligence, navigation and targeting functions that underpin contemporary warfare. Space technology is also a force multiplier as new opportunities and threats emerge for land operations with innovations in space-based weapons and anti-space operations. Disruptive technologies that are likely to shape the future of space warfare represent both an opportunity and a challenge.

Space-based capabilities have always been at the core of modern military operations, evidently and clearly demonstrated by historical lessons from the Gulf War to current regional conflicts. The battlefield extends far-off Earth now due to the rapid progress in ASAT systems, directed-energy weapons, hypervelocity platforms enabled by AI-driven control networks.

As countries rebuild their arsenals and reorient doctrines, space becomes a major domain the domination of which will determine global strategic results. Resilient, integrated and compliant operations in this domain are key to deterrence and prevent proliferation of anti-satellite weapons. The future of warfare is set to be defined by our ability to harness disruptive technologies while addressing the complex challenges they bring. The co-evolution of space and land warfare dictates a general military strategy where the Army's reliance on space-based capabilities is fully integrated into operational planning and execution. With the Army preparing for future fights, it must invest in both protecting and leveraging space to secure its ability to fight in a contested, more complex space environment. This chapter underscores the importance of discussing space technology in the same breath as land warfare. The interconnectedness of these domains means that the outcome of every Army operation is, to some extent, dependent on the capabilities provided by space. As new technologies emerge to radically transform the battlespace, the ability for the Army to harness and defend space technology will be a decisive factor in its operational success.

Notes

1   Dr. Jeffrey M. Reilly, Multidomain Operations, A Subtle but Significant Transition in Military Thought, *Air and Space Power Journal* 30, https://www.airuniversity.af.edu/portals/10/aspj/journals/volume-30_issue-1/v-reilly.pdf

2   Ionela Cătălina Manolache, Headquarters Multinational Division South-East Bucharest, Romania, The Role of Multi-Domain Operations in Modern Warfare, *Land Forces Academy Review*, Vol. XXVIII, No. 3(111), 2023 https://sciendo-parsed.s3.eu-central-1.amazonaws.com/64f05101774f3e139fa9ac15/10.2478_raft-2023-0020.pdf

3   Mark Smith, "Anti-satellite weapons: History, types and purpose," August 10, 2022 https://www.space.com/anti-satellite-weapons-asats

4   Paul Glenshaw, "The First Space Ace, F-15 vs. Satellite," April 2018, https://www.smithsonianmag.com/air-space-magazine/first-space-ace-180968349/

5   Juliana Suess, Jamming and Cyber Attacks: How Space is Being Targeted in Ukraine, *RUSI*, 5 April 2022 https://www.rusi.org/explore-our-research/publications/commentary/jamming-and-cyber-attacks-how-space-being-targeted-ukraine

6   Amitav Mallik, High Power Lasers–Directed Energy Weapons, Impact on Defence and Security, Defence Research and Development Organisation, Ministry of Defence, https://drdo.gov.in/drdo/sites/default/files/monographs-documents/28-highpower-lasers.pdf

7   Doris, Lucas. (2025). AI in Space Exploration: Autonomous Systems for Planetary Rovers and Satellite Data Analysis. https://www.researchgate.net/publication/390056252_AI_in_Space_Exploration_Autonomous_Systems_for_Planetary_Rovers_and_Satellite_Data_Analysis

8   Marc Carbone, NASA GRC, Autonomous Power Control Lead, AI-enabled Autonomous

Systems: Space Power Applications, 10-12 March 2024 https://ntrs.nasa.gov/api/citations/20240002420/downloads/IAPG_2024_Final.pdf

9 Larry Greenemeier, "GPS and the World's First "Space War", Scientific American," February 8, 2016, https://www.scientificamerican.com/article/gps-and-the-world-s-first-space-war/

10 Missile Defense Project, "Tomahawk," Missile Threat, Center for Strategic and International Studies, September 19, 2016, last modified April 23, 2024, https://missilethreat.csis.org/missile/tomahawk/.

11 Hülya Kinik, Sinem Çelik, The Role of Turkish Drones in Azerbaijan's Increasing Military Effectiveness: An Assessment of the Second Nagorno-Karabakh War, Insight Turkey Fall 2021, December 14, 2021 https://www.insightturkey.com/articles/the-role-of-turkish-drones-in-azerbaijans-increasing-military-effectiveness-an-assessment-of-the-second-nagorno-karabakh-war

12 Gregory Gagnon, Why Military Space Matters, *Joint Force Quarterly* 110, July 7, 2023 https://ndupress.ndu.edu/Media/News/News-Article-View/Article/3450009/why-military-space-matters/#:~:text=Over%20the%20past%20two%2Dplus,the%20challenge%20of%20the%20past.

13 Iron Dome Air Defence Missile System, Army Technology, October 20 2023 https://www.army-technology.com/projects/iron-dome/?cf-view

14 Kingston Reif, Missile Defense Systems at a Glance, Arms Control Association, August 2019, https://www.armscontrol.org/factsheets/missile-defense-systems-glance#:~:text=Satellite%20Sensors%20and%20Ground%2D%20or,it%20and%20eliminate%20the%20threat).

15 Marko Höyhtyä and Sari Uusipaavalniemi, The space domain and the Russo-Ukrainian war: Actors, tools, and impact, The European Centre of Excellence for Countering Hybrid Threats, January 2023, https://www.hybridcoe.fi/wp-content/uploads/2023/01/20230109-Hybrid-CoE-Working-Paper-21-Space-and-the-Ukraine-war-WEB.pdf

16 Tim Zadorozhnyy, Ukraine receives 5,000 more Starlink terminals from Poland, minister says, The Kiev Independent, April 3, 2025 https://kyivindependent.com/ukraine-receives-5-000-more-starlink-terminals-from-poland-minister-says/

17 Thomas Withington, Ukraine's Favourite Dish, European Security & Defence, 30 May 2023 https://euro-sd.com/2023/05/articles/30035/ukraines-favourite-dish/

18 Greg Hadley, Saltzman: China's ASAT Test Was 'Pivot Point' in Space Operations, Jan 13, 2023 https://www.airandspaceforces.com/saltzman-chinas-asat-test-was-pivot-point-in-space-operations/

19 Anti-Satellite Missile, DRDO, ASAT Ebook, Ministry of Defence, 2020 https://www.drdo.gov.in/drdo/sites/default/files/inline-files/ASAT_book_English.pdf

20 Lisa Sodders and Brad Smith, Space Systems Command Public Affairs Focused on the Threat: Directed Energy Weapons (Part 3 of 6), Sept. 20, 2024 https://www.ssc.spaceforce.mil/Newsroom/Article-Display/Article/3913339/focused-on-the-threat-directed-energy-weapons-part-3-of-6

21 Lisa Sodders and Brad Smith, Space Systems Command Public Affairs Focused on the Threat: Directed Energy Weapons (Part 3 of 6), Sept. 20, 2024 https://www.ssc.spaceforce.mil/Newsroom/Article-Display/Article/3913339/focused-on-the-threat-directed-energy-weapons-part-3-of-6

22 Case Study Viasat, Cyber Peace Institute, June 2022 https://cyberconflicts.cyberpeaceinstitute.org/law-and-policy/cases/viasat

23 Outer Space Treaty, United Nations Office for Outer Space Affairs, https://www.unoosa.org/oosa/en/ourwork/spacelaw/treaties/outerspacetreaty.html

24 "Orbital Debris Quarterly News", Vol 14, Issue 4, October 2010, NASA Orbital Debris Program Office, http://www.orbitaldebris.jsc.nasa.gov/newsletter/newsletter.html

25 Sankaran, J. (2022). Russia's anti-satellite weapons: A hedging and offsetting strategy to deter Western aerospace forces. *Contemporary Security Policy*, 43(3), 436–463. https://doi.org/10.1080/

13523260.2022.2090070

26 Secretary of the Air Force Public Affairs, X-37B Orbital Test Vehicle concludes seventh successful mission, US Space Force, March 13, 2025, https://www.spaceforce.mil/News/Article-Display/Article/4112259/x-37b-orbital-test-vehicle-concludes-seventh-successful-mission

27 Joseph Trevithick, "Critical to Defending U.S. Citizens: Space Force Boss," Mar 20, 2025 https://www.twz.com/space/putting-missile-interceptors-in-space-critical-to-defending-u-s-citizens-space-force-boss

28 PAROS Treaty, Proposed Prevention of an Arms Race in Space (PAROS) Treaty, Nuclear Threat Initiative, https://www.nti.org/education-center/treaties-and-regimes/proposed-prevention-arms-race-space-paros-treaty/#:~:text=PAROS%20Treaty,Background

29 Victoria Samson, "Russian Coorbital Antisatellite Testing, Secure World Foundation," December 2024 https://swfound.org/media/207997/fs24-05_russian-co-orbital-anti-satellite-testing.pdf

30 Dylan Malyasov, "New details emerge on Russia's secret laser weapon system," Aug 3, 2024 https://defence-blog.com/new-details-emerge-on-russias-secret-laser-weapon-system/

31 Mike Wall, "US and UK militaries pick Rocket Lab's HASTE launcher to help test hypersonic tech," Apr 16, 2025, https://www.space.com/space-exploration/tech/us-and-uk-militaries-pick-rocket-labs-haste-launcher-to-help-test-hypersonic-tech

32 Insight, "The core components of space situational awareness," https://www.riskaware.co.uk/insight/spaceaware/the-core-components-of-space-situational-awareness/#:~:text=Monitoring%20and%20tracking%20space%20objects&text=For%20instance%2C%20the%20 U.S.%20Space,evasive%20actions%20by%20satellite%20operators.&text= Monitoring%20also%20extends%20to%20understanding%20the%20status%20of%20these%20objects.

33 Vice Admiral Brian Brown, U.S. Navy (Retired), "The Challenge of Joint Space Operations," January 2024, https://www.usni.org/magazines/proceedings/2024/january/challenge-joint-space-operations

34 Deepak Bhatt, "The New Era of Space Exploration and Commercial Space Technology," Nov 1, 2024, https://www.globaltechnologyreview.com/post/the-new-era-of-space-exploration-and-commercial-space-technology#:~:text=Commercial%20space%20technology%20is%20pushing,to%20unlock %20the %20final%20frontier.

35 Tereza Pultarova, "Starlink satellites: Facts, tracking and impact on astronomy," March 28, 2025, https://www.space.com/spacex-starlink-satellites.html

36 Talia M. Blatt, "Anti-Satellite Weapons and the Emerging Space Arms Race," *Harvard International Review*, May 26, 2020, https://hir.harvard.edu/anti-satellite-weapons-and-the-emerging-space-arms-race/

37 Michelle Starr, "Earth's Space Debris Problem is Getting Worse, and There's an Explosive Component SPACE," October 13, 2020, https://www.sciencealert.com/the-space-debris-problem-is-getting-worse-not-better

38 Mike Wall, "Kessler Syndrome and the space debris problem," July 15, 2022 https://www.space.com/kessler-syndrome-space-debris

39 Harun Siljak, Down to Earth. "China's Quantum Satellite Enables First Totally Secure Long-Range Messages," Jun 18, 2020. https://www.downtoearth.org.in/science-technology/china-s-quantum-satellite-enables-first-totally-secure-long-range-messages-71831.

15

Tryst with Technology: Indian Army's Path to Disruptive Innovation

"Transformation of Indian armed forces has become a prerequisite to stay relevant in the fast changing geopolitical environment."

—General Bipin Rawat

As the Indian Army travels through the rapidly progressing landscape of multi-domain war, it must adapt its methods and skills to exploit the potential of emerging disruptive innovations. This concluding chapter aims to provide a set of reasoned recommendations based on detailed analysis of armoured warfare, infantry modernisation, artillery precision, progressive aviation, air defence and space-based capabilities. These strategies are designed not only to combat current threats but also to provide the Indian Army with a sustained competitive edge throughout the spectrum of future disputes. In addition, to successfully integrate such progress into regular procedures, a resilient system of technological absorption, including personnel, education, doctrine development, and organisational restructuring, are important. The rapid adaptation of disruptive tools has brought about a new paradigm in warfare, demanding that the Indian Army evolves swiftly and decisively.

As we seek to transform our forces, one also needs to have a look at the complete ecosystem of defence R&D, production and in service absorption of these technologies through the prism of indigenous capacities such as funding, technical skill, scientific talent pool, strategic thought and national will. The current concluding chapter, drawing on the analysis of the preceding chapters, outlines a series of recommendations for each of the combat arms and for the broader institutional framework. These recommendations seek to ensure that the Indian Army is well equipped to deal with emerging threats, exploit technological advantages, and maintain active superiority in future multi-domain wars.

Mechanised Forces

The Mechanised Forces lend themselves for rapid deployment and heavy volume of firepower engagements, ideal for offensive and defensive operations over vast spaces. Currently, the Indian Army relies on aging BMP-2 vehicles, which have been in service since the 1980s.[1] The Futuristic Infantry Combat Vehicle (FICV) project, approved under the 'Make in India' initiative, aims to induct 1,750 tracked vehicles to replace these, with recent RFIs issued in June 2021 for development.[2] The FICV is a future combat system expected to be amphibious, equipped with a manned turret, fire-and-forget top-attack anti-tank guided missiles (ATGMs), at least 30 mm automatic cannon, co-axial machine gun, and a stabilized remote control weapon station (RCWS) with a 12.7 mm machine gun. Certain steps which need to be taken in order to provide impetus to the ongoing transformation of the Armoured and Mechanised forces are:

Modern and Enhanced Production Capacities: Recently the overhaul of the T 72 tanks has started at the Vehicle Factory Jabalpur (VFJ), a key unit of the Armoured Vehicles Nigam Ltd (AVNL).[3] The overhaul involves complete refurbishment of the engine, fire control systems, armour protection with upgrades for night vision, thermal imaging and modern electronics. However, with one unit getting overhauled in 6 to 7 months, it would take almost 5 years to overhaul the complete inventory of T 72.[4] With every alternate year, India getting embroiled into a conflict with either of the adversaries, it is imperative to step up our production capacities and modernize our processes.

Expedite FICV Procurement: Accelerate the development and induction of FICVs, ensuring they incorporate AI for autonomous operations, cyber protection for vehicle control systems and advanced communication systems for better coordination within the combat team /group. The project, under the 'Atmanirbhar Bharat' program, should leverage private sector partnerships with Indian companies in defence sector like Larsen & Toubro, which has selected Allison Transmission's propulsion solution for prototypes.[5]

Light Armoured Vehicles and Tanks: Given the vast expanses at our northern borders, it may be prudent to induct lighter version of tanks and AFVs to exploit the advantages of manoeuvre warfare at high altitude battlefields. This however, must be done without compromising on the other two critical factors of fire power and protection, further fortified with emerging technologies such as AI, Active Armour Protection System, loitering munitions and integrated drones. Once PLA fielded the light tanks and deployed them in Doklam in 2017, the Indian Army projected the requirement for 364 light tanks on fast-track basis, 59 to be manufactured by DRDO and the rest 295 to be manufactured under the Make-I category.[6] Consequently, within three years, DRDO rolled-out Armoured Fighting

Vehicle-Indian Light Tank (AFV-ILT) called Zorawar, which under trail as of now, with likely delivery to Indian Army by 2027.[7]

Efficiency and Stealth: India like South Korea must embark on equipping future tanks with alternate power trains such as the K3 Tank which utilizes hydrogen fuel cells, advanced electric motors, and rechargeable batteries, providing an environmentally friendly alternative to diesel engines.[8] This innovative system not only reduces emissions but also allows the tank to operate in near silence, greatly enhancing its stealth capabilities, providing an essential strategic advantage in modern warfare.

Integrate Unmanned Systems: Use drones and unmanned land vehicles for reconnaissance and support missions, acquiring situational awareness, and reducing vulnerability to force. The army's focus on disruptive tools must manifest in large scale procurement of multi capacity drones and UGVs for enhance battle field intelligence and domination.

Enhance Communication Systems: Improve communication systems by retrofitting advanced communication systems, such as Software Defined Radios (SDRs), to ensure seamless integration and real-time information sharing with other arms, thus supporting fast paced network centric operations.

Integration of Advanced Sensors and Networking: Future armoured vehicles should be equipped with advanced sensors & detectors, including thermal, infrared and LIDAR systems integrated into a real-time command, control, communications and intelligence (C3I) network. This will provide enhanced situational awareness and facilitate coordinated manoeuvres on the battlefield.

Incorporation of Unmanned Systems: The deployment of unmanned ground vehicles (UGVs) as reconnaissance platforms or even as unmanned turret systems can reduce the risks to own personnel, ensure relentless operations and increase redundancies by allowing the capability to operate in intense hostile and dirty environments.

Enhanced Survivability and Counter-EW Measures: Given the prevalence of electronic warfare and cyber threats, armoured platforms should incorporate latest hardened communication systems and electronic counter-countermeasures (ECCM) to ensure they remain effective in contested electromagnetic environments. The vintage ECCM suites may not withstand the integrated onslaught of modern day electronic and cyber warfare tools being used in tandem by the enemy forces.

Upgraded Firepower and Precision Weapons: Integration of smart munitions and advanced targeting systems, leveraging artificial intelligence for real-time decision support, will improve accuracy, response time and lethality against modern, agile adversaries. Systems such as the US Advanced Targeting and Lethality Aided System (ATLAS) on M1 Abrams can be developed for the new generation of tanks and AFVs[9] and the existing fleet can be retrofitted with systems akin to the 'Ground Warden' which drastically hastens the targeting cycle.[10]

Enhanced Armour and Protection: Invest in advanced materials, such as nanotechnology-based or reactive armour, and stealth technologies to combat emerging hazards such as hypersonic projectiles and smart munitions.

Cybersecurity Measures: Equip armoured vehicles with robust cybersecurity protocols to prevent hacking and protect control systems from cyber threats. Train troops to revert to basic in case of down time of electronic and autonomous systems due to cyber-attacks and try to survive in denied environments.

Energy Weapons: Explore directed energy weapons for point defence against drones, missiles, and other threats. DRDO has already made certain breakthroughs in DEWs and it goes without saying that certain versions for the mechanised forces are an absolute imperative.

Infantry

The Infantry is the backbone of the Army, continuously employed in counter-terrorist and counter-insurgency operations, requiring modern equipment for enhanced combat effectiveness. Current modernization efforts include procuring new generation lightweight assault rifles, bulletproof jackets, helmets, hand-held thermal imagers. The Futuristic Infantry Soldier As a System (F-INSAS) program aims to equip soldiers with modular weapon systems, integrated communication, and health monitoring, with bulletproof helmets featuring flashlight, thermal sensors, and night vision.[11]

Modernize Personal Equipment: Provide lightweight assault rifles, lighter protective equipment, and light-enhanced night vision devices to help increase lethality and survivability of soldiers, while ensuring indigenization to mitigate import dependency. Better ergonomic gear, with communication kit based on latest radio sets must be given as personal life time issue to every infantryman.

Implement F-INSAS: Full implementation of the F-INSAS or an equivalent program so that soldiers can be provided integrated systems for communicating, navigating, and including sensors to monitor health parameters and provide quick medical relief, enhancing operational readiness.

Increase Specialist Roles: Convert more and more infantry battalions into Special Forces or Special Operations forces, in order to maintain adequate punch across the diverse, complicated, and challenging threat landscape we are facing without increasing the number of general duty troops but creating more specialists in every unit.

Wearable Technologies and Augmented Reality: Each soldier should be equipped with wearable sensor arrays, and augmented reality (AR) headsets, to provide them real time operational intelligence, terrain data and threat overlays. Once it becomes an integral part of their training, soldiers will be able to leverage the same equipment in field and benefit from increased situational awareness, assisting better decision making.

Integrating Unmanned Ground Systems and Unmanned Aerial Systems: Though ever increasing numbers of lightweight unmanned systems, for example micro-UAVs for information gathering, unmanned ground systems for logistics, robots for mapping minefields are available off the shelf and are being inducted also but the Army needs to approach this technology as separate domain altogether. The invasive pace at which this technology is transforming tactical operations, requires us to understand, drive and control the entire ecosystem and life cycle of these platforms.

Improved communications and cyber capabilities: Strong, secure communications networks that are jamming and cyber-resilient will be necessary. Soldiers will need to be trained for an environment where information is processed quickly by AI systems, they must learn to leverage that information and operate with minimum disruption in case such information is denied to them by the adversary.

Modernized Infantry weapons: Emerging technologies will in no time become common. Therefore, our procurement and equipping cycles must be able to keep pace with the technology and outpace our adversaries. This underscores the need to keep our soldiers always equipped with next-generation personal weapons, smart munitions, and precision targeting so that infantry engages evolving threats with speed and efficiency.

Smart weapons: Allow precision targeting via AI-assisted targeting systems incorporated in smart munitions to include non-lethal weapons minimizing collateral damage. Also keep working on putting a DEW into an infantryman's hands, which may be a game changer in a highly digitized battlefield.

Biotechnology interventions: To improve soldier performance in extreme conditions, explore biotechnologies directly modifying human capabilities on the battlefield.

Artillery

Artillery modernization is critical for providing fire support in various terrains, especially given tensions with China along the Line of Actual Control (LAC). The Field Artillery Rationalization Plan (FARP), proposed in 2000, aims to procure around 3,000 to 3,600 pieces of artillery, primarily 155mm guns, including towed, mounted, self-propelled, and ultra-light howitzers.[12] Recent inductions include 145 M777 Ultra-Light Howitzers, K-9 Vajra-T self-propelled guns, and indigenous developments like Dhanush and ATAGS, with a focus on indigenization.[13]

Accelerate Howitzer Procurement: Timely modernization of existing howitzer systems should be ensured, and modifications made with specific emphasis on indigenous production avenues with through DRDO and private sector collaborations with companies such as L&T and Bharat Forge.[14]

Move to smart munitions: The investment in smart or precision munitions systems will enable better accuracy and reduce collateral damage, increasing efficiency in the execution of precision strikes, especially in operations in mountainous terrain.

Embrace artificial intelligence and automation: AI can better assist target acquisition, fire control, predictive maintenance and overall operational utility. The Army Technology Board had started working with IITs and IISc towards faster absorption of advanced AI-enabled systems, however, we need to hasten their induction as the adversary has been upgrading his systems at a staggering pace.[15]

Enhance Surveillance: Use sophisticated set of surveillance radars such as Swathi Weapon Locating Radar and Unmanned Aerial Vehicles as part of combat ISTAR (intelligence, surveillance, target acquisition, reconnaissance) framework to maximize the possibility of target acquisition and battlefield awareness, while ensuring modernization of Artillery's Surveillance and Target Acquisition Regiments.[16]

Digital Fire Control and Targeting: Leveraging AI and machine learning for fire control systems will allow artillery units to process battlefield data swiftly, adjust fire plans in real time, and engage targets with greater precision.

Integration with Sensor Networks: Artillery batteries need to be connected with aerial, ground, and space-based sensor networks across all services and agencies, to create a single operational picture enabling rapid recalibration of fire missions.

Long-Range Precision Munitions: Investment in advanced, precision-guided munitions will extend the reach and effectiveness of artillery, ensuring that critical targets can be neutralized before adversaries can mount a counter-offensive. Increase reliance on GPS and laser-guided artillery shells to achieve higher accuracy and effectiveness in strikes. Develop or procure hypersonic artillery systems to extend range and reduce response times on the battlefield.

Enhanced Mobility and Survivability: Mobile, self-propelled artillery systems with advanced counter-battery capabilities will improve responsiveness and reduce vulnerability to enemy reconnaissance and strikes. Matching mobility is of paramount importance in deserts and plains where artillery has to provide support to armoured and mechanized columns.

Automated and Autonomous Systems: Implement robotic systems for faster, safer, and more efficient ammunition handling. At the same time autonomous light guns may be of operational advantage in the mountains and along the LC/ LAC where moving and operating artillery pieces across narrow rugged valleys, treacherous slopes and glaciated peaks demands superhuman will and physical capabilities.

Counter-Battery Radar: Enhance detection and neutralization of enemy artillery with advanced radar systems for improved survivability of own artillery units.

Army Aviation

India's Army Aviation Corps provides crucial support through helicopters and UAVs, essential for reconnaissance, observation and logistical support in remote areas like the northern and eastern borders. The Corps has been ALH, Chetak and Cheetah helicopters, with plans to replace them with 190 Light Utility Helicopters (LUH).[17] In the recent past, acquisition of modern platforms such as the Apache and Chinook has been undertaken by the Corps to bolsters its capability but in limited numbers.

Replace Aging Fleet: Accelerate the induction of LUHs and LCHs to improve operational capability, while being cognizant that the Cheetah and Chetak will begin their phase out process in 2027 with 50+ units exceeding their Total Technical Life (TTL) to retire in that period.[18]

Expanding Indian Army Aviation's Capabilities: A narrow focus on traditional roles may limit the AAC's flexibility and its ability to respond to emerging challenges on the battlefield. Additionally, the reliance on other branches for critical capabilities such as attack helicopters and UAVs could create operational dependencies that may hinder the AAC's effectiveness in high-intensity conflicts. There the Aviation Corps must work towards expanding organic capabilities and assume responsibility for all UAV, attack helicopters and light and medium lift utility helicopters while continuing to rely on the IAF for fixed-wing assets. This approach would allow the AAC to enhance its operational effectiveness without overextending its resources.

Invest in UAVs: Procure and develop UAVs for surveillance/reconnaissance/combat roles consistent with Army's emphasis on multi-capacity drones for battlefield intelligence and close support.

Integrate Advanced Technologies: Implement AI for flight control, mission planning, and maintenance to improve efficiency and safety, leveraging the private industry with oversight from the AI Centre for Excellence at MCTE for research.

Acquire Medium and Heavy Lift Helicopters: Medium and Heavy lift helicopters need to be procured in sufficient numbers, addressing the tactical lift and logistic deficiencies and enhance operational flexibility.

Modernization of Rotorcraft and UAVs: The helicopters will need upgrades with highly advanced avionics, sensors, and communication and EW suites. Other important step to bolster the overall capability of the corps will be induction of autonomous and semi-autonomous unmanned aerial vehicles, which can provide continuous surveillance and rapid strike capabilities.

Digital Cockpit and Cross Domain Networking: The future aircraft should incorporate a digital cockpit that encompasses Augmented Reality displays and AI-based, decision support systems that allows pilots to seamlessly monitor, prioritize and make decisions from a sea of complex data being fed from multiple platforms in real time.

Improved Maintenance and Sustainment: AI assisted predictive maintenance through data analytics and IoT sensors will enhance readiness and minimize downtime during high-tempo operations.

Interoperability: Army aviation capabilities must seamlessly mesh into the broader joint all-domain command and control (JADC2) architecture, ensuring that aerial assets contribute effectively to the overall battlefield picture.

Vertical Lift: Invest into next generation vertical lift capabilities that can fly faster, longer, and carry more payload.

Air Defence

Army Air Defence (AAD) is responsible for protecting against aerial threats, particularly below 5,000 feet, with a history of active operations dating back to 1940.[19] Modernization has stagnated post-1996, but recent efforts to build a comprehensive air defence picture include Project Akashteer, a networking and automation project worth nearly Rs 2,000 crore, was expected to be completed by March 2024.[20] Current systems include Akash SAM, S-125 Pechora and Spyder Air Defence System, with a focus on countering drone threats, especially given lessons from the Ukraine conflict.[21]

Implement Project Akashteer: Complete the networking and automation project to create a comprehensive air defence picture, similar to the IAF's Integrated Air Command and Control System, enhancing coordination and response.

Procurement of Advanced Systems: Induction of lighter radars and mobile weapon systems for use in mountainous terrain. This is particularly important due to the recent shift in focus from our western borders to northern borders, with speedy induction of indigenous systems like Akash SAM leading the way.

Development of Counter-Drone Capabilities: Procurement of technologies to detect and neutralize UAV threats is expedient given the attractiveness of their use across varied operational scenarios, leveraging insights from global experiences.

Embracing AI: Induct AI enabled systems for real-time threat assessment and decision-making to shorten response time and efficacy while ensuring their integration with other branches of the military to create a layered defence.

Multi-Layered Defence Systems: Ground-based air defence systems should be integrated into a multi-layered grid that combines guns, short-range missile systems, directed energy weapons and electronic warfare capabilities to create overlapping fields of protection.

Advanced Radars and Sensor Fusion: Use of advanced radar systems with active electronically scanned arrays (AESAs) and multi-sensor fusion will improve target detection and tracking, even in cluttered or jamming environments.

AI-Enhanced Threat Analysis: The integration of artificial intelligence and machine learning in air defence networks will assist with improved threat assessment and identification enabling quicker decision making and appropriate response times to incoming missiles, drones and hypersonic threats.

Mobility and Rapid Deployment: Air defence assets must be highly mobile, with systems that can be quickly repositioned to respond to evolving threat trajectories, ensuring that gaps in the protective shield are minimized.

Hypersonic Missile Defence: Invest in AI-based countermeasures that proactively predict and intercept hypersonic threats.

Directed Energy Weapons: Integrate laser systems on aircraft, tanks, ICVs and vehicles for missile defence and countering aerial threats.

Cyber and Electronic Warfare: Develop capabilities to disrupt enemy air defence suppression through cyber and electronic warfare.

Space

Space capabilities are essential for communication, navigation, and intelligence, with the Indian Army utilizing satellites like GSAT-7A for communication and Cartosat series for reconnaissance.[22] The Defence Space Agency (DSA), established in 2018, oversees space-warfare and satellite intelligence assets, consolidating capabilities from all three services. India has launched several military satellites since 2014, including EMISAT for electronic intelligence and RISAT series for surveillance, with ISRO supporting these efforts.[23] Though a lot needs to be done as far as developing military applications of space based assets is concerned and most of it doesn't even lie in the gambit of the Armed Forces, but certain recommendations where the armed forces and the army in particular needs to contribute are in succeeding paras.

Optimizing the Use of Satellites: Maximise satellite applications, communication, navigation, and satellite imagery, for operational planning and execution, making use of satellite capabilities such as GSAT-7A for secure, real-time data transfer and Cartosat for near-real time satellite imagery.[24]

Training of Personnel: The Indian Army needs to educate its personnel on the impact, utility and future development in the space domain, so that there is no inhibition in leveraging present and future systems and platforms which will leverage the spaced based assets in an all-encompassing manner.

Dedicated Military Satellites: Work with ISRO to launch more satellites that meet military requirements, such as dedicated reconnaissance satellites, or communication satellites to give greater operational autonomy.

Investing in Space Based Surveillance and Communication: The Indian Army should prioritize the development and acquisition of space-based sensors for early warning, ISR (intelligence, surveillance, and reconnaissance), and secure communications.

Integration with Multi-Domain Operations: Space assets must be fully integrated into the broader air defense and Joint Operational Networks to contribute to real-time data fusion and enhance overall situational awareness.

Defending Against Space Threats: As adversaries are increasingly challenging space systems, resilient counter-space options, such as hardening of satellites and concepts for rapid replacement, are necessary.

Encouraging Indigenous Capabilities for Space: Though Indian Space Program is highly indigenous but enhancing domestic research and development in military applications will lessen reliance on foreign systems and preserve some measure of strategic autonomy in this important area.

Developing Anti-Satellite Capabilities: Develop technologies to protect Indian space capabilities and, where necessary, to neutralize adversary satellites.

Exploring Quantum Communication: Investigate potential of quantum technology for secure, instantaneous communication to support battlefield operations.

ABSORPTION OF EMERGING TECHNOLOGIES

The Indian Army must carefully absorb, develop and apply, among other things, the new emerging technologies in the current era of warfare, such as Electronic Warfare, Cyber Warfare, Artificial Intelligence and Quantum Technologies and realistically, incorporate them into operations in order to help India realize its vision of being a superpower like the USA and China. These technologies are pivotal in modern warfare, offering critical advantages in intelligence, communication, and operational effectiveness. The USA, for instance, has initiatives like the Third Offset Strategy to exploit these technologies, while China leads in state-driven AI and quantum research.[25] For India, maintaining a technological edge is essential for regional security and global standing, particularly given threats from neighbors and the need to assert itself as a "Net Security Provider." The Indian Army, as a key component of this effort, must adapt to these changes to ensure operational superiority and strategic autonomy.

Electronic Warfare

EW involves the use of the electromagnetic spectrum for sensing, protecting, and communicating, as well as disrupting enemy operations. The Indian Army currently employs systems like Samyukta, a mobile integrated EW system developed jointly by DRDO, Bharat Electronics Limited (BEL), Electronics Corporation of India Limited (ECIL), and the Corps of Signals. Samyukta operates on 145 ground mobile vehicles, covering an area of 150 km by 70 km, with capabilities for surveillance, analysis, interception, direction finding, and jamming of communication and radar signals from HF to MMW.[26] Additionally, the Himshakti EWS, developed by BEL, is designed for mountainous regions and can jam frequencies over 10,000 square kilometers, ideal for mechanized forces.[27] However, reports indicate that the Army's approach to EW is not as evolved as China's People's Liberation Army (PLA), with fragmented command structures and a need for better integration.

Recommendations for Electronic Warfare

Upgrade Existing Systems: Continuously modernize systems like Samyukta and Himshakti to counter evolving threats, focusing on cognitive EW and AI integration for adaptive jamming.

Invest in R&D: Allocate resources for developing next-generation EW technologies, addressing gaps identified in comparisons with China's PLA, such as advanced signal intelligence and countermeasures.

Training and Skill Development: Enhance training programs at institutions like MCTE to ensure personnel can operate and maintain advanced EW systems, emphasizing man-portable and terrain-specific solutions.

Inter-Service Collaboration: Foster collaboration with the Navy and Air Force to share knowledge and resources, ensuring a unified EW doctrine and capability across services.

Cyber Warfare

Cyber Warfare encompasses actions to attack or defend information systems, crucial for modern network-centric operations. The Indian Government established the the National Cyber Coordination Centre (NCCC) in 2014 and Defence Cyber Agency in 2021, indicating growing awareness of these threats.[28] However, efforts are still evolving, with a need for coherent strategies and offensive capabilities. The Army due to its vast manpower, equipment profile and dispersed locations is investing considerable resources in cyber technology adoption, with initiatives like the Military College of Telecommunication Engineering (MCTE) training personnel in cyber warfare. Challenges include legacy voice-centric communications and the need for better integration into military strategy.

Recommendations for Cyber Warfare

Develop Offensive Capabilities: Build capabilities to conduct cyber operations that can disrupt adversary networks, focusing on industrial espionage and softening defences, particularly against China and Pakistan.

Strengthen Cybersecurity: Implement robust measures to protect critical infrastructure and military networks, utilizing the Defence Cyber Agency and NCCC, and addressing fragmentation under the 2008 IT Amendment Act.[29]

Public-Private Partnership: For both cyber defence and cyber offence consider participation of key stakeholder in the private sector for expertise and technology, with the caveat that only domestic companies will execute and create these cyber defence and defence solutions.

International Cooperation: Seek partnerships with countries that are likely to share intelligence and best practices. Consider relations with the US, Israel, and Japan in order to enhance ISR capabilities and migrate towards a more aggressive cyber strategy.

Artificial Intelligence

AI offers transformative applications in military operations, from autonomous weapons to decision-making support. On the recommendation of a task force set up by DDP under the Chairmanship of Shri N Chandrasekaran, Chairman Tata Sons, and in consultation with all stakeholders, Defence Artificial Intelligence Council (DAIC) has been set up under the Chairmanship of Raksha Mantri to provide necessary guidance and structural support. Further, Defence AI Project

Agency (DAIPA) has been created under the Chairmanship of Secretary DDP for enabling AI based processes in defence Organisations.[30] The Indian Army in the early stages of AI integration, with the Army Technology Board (ATB) collaborating with IITs, IISc, and other R&D organizations. The AI Centre for Excellence at MCTE in Mhow, Madhya Pradesh, focuses on areas like situational awareness, sensor fusion, faster decision-making, and autonomous weapons systems. The Indian Army in collaboration with Bharat Electronics Limited (BEL) has launched an AI Incubation Centre (IAAIIC) in Bengaluru. This initiative aims to accelerate the development and adoption of AI technologies within the Indian Army. DRDO has developed AI-enabled systems for the Army, detailed in reports, but there is little success in incubating AI-based research internally. Challenges include a deficit in data science talent and reliance on external professionals, with the Corps of Signals and Directorate General of Information System (DGIS) needing specialist officers.

Recommendations for Artificial Intelligence

AI Task Force: Since AI has become an enabler across various arms and services and there will be a rush for inducting systems into the Army, an apex level AI Task Force must be created which can coordinate and synergise the efforts of various stake holders and procurement verticals. This is of utmost importance and must be done for other disruptive technologies too, in order to have effective control over the entire life cycle of niche equipment which may come in to the army over the years.

Increase Investment: Allocate more funds for AI research and development, with a budget like the Defence AI Centre (DAIC) at INR 1,000 crore annually. Further, an AI roadmap has also been finalised for each DPSU under which 61 defence specific AI projects have been identified for development. Out of these 61 projects, 26 projects have been completed by the DPSUs.[31]

Talent Acquisition and Retention: Attract and retain data science and AI talent within the Army, potentially through specialist officer programs and civil-military infusion, addressing the deficit in organic talent.

Collaborate with Academia and Industry: Partner with universities like IITs and IISc, and tech companies like IBM, Google, Infosys, and TCS, to leverage cutting-edge AI technologies for military applications.

Ethical and Legal Frameworks: Develop policies and guidelines for the ethical use of AI, addressing the lack of specific laws and ensuring alignment with democratic values, as seen in NATO strategies.

Quantum Technologies

Quantum technologies, including quantum computing, communication, and cryptography, promise revolutionary capabilities in secure communication and computational power. The Indian Army has established a quantum laboratory at MCTE, supported by the National Security Council Secretariat, focusing on Quantum Key Distribution, Quantum Communication, Quantum Computing, and Post-Quantum Cryptography.[32] Efforts are nascent, with evaluations of impact on military operations premature, but steady progress is being made to address deficits.

Recommendations for Quantum Technologies

Accelerate Research: Invest in quantum research to develop practical applications for military use, focusing on quantum key distribution and communication to enhance secure operations.

Build Expertise: Train personnel in quantum science and technology, leveraging the quantum laboratory at MCTE and collaborating with national research institutions.

Engage with global scientific community: Work with international quantum research organizations to remain abreast with latest developments and innovations, to ensure the Army benefits from the advances in the field globally.

Quantum-Safe Cryptography: Preparing for quantum computing includes developing quantum-resistant encryption methods, so the military can mitigate risks to present cryptographic systems.

TECHNOLOGICAL ABSORPTION: HR, TRAINING, DOCTRINE, STRUCTURE AND ORGANIZATION

The point that whether technology drives warfare or the other way round has been discussed in the very first chapter. Similar is the case with modernisation and technological absorption in an organisation. Till now the Indian Armed Forces have been importing sophisticated weapons and systems, which largely channelised our energies towards Technology led transformation. However, over the years, since our Defence Industrial Complex has developed indigenous capacities to produce world class systems and India now is exporting military equipment to friendly countries. As our indigenous capacities grow further fuelled by healthy economic growth rates, the time has come to adopt the Tactics-led approach to transformation, emphasizing process redesign, workforce training and cultural adaptation, with technology selected based on its ability to meet tactical goals keeping the combatants as focus and not the technology.[33]

The transformative potential of emerging disruptive technologies will only be effectively and positively leveraged if the Indian Army possesses the ability to absorb and institutionalize them. This requires realigning of our human resource management, institutional training, doctrinal guidance and organizational structures and processes. The focus must be on recruiting and retaining technical talent in areas such as artificial intelligence, cyber warfare, sensors integration and advanced engineering with career paths and incentive programs tailored to both military and civilian to attract and retain the best talent. At the same time, it is imperative to modernise the training curriculum by incorporating immersive simulation-based experience modules replicating the pressures and challenges of multi-domain operations. Technologies such as AR and VR can enable realistic, scenario-based training that prepares soldiers to effectively exploit advanced systems on the battlefield. Doctrines also need to be updated in-line with network-centric warfare, with a focus on agility, fast-paced decision-making and multi-domain integration.

The doctrinal reform must be matched by necessary organizational transformation that fosters decentralization and empowers units to make real-time decisions based on comprehensive, shared intelligence. Joint technology integration organizations that span combat arms will assist with integration between research and development projects all the way to operational field units. The creation of technology integration units that span across combat arms will be able to facilitate seamless communication between research and development, field operations and strategic planning, ensuring that technological innovations are rapidly and effectively integrated into operational frameworks. Ultimately, a flexible and adaptive organisational structure, one that embraces continuous learning and iterative improvement, will be essential for the Indian Army to remain at the forefront of modern warfare.

HR

The backbone of technological adoption lies in the Army's personnel, necessitating a strategic overhaul of its human resources approach. STEM recruitment must be a top priority, focusing on attracting and retaining individuals with expertise in Science, Technology, Engineering, and Mathematics (STEM) as is being done through the TES entry scheme.[34] These professionals whether engineers, data scientists, or cybersecurity experts bring the technical acumen needed to operate, maintain and innovate with advanced systems like AI-driven UGVs, quantum communication networks, or hypersonic weaponry. Recruitment campaigns should target universities and tech hubs, offering incentives like specialised career paths or advanced training opportunities, while retention efforts must include competitive pay, continuous education, and assured promotions to a certain level in the hierarchy, commensurate to growth prospects of similar talent in the civil technological ecosystem. By building a tech-savvy workforce, the Army ensures it has the human capital to exploit disruptive technologies effectively, bridging the gap between concept and battlefield application.

In addition to this, I believe the creation of specialized tech units is critical to support innovation and absorption of emerging technologies. Such units are designed to focus on artificial intelligence, cyber warfare, space operations, or such fields to serve as incubators for new technology, testing systems like autonomous drones or electronic jamming tools in real-world conditions. Having a mixed team of army men and civilian experts will allow these units to collaborate on creating advanced combat capabilities and validate their operational utility, keeping the Army well ahead of its potential competitors. As an example, a cyber-warfare unit could assist in improving the EW suite of armoured vehicles and a space unit could collaborate with ISRO to enhance satellite capabilities. These specialized teams not only enhance combat readiness but also foster a culture of innovation, positioning the Indian Army as a leader in 21st-century warfare.

Leadership Commitment: The Chief of Army Staff, General Manoj Pande, has emphasized leveraging technology as a catalyst for transformative change, aligning with the "Indianise to Modernise" slogan. This top-down commitment sets the tone for organizational culture.

Recruitment of Technically Skilled Personnel: The Agnipath scheme, introduced in June 2022 and implemented from September 2022, recruits young soldiers called Agniveers for a four-year tenure, with a specific category, Agniveer Technical, targeting youth with pre-existing technical qualifications from ITIs and polytechnics. The scheme aims to build a technically adept military, with Agniveers receiving advanced training and certifications in areas like engineering, medicine, computer technology, aviation, and communications, enhancing their role in maintaining and repairing military equipment and managing IT systems.

Incentivizing Innovation: The "Inno-Yoddha" competition, rebranded as an annual platform, allows soldiers to present ideas, which are tested at formation levels and assessed for relevance and uniqueness. Innovators receive recognition for exceptional contributions. Career progression support and extended deputation for technical roles further motivate personnel.

Collaboration and Partnerships: The Army is expanding collaborations with academia, start-ups, and industry to fast-track advancements. The Foundation for Innovation and Technology Transfer at IIT Delhi partners for production agency selection, aligning with the defence ministry's reform agenda.

Training

Training is the key to technology adoption, and the Indian Army must explore innovative ways to prepare their soldiers for battle environment which is technology-enabled. Continuous education is essential for army personnel to remain current in regards to emerging disruptive technologies and their use for military purposes. Continuous education should move beyond the use of simple workshops and lectures to establish a series of workshops, seminars and courses from academic and industry partners as a robust framework of what continuous education may look like beginning with the basics such as AI programming, operational drone use or the ethics of biotech. For instance, junior officers might attend a seminar on how a smart munition can be integrated into their infantry tactics, while their senior commanders may attend a seminar on the use of space-based ISR and how it will alter the current paradigm. By creating an environment of continuous learning, the Army prepares its personnel to be flexible and competent, when technology becomes available and it can be used in conflict.

To translate knowledge into practical skills, simulation-based training must become a cornerstone of Army preparedness, leveraging advanced simulations, war games, and VR/AR platforms. These systems fully immerse soldiers in a simulated realistic scenario combined with technology such as being in command and control of a fleet of Unmanned Ground Vehicle (UGV), or executing air defense against an unseen hypersonic missile strike. With each exercise, soldiers will gain experience of making time bound decisions and coordinating within a virtual world that is as dynamic as the real battle field, without being exposed to danger. VR can easily replicate the chaos and dangers of urban combat, complete with smart weapon interfaces, while AR overlays tactical data during field exercises, enhancing real-world training. Such realistic training, not only saves crucial resources while building technical proficiency but also fosters teamwork and adaptability, preparing units to operate seamlessly with autonomous systems, networked sensors and energy weapons. By scaling these simulators across all ranks and roles, the Army can maintain a high state of readiness for the complexities of future conflicts.

Creating the Mindset: The senior hierarchy of the Army needs to be trained first on the overall technological ecosystem and how the emerging technologies are fast transforming the landscape of modern warfare. Special courses designed to inculcate the culture of technological advancement in the upper echelons of the Army will have an overwhelming effect on all facets of training, recruitment, planning and procurement. Once the hierarchy's mindset is aligned with modern warfare, it would be easy to hasten the processes.

Reimagined Training Programs: Revise training courses to include simulation and virtual reality training events to model multi-domain operations. This should include technical training and doctrinal adaptation so that soldiers can operate with advanced systems while they are training.

Continuous Education: Implement future warfare course and training programs to keep personnel updated on advancements in disruptive technologies and their military applications.

Collaborate with SMEs: Incorporate the best minds in the technical field to train personnel. Invite, hire and recruit subject matter experts to impart training to all ranks during courses and training exercises.

Doctrine

The Army's doctrinal framework must keep pace and ideally evolve to reflect the realities of modern warfare, ensuring that strategies and tactics align with technological advancements. Updated warfare concepts are essential, incorporating hybrid warfare, blending conventional and irregular tactics alongside autonomous systems, AI, and space-based operations. This requires revising pamphlets and manuals and modifying operational plans to integrate modern platforms and tools, like UAVs into reconnaissance doctrines, AI into targeting protocols and satellites into communication strategies. For instance, a new doctrine might outline how infantry units coordinate with drone swarms or how artillery leverages hypersonic munitions for deep strikes. By embedding these concepts into its intellectual foundation, the Army ensures that its approach to conflict remains relevant, leveraging technology to outmanoeuvre adversaries across multiple domains.

The significance of agility and adaptation is equally substantial, as technological change traverses the globe at a rapid pace, and calls for an agile and adaptive force. Doctrines must underscore rapid adaptation, at a pace that compels units to integrate new systems like laser defences or quantum networks within weeks, not years – while empowering decentralized decision-making to exploit fleeting opportunities on the battlefield. This might involve training platoon leaders to adjust tactics on the fly when supported by autonomous UGVs or enabling air defence units to counter novel threats without awaiting higher approval. By fostering an agile mindset, supported by streamlined procurement and experimentation processes, the Army can stay ahead of adversaries who may deploy unexpected innovations, maintaining operational flexibility in an unpredictable global security landscape.

Doctrinal Change: Revisit military doctrines to embrace contemporary network-centric, multi-domain warfare. This includes creating new tactics, techniques, and procedures (TTP) for integrating new technologies into conventional operations.

Modern Warfare Challenges: Revise military doctrines to include hybrid warfare, autonomous systems, artificial intelligence, and space operations that reflect the changes and developments in war fighting.

Adapting to Change: Focus on the need for military organization to adapt rapidly to changing technologies and empower decentralized decision-making to achieve agility in operations.

Structure and Organisation

Organisational reform is critical to institutionalise technology absorption, ensuring that the Army's structure supports innovation and integration. We must create cross-functional teams, whereby the military personnel from various arms and services can form collaborative teams with technology experts and think tanks who are in the business of development and innovation of emerging technologies. This shall allow not only boost the development of contemporary military platforms but would allow for insights into the doctrinal aspects right from the conceptual stage of their development. These teams would collaborate on projects like retrofitting helicopters with stealth systems or integrating counter-battery radar into artillery units, fostering a synergy of operational insight and technical expertise. The Army Design Bureau has been incorporating the stakeholders from the private defence industry and the results have been encouraging. As the interaction with industry increases, a complete defence ecosystem will get created to cater for the needs of the armed forces. By creating cross-functional teams to steer important projects, the Army will be able to drive practical innovation, ensuring that new technologies are tailored to the Army's unique needs and seamlessly embedded into its operations.

A centralized apex authority may be needed to coordinate large number of projects and to guide the integration of new technologies for all combat arms and services. This central authority can focus on priorities, allocation of resources, tracking progress and ensuring an achievable and unified approach towards adopting a new system/technology like hypersonic defences or space-based ISR. It could standardise protocols for cybersecurity across all smart platforms, streamline training for smart weapons and evaluate the battlefield impact of niche systems. By providing top-down direction, this body prevents compartmentalisation and duplication, aligning technological advances with the Army's strategic goals and maximizing their operational impact.

Finally, collaborative ecosystems should be cultivated to speed up the development and adoption of defence innovations by accessing external expertise and resources. Partnerships with academia such as IITs or IISc provide opportunities for research into niche domains like quantum communication or nanotechnology protective armour, while partnerships with major technical organisations/ institutes such as DRDO and ISRO will allow for tailor made platforms at comparatively low cost with government oversight. Commercial entities and major defence manufacturers with adequate financial clout will be able to turn prototype defence innovations into scalable solutions. Collaborations with offshore partners, particularly in the US, Russia or Israel, can provide access to existing defence innovation systems, the best practices, and options for joint production and co-development of technology as in case of the Brahmos family of missiles. These ecosystems accelerate innovation by transforming ideas into technologies ready for use in the field, and ensuring the Army is able to leverage global defence innovation and enhancing its capabilities through shared knowledge and collective effort.

Organisational Restructuring: If possible, develop joint coordinated technology integration units that cut across combat arms and services. These units should have the mandate to coordinate R&D, facilitate rapid prototyping, and ensure interoperability among diverse platforms.

Cross-Functional Teaming: Create teams that are a combination of military personnel and technologists to enhance collaboration with defence industry and advance innovation.

Centralized Coordination: Establish a centralised apex body for coordination, oversight and integration of developing technology across combat arms.

Collaborative Ecosystem: Collaborate with academia, industry, and international allies in order to speed technology development through the sharing of experiences and best practices.

Flexible and Adaptive Structure: The Army has been actively creating more flexible and adaptive structures but there is a lot which is required to create a commensurate defence oriented culture in the industry which supports technological innovation and strategic vision. Even our major defence producers have to thrive in the same heavily bureaucratised commercial ecosystem and thus find it difficult to keep pace with the emergent needs of the armed forces. There is an urgent need to infuse flexibility and ease of business in our overall industrial set up so that the complete supply chain is energized to deliver timely defence solutions.

Borrowing Brilliance

The USA and China provide benchmarks for technology absorption. The US military employs strategies like the Third Offset Strategy, leveraging DARPA for breakthrough technologies, and invests heavily in R&D, collaborating with the private sector and academia. China, under Xi Jinping, has consolidated technical functions under the PLA Strategic Support Force, focusing on AI and quantum technologies with state-led initiatives. Both nations emphasize strategic planning, organisational restructuring, and international cooperation, lessons the Indian Army can adopt. To effectively absorb and integrate these technologies, the Indian Army should consider the following, tailored to each domain and supported by general strategies.

Strategic Planning: Formulate a comprehensive strategy for technology absorption, aligned with national security objectives and the "Year of Technology Absorption" initiative in 2024, emphasizing *Atmanirbharta*.[35] Have a long term plan on emerging technologies with permanent staff and leadership to carry through the development through complete cycle of fielding these products.

Organisational Restructuring: Consider creating dedicated units or command for emerging technologies, similar to the US DARPA model, to streamline integration and innovation. Do not rely on DRDO solely for technological feats.

Budget Allocation: Ensure sufficient funding, addressing low expenditure on defence R&D (e.g. INR 18,669.66 crore in 2021-22 against a budget estimate of INR 20,757.44 crore),[36] and support private sector R&D capabilities through adequate handholding and assured business.

Continuous Assessment: Regularly evaluate technological advancements and refine strategies, learning from employment of technologies in recent conflicts and their impact on warfare to maintain a technological edge.

Doctrine Development: Update military doctrines simultaneously with induction of niche technologies, ensuring seamless integration into operations and training. Technologies such as autonomous platforms have wide ranging impact on tactics, doctrine, logistics as well as manpower management. Therefore, induction of such platforms require a holistic adjustment to existing processes.

Training and Exercises: Conduct simulations and exercises that simulate the use of EW, cyber, AI and quantum technologies to keep the troops abreast with the latest developments on the battlefield, keep them tech savvy and enhancing operational readiness.

Agile Procurement: Streamline procurement processes to quickly adopt new technologies, reducing dependency on imports and push for special provisions for procurement of niche systems.

Culture of Innovation: Foster an environment that encourages innovation within the Army, leveraging initiatives like Innovations for Defence Excellence and MSMEs for technology development.

"Soldier Scientist and Scientist Soldier" Concept

To maintain a competitive edge in modern warfare, the Indian Army should champion the innovative concept of Soldier Scientist and Scientist Soldier, fostering a culture where military personnel and scientific minds collaborate closely to drive technological advancements. The Soldier Scientist approach involves inspiring soldiers to contribute to research and development, converting them from mere users of technology into active innovators. This can be achieved by introducing specialised training programs that teach soldiers foundational skills in areas like artificial intelligence, robotics, or materials science, enabling them to identify operational gaps and propose practical solutions. Our premier institutes like College of Military Engineering, Pune, Military College of Telecommunications & Engineering, Mhow, Army Institute of Technology, Pune are fine examples of this approach. Given a chance, a soldier on the frontlines could join forces with engineers to design portable, solar-powered communication devices suited for remote terrains.

Conversely, the Scientist Soldier initiative requires entrenching civil scientists within army units to experience first-hand the challenges of combat environment, such as the need for lightweight equipment or robust systems in extreme weather. This engagement would ensure that scientific research is tailored to the Army's real-world needs, producing technologies like AI-enhanced targeting systems or biodegradable field supplies that are both innovative and deployable. To implement this vision, the Army could establish innovation hubs or "Tech Battalions," where soldiers and scientists work together, supported by adequate funding for experimental projects and have access to cutting-edge facilities. If we can have an Ecological TA Battalion, might as well have a Technological TA Battalion too. Regular competitions, hackathons, and joint workshops could further encourage this synergy, creating a vibrant ecosystem where ideas flow freely between the battlefield and the laboratory. By promoting this dual concept, not only will the Indian Army be able to accelerate the development of relevant technologies but also nurture a progressive workforce ready to adapt to the complexities of future conflicts.

This "Soldier Scientist & Scientist Soldier" approach is exemplified by several international models and recent Indian ones too. For instance, the Israeli Defence Forces actively encourage soldiers to pursue higher education in STEM fields and later contribute to defence research and development, a strategy that has led to ground breaking innovations in unmanned systems and cyber capabilities. Similarly, the Canadian Army's Integrated Soldier System program integrates advanced technological training with field operations, thereby transforming ordinary troops into agile operators capable of managing complex digital networks and sensor systems.[37] Armies world over have begun to incorporate elements of augmented reality, biometric monitoring, and real-time data feeds, effectively blurring the lines between frontline operations and technological innovation.

In a remarkable example of this concept, a young Indian Army officer, Colonel Prasad Bansod, has emerged as the driving force behind the design and development of India's first-ever 9mm machine pistol – the ASMI. The Colonel, who actually doesn't come from a technical background was able to achieve this feat on the basis of rich operational experience which enabled the DRDO to create a prototype in a very short time. 550 pieces of this weapon have already been ordered by the Army.[38] Similarly, the Ten AI Weapon System (TAIWS) developed by Col Ashish Dogra and Lt Col Prashanth Agrawal of EME in association with IIT Bombay and students of MIET, Jammu, has completed field trials and after certain modifications is expected to be deployed on the Line of Control.[39]

By institutionalizing joint training modules, inter-service exchanges, and immersive field exercises where soldiers collaborate with scientists and engineers, the Indian Army can foster a culture in which every soldier becomes a potential innovator and every scientist gains the experiential insights necessary to develop truly operational solutions. On the other hand, the Scientist Soldier approach integrates scientists into military environments to align research with battlefield realities. For instance, scientists from the Defence Food Research Laboratory (DFRL) worked with Army units in high-altitude regions, resulting in a lightweight, high-calorie ration pack tailored for sub-zero conditions, improving troop nutrition and mobility.[40]

These examples highlight how soldiers and scientists, when collaborating closely, can produce deployable technologies that enhance operational effectiveness. The Indian Army's Army Design Bureau, the directorate which has been spearheading the development of niche products, has filed for the registration of 75 Intellectual Property Rights (IPR) of which 12 have already been granted. The products have been designed and developed in-house by "service innovators" or funded by the Army Technology Board. Currently, the board is funding 100 projects of which 75 are in final stages.[41] This symbiotic relationship will ensure that battlefield innovations are not only technologically advanced but also practically viable, tailored to the unique challenges of modern warfare.

Leveraging DRDO and ISRO

The Indian Army must strengthen its collaboration with the Defence Research and Development Organisation (DRDO) and the Indian Space Research Organisation (ISRO) to harness their expertise and bolster its technological array. These premier institutions offer unparalleled capabilities in critical domains, DRDO with its specialisation in defence systems like missiles and drones, and ISRO with its advancements in satellite technology and space-based applications. At the same time, several projects which have been delayed must be revitalised and fixing accountability for the delays is one of the best measures to accelerate development.

To benefit from the strengths of these two organisations, the Army should pursue joint development programs, pool resources to create army specific solutions such as hypersonic missiles for precision strikes or satellite constellations for real-time battlefield intelligence and navigation.

For example, integrating DRDO's missile guidance systems with ISRO's navigation satellites could enhance the Army's ability to conduct long-range operations with pinpoint accuracy. This collaboration can work through regular coordination mechanisms, including joint task forces, shared testing facilities, and technology transfer agreements that ensure seamless knowledge exchange. Posting or subsuming Army officers within DRDO and ISRO project teams during the R&D phase would further align innovations with operational requirements, reducing the gap between prototype and deployment. Additionally, conducting joint field exercises where technologies like quantum communication devices or autonomous drones are tested in simulated combat environment can refine these systems for practical use. All these steps will only bear results if there is a system of time bound results being monitored at the apex level. Since, it's a matter of national security, sluggishness by any of the stake holder must not be tolerated at any stage of development. By fostering this synergy, the Army can tap into cutting-edge advancements tailored to India's unique security challenges, such as securing high-altitude borders or countering threats in dense urban areas. This collaborative approach not only strengthens the Army's warfighting capabilities but also reinforces India's self-reliance in defence technology, ensuring that national scientific prowess directly translates into military readiness.

DRDO and ISRO, the twin pillars of India's indigenous defence innovation, offer a formidable foundation for transforming the operational capabilities of the Indian Army. DRDO has attained significant success in the field of advanced missile systems, radars, and directed energy weapons, and the Akash and Shaurya missile systems are prime examples of its technological prowess.[42] Another prominent example is the BrahMos supersonic cruise missile, a precision-strike weapon developed in partnership with Russia's NPO Mashinostroyenia and inducted into the Army, enhancing its firepower.[43] The "Integrated Guided Missile Development Programme (IGMDP)" is also a notable achievement, wherein DRDO and the Army jointly developed the Agni and Prithvi missile series, strengthening India's deterrence capabilities.[44] ISRO offers the advantage of satellite technology as demonstrated by its GSAT series and IRNSS (NavIC) navigation system, providing an extraordinary advantage of space-based surveillance and secure communications to armed forces.[45] The RISAT-2B satellite, with its synthetic aperture radar (SAR), provides all-weather surveillance critical for the Army's intelligence, surveillance, and reconnaissance (ISR) operations along borders.[46]

Additionally, ISRO's NavIC navigation system offers precise positioning data, improving the accuracy of artillery and missile systems. These examples showcase how DRDO and ISRO's expertise in defence and space technologies directly manifests into operational readiness for the Indian Army. The synergy between these organisations can accomplish complex systems, such as the DRDO's Communication-Centric Intelligence Satellite (CCI-Sat) project, which leverages ISRO's launch capabilities, is poised to provide enhanced real-time intelligence, thereby bridging the gap between ground operations and space-based assets.[47] In addition, new joint projects including one recently proposed to counterweight the integration of DRDO sensor technologies with ISRO satellite systems reveal the benefits of a holistic, multi-domain solution. These examples showcase how DRDO and ISRO's expertise in defence and space technologies directly translates into military readiness for the Indian Army.

Handholding Startups, MSMEs and Industrial Complex

The Indian Army should actively nurture startups, Micro, Small, and Medium Enterprises (MSMEs), and the broader industrial complex to unlock a wealth of innovative solutions adapted to its operational needs. Startups and MSMEs, with their agility and creativity, are well-positioned to address niche challenges, such as developing low-cost surveillance drones or wearable sensors for soldier safety, that larger firms might not find lucrative. To support such entities, the Army could establish defence innovation incubators within military stations, providing access to testing ranges, technical expertise, and mentorship from seasoned personnel. For example, a startup designing AI-based logistics optimisation tools could refine its product with Army input and field trials with a logistic unit. Alongside this, offering financial assistance, such as grants, seed funding, or assured procurement contracts would enable these small players to scale their ideas from concept to reality. The Army Design Bureau does provide handholding support, guiding startups and MSMEs through the complexities of defence standards, certifications, and supply chain integration via workshops and dedicated liaison officers. Simultaneously, partnering with the industrial complex comprising established manufacturers with kind of monopoly in their respective fields, ensures that successful prototypes can be mass-produced efficiently. For example, a startup's breakthrough in making ruggedised batteries, could go in for mass production through collaboration with a major industrial entity, delivering reliable power solutions to troops in the field. By promoting such an ecosystem, the Army can create a stream of home-grown innovations, reducing dependence on foreign technology while boosting India's defence economy. This strategy not only enhances military capabilities with bespoke, cost-effective solutions but also positions the Army as a catalyst for technological entrepreneurship, aligning national security with economic growth.

It is imperative for the Indian Army to bolster this ecosystem to be more agile and innovative. Successful models of such arrangements exist both domestically and internationally. For instance, the US defence sector's collaboration with Silicon Valley startups through initiatives like the Defence Innovation Unit (DIU) has accelerated the development and fielding of novel technologies such as AI-driven analytics and unmanned systems.[48] In India, the Ministry of Defence's Innovations for Defence Excellence (iDEX) program is already a prime example of how startups and MSMEs are being engaged to develop cutting-edge systems, from advanced counter-drone systems to lightweight, high-performance materials.[49] Companies like NewSpace Research & Technologies have partnered with HAL on projects such as the HAL Combat Air Teaming System, demonstrating the potential for small, agile firms to contribute significantly in realising state of the art defence projects.[50]

Additionally, numerous MSMEs across India have been instrumental in supplying specialised components for indigenous platforms like the Tejas fighter and the Akash missile system. By establishing dedicated mentoring mechanisms, providing technical guidance and streamlining regulatory processes, the Indian Army can create an environment where innovation thrives. Encouraging public-private partnerships through joint ventures and preferential procurement policies will further integrate these enterprises into the national defence ecosystem, ensuring that ground-breaking ideas are rapidly developed, tested and scaled to meet long term operational needs. Such a strategy not only enhances technological self-reliance but also reinforces India's larger "Make in India" initiative, thereby bolstering economic growth and industrial competitiveness in the defence sector.

One prominent example is the partnership with Tonbo Imaging, a startup based in Bengaluru that has produced the "Eye of God," a lightweight thermal imaging system enhancing night vision for infantry soldiers, now widely used by the Army.[51] Under the "Innovations for Defence Excellence (iDEX)" initiative, startups like Big Bang Boom Solutions have created anti-drone systems currently under trials by the Army, illustrating how small firms address niche requirements of the forces.[52] The Army Technology Board (ATB) also funds MSME projects, such Innovative Simulator for Unmanned Vehicles by Combat Robotics, a Pune-based startup.[53] On the industrial front, the Army partnered with Tata Advanced Systems to mass-produce wheeled armoured vehicles originally designed by a smaller firm, boosting mobility solutions.[54] These examples demonstrate how the Army's handholding and partnerships enable startups, MSMEs, and larger industries to deliver tailored, cost-effective technologies for military use.

Understanding Atmanirbharta in Defence

The debate surrounding *Atmanirbharta* in Defence, or self-reliance in defense production, reflects India's ambition to reduce dependency on imports and bolster indigenous capabilities, especially given its historical position as the second-largest arms importer globally, as noted in recent reports. *Atmanirbharta* aims at achieving complete sovereignty over the life of a weapon, platform or a system, right from the drawing board stage to its in service exploitation and finally replacement by a more capable and latest version. Thus, it seeks for singular focus on the inception, conception, validation, scaling and action (ICVSA) cycle.[55]

Atmanirbharta in the manufacturing sector is a government initiative aimed at enhancing indigenous production, as proven by recent contracts worth Rs 39,125.39 crore signed by the Ministry of Defence, indicating a significant boost to the Make in India initiative.[56] This strategy seeks to mitigate risks from global supply chain disruptions, such as those seen during the COVID-19 pandemic and the Russia-Ukraine conflict, where nearly 60% of India's equipment imports from Russia were affected. Indigenous production, like expensive emerging technology based platforms, is expected to lower costs in the long term compared to continuous imports, despite initial expenses in R&D and production. Once the product hits the assembly line and the complete ecosystem of maintenance, repair and replacement comes up, the initial cost gets easily recovered.

Balancing Indigenous Production in a Global Context

Given the interconnected nature of the global defense industry, the Indian Army must balance self-reliance with international collaboration. Research suggests prioritizing critical technologies for indigenous development, such as AI, cyber security, and advanced materials, which are essential for national security. For instance, the Army has established a quantum laboratory at the Military College of Telecommunications Engineering (MCTE) in Mhow, focusing on Quantum Key Distribution and Quantum Computing, which surely will get support from the national allocation of INR 8,000 crore over five years to this field under the National Mission on Quantum Technologies & Applications.[57]

However, for technologies where domestic expertise is limited, leveraging global partnerships is crucial. This can involve joint ventures, technology transfers, or licensed production, as seen in the C-295 aircraft project, where the first 16 units are imported from Airbus, and the remaining 40 are produced domestically at the TATA Aircraft Complex in Vadodara.[58] This approach ensures access to cutting-edge technology while building local capabilities.

Impact on Modernization and Absorption of Emerging Technologies

The push for *Atmanirbharta* is likely to impact the Indian Army's modernization, particularly in absorbing emerging disruptive technologies such as AI, cyber warfare, drones, and quantum computing. The Army is investing in AI through the Army Technology Board, collaborating with Indian Institutes of Technology (IITs) and Indian Institutes of Sciences (IISc), and has an AI Centre of Excellence at MCTE. Applications include intelligence, surveillance, reconnaissance, targeting, logistics, and planning, as highlighted in recent analyses.

However, challenges exist, such as reliance on legacy voice-centric communications rather than data-centric systems, and limited success in incubating AI-based research due to non-specialist officers with short tenures. To address this, the Army is acquiring Software Defined Radios (SDRs) for secure communication, with tenders issued to enhance infantry mobility. An unexpected detail is the growing role of startups and MSMEs, supported by the Innovations for Defence Excellence (iDEX) scheme, which could drive innovation and reduce reliance on foreign suppliers, as seen in the development of secure Army Mobile Bharat Version (SAMBHAV) handsets.[59] To effectively balance *Atmanirbharta* with global interdependence, the following recommendations are proposed:

- **Enhance Training and Education:** Funding must continue to improve advanced training facilities at MCTE, AIT, CME and other training centres to ensure that personnel can cope with the rapidly developing technologies. An example of this is the training for cyber warfare through the use of cyber ranges and labs.

- **Accelerate R&D Efforts:** Increase funding for R&D, particularly in AI, cyber, and quantum technologies, and encourage innovation through competitions and collaborations with academia and industry. For example, the Army's collaboration with IITs and IISc for AI development is a step in the right direction.

- **Upgrade Infrastructure:** Modernize communication and information systems to support data-centric operations, moving away from legacy systems. The acquisition of SDRs is a critical step, with a focus on man-portability for infantry mobility.

- **Develop Indigenous Solutions:** Prioritize indigenous development of critical systems, such as modern infantry rifles (70,000 AK-203s for the Army)[60] and unmanned ground vehicles like DRDO's Daksh,[61] to reduce import dependency. The Union Budget 2025-26 has allocated a record Rs 6.81 lakh crore to the Ministry of Defence marking a 9.53%increase from the previous fiscal year and constituting 13.45% of the total Union Budget, the highest among all ministries.[62] This comprehensive allocation underscores the government's commitment to developing technologically advanced and self-reliant armed forces, aligning with the vision of 'Viksit Bharat 2047'.

- **Integrate Technologies into Doctrine:** Update military doctrines to incorporate the absorption of emerging technologies, ensuring rapid adaptation to modern warfare, as underscored by the Defence Minister on the 75th Army Day in 2023.[63]
- **Strengthen Public-Private Partnerships:** Foster closer collaboration between government, defense public sector undertakings (DPSUs), private companies, and academia. The private sector already contributes 21% to production,[64] as seen with TATA's role in the C-295 project.
- **Develop Family of Systems and Platforms:** A family of weapons and systems can be developed as a cohesive ecosystem where individual platforms share components, logistics, and operational frameworks. Develop common standards for ammunition, power systems, communication protocols, and software to ensure compatibility across the family. This concept, often referred to as a "family of systems" (FoS) or "system of systems" (SoS), is widely used in aviation and aerospace and will not only ensure economy of resources but also standardisation, interoperability, maintainability, modularity, multi-utility and future proof our systems.[65]
- **Streamline Procurement Processes:** Simplify and expedite procurement procedures to encourage domestic production and reduce delays, ensuring timely acquisition of modern technologies.
- **Support MSMEs and Startups:** Provide financial and technical support through schemes like iDEX and the Technology Development Fund, targeting innovation in areas like AI and smart weapons.
- **Establish Testing and Certification Facilities:** Set up advanced testing and certification centers to ensure indigenous products meet international standards, supporting quality assurance for exports.
- **Promote Export Opportunities:** India must target international markets, with defence exports growing from Rs 686 crore in FY 2013-14 to Rs 21,083 crore in FY 2023-24, and aim for Rs 50,000 crore by 2029.[66] This not only makes commercial sense but also establishes India as a defence power house which certainly reflects well on its Armed Forces. Key exports such as BrahMos missiles and Tejas aircraft, boost geopolitical strategy too.
- **Develop Comprehensive Policies:** Formulate clear policies for the use of emerging technologies, addressing legal gaps in the Information Technology Act 2000, and ensuring ethical considerations and cybersecurity.

- **Institutional and Legal Considerations:** The Indian Army faces challenges due to the lack of specific laws for AI and quantum technologies, with the current Information Technology Act 2000 lacking provisions for cyber war and defense.[67] There is a need for a Cyber Security Act and updates to address emerging technologies, ensuring a robust legal framework.
- **Economic and Strategic Implications:** The focus on *Atmanirbharta* is expected to boost the economy through job creation and reduced import bills, with defense production reaching Rs 1,27,265 crore in FY 2023-24, up 174% from FY 2014-15.[68] Exports to over 100 nations, including the USA, France, and Armenia, position India as a burgeoning hub for indigenous production, with potential geopolitical benefits in the Asia-Pacific region.

With the Indian Army looking at future war fighting being shaped by new disruptive technologies, it will need to adopt a multi-pronged approach that entails modernisation of every aspect of combat and transformation of its institutional architecture. Each arm, from the Infantry to Armoured Corps, Artillery, Army Aviation, Air Defence, and Space, would require tailor made unique innovations and the adoption of novel systems such as autonomous vehicles, smart munitions, hypersonic artillery, combat drones, integrated air defences, and space-based surveillance.

Eventually, the Army would be able to enhance its lethality, survivability and adaptability across all domains of warfare. Simultaneously, the broader process of technological absorption through reforms in human resource, training, doctrine, and organisational structure, will be crucial to ensuring that these advancements translate into real-world operational superiority. Operational effectiveness will not be solely achieved through procurement of niche systems but through building an all-encompassing ecosystem where all elements of warfare are aligned with blistering pace of technology. By embracing these comprehensive recommendations, the Indian Army can create a decisive edge in multi-domain warfare, securing India's sovereignty and influence in an era defined by technological disruption and strategic competition.

From the Trenches to Tomorrow

As I draw this work to a close, I must admit that, after 25 years of dedicated service as an infantry officer, where thinking from the knees has been the heartbeat of every decision, the author has stepped boldly into the thrilling, vast, and often bewildering realm of emerging disruptive technologies and their profound impact on the future of warfare. With a career forged in the crucible of the frontlines, I have approached this vast and ever-evolving topic with a deep sense of humility, striving to do justice to its complexity and captivate readers with its possibilities. It is my heartfelt hope that you, whether a seasoned military professional or a curious explorer, have found this journey through these pages both enjoyable and illuminating, uncovering insights into the dynamic interplay of technology and conflict. As you turn this final page, may you feel the weight of an infantry officer's courage to venture beyond the familiar, and may the knowledge shared here linger, sparking your own reflections on the challenges and opportunities that await.

Notes

1   IDRW. "Indian Army Bolsters BMP-2 2K Fleet with RFI for Counter Unmanned Aerial Systems (C-UAS) under Make in India." April 30, 2025. https://idrw.org/indian-army-bolsters-bmp-2-2k-fleet-with-rfi-for-counter-unmanned-aerial-systems-c-uas-under-make-in-india/#google_vignette.

2   IDRW. "Indian Army Gears Up for Futuristic Infantry Combat Vehicles." April 30, 2025. https://idrw.org/indian-army-gears-up-for-futuristic-infantry-combat-vehicles/#:~:text=The%20Indian%20Army%20is%20poised,guided%20missiles%20for%20superior%20firepower.

3   The Hitavada. "Achievement: VFJ Embarks on New Overhauling Mission of T-72 Tanks." April 16, 2025. https://www.thehitavada.com/Encyc/2025/4/16/Achievement-VFJ-embarks-on-new-overhauling-mission-of-T-72-tanks.html.

4   IDRW. "Vehicle Factory Jabalpur Begins Overhauling T-72 Tanks, Strengthening Indian Army's Armored Might." April 30, 2025. https://idrw.org/vehicle-factory-jabalpur-begins-overhauling-t-72-tanks-strengthening-indian-armys-armored-might/.

5   Motor India. "Allison Transmission Selected by L&T for India's New Futuristic Infantry Combat Vehicle." 20 Feb 2023. https://www.motorindiaonline.in/allison-transmission-selected-by-lt-for-indias-new-futuristic-infantry-combat-vehicle/.

6   Shivani Sharma, "Zorawar: Advanced Light Tank for Indian Army Near China Border." *India Today*, March 5, 2025. https://www.indiatoday.in/india/story/zorawar-advanced-light-tank-indian-army-china-border-belgium-2689324-2025-03-05.

7   LT Gen KJ Singh, Chanakya Forum. "Zorawar: India's Light Tank – Is It Enough?" 22 May 2024. https://chanakyaforum.com/zorawar-indias-light-tank-is-it-enough/.

8   Army Recognition. "Hyundai Rotem Develops Hydrogen-Powered Next-Gen K3 Tank with Enhanced Stealth and Longer Range." 31 Oct 2024, https://armyrecognition.com/focus-analysis-conflicts/army/defence-security-industry-technology/hyundai-rotem-develops-hydrogen-powered-next-gen-k3-tank-with-enhanced-stealth-and-longer-range#:~:text=On%20October%202023%2C%202023%2C%20Hyundai,friendly%20alternative%20to%20diesel%20engines.

9   Nathan Strout, How the Army plans to revolutionize tanks with artificial intelligence, 30 Nov 2020, https://www.c4isrnet.com/artificial-intelligence/2020/10/29/how-the-army-plans-to-revolutionize-tanks-with-artificial-intelligence/

10 Joe Saballa, MBDA Unveils AI-Enabled Tech to Find Hidden Targets, The Defense Post, June 21, 2024, https://thedefensepost.com/2024/06/21/mbda-ai-find-targets/

11 Indian Express. "Rajnath Singh Hands Over F-INSAS to Indian Army." *Indian Express*, Aug 17, 2022, https://indianexpress.com/article/cities/pune/rajnath-singh-hands-over-f-insas-to-indian-army-8093387/.

12 Lt General P.C. Katoch (Retd) "Indian Artillery Modernisation." SP's Land Forces, Issue 4, 2021, https://www.spslandforces.com/story/?id=766&h=Indian-Artillery-Modernisation#:~:text=Field%20Artillery%20Rationalisation%20Plan%20(FARP,it%20comes%20to%20artillery%20modernisation.

13 Pawar, Balli. "More Power to Fire." Force India, April 30, 2025, https://forceindia.net/guest-column/guest-column-balli-pawar/more-power-to-fire/.

14   Laxman Kumar Behera, "India's Defence Industry: Achievements and Challenges." Observer Research Foundation, May 06, 2024. https://www.orfonline.org/research/india-s-defence-industry-achievements-and-challenges#:~:text=Established%20in%201958 %2C%20the%20 DRDO%20is%20i nvolved,technologies%20for%20both%20strategic%20and %20convention al%20weapons.&text=Some%20of%20the%20big%2Dticket%20items%20being%20manufactured, Rocket%20launchers%20(Tata%2 0and%20L&T)%2C%20among%20others.

15   Rajat Pandit, "Army Steps on the Gas for High-Tech Infusion for Futuristic Warfare, Plans to Induct Domain Specialists." *Times of India*, Nov 22, 2024. https://timesofindia.india times. com/india/army-steps-on-the-gas-for-high-tech-infusion-for-futuristic-warfare-plans-to-induct-domain-specialists/articleshow/115574996.cms#:~:text=We%20are%20 proactively%20 engaging%20industry,benchmarks%2C%20 timelines%20and%20monthly%20reviews.

16   Eletimes. "Swathi Weapon Locating Radar: A Beacon of Indian Defense Technology." Jan 18, 2025, https://www.eletimes.com/swathi-weapon-locating-radar-a-beacon-of-indian-defense-technology#:~:text=Technical%20Features%20and%20Capabilities,uninter rupted%20 operation%20during%20critical%20missions.

17   Business Standard. "Army Aviation Looking at Phasing Out Cheetah, Chetak Helicopters from 2027." Nov 07, 2023. https://www.business-standard.com/india-news/army-aviation-looking-at-phasing-out-cheetah-chetak-helicopters-from-2027-123110701213_1.html.

18   Dinakar Peri, Business Standard. "Army Aviation Looking at Phasing Out Cheetah, Chetak Helicopters from 2027," Oct 23, 2023. https://www.business-standard.com/india-news/army-aviation-looking-at-phasing-out-cheetah-chetak-helicopters-from-2027-123110701213_1.html.

19   Wikipedia. "Corps of Army Air Defence." Viewed April 05, 2025. https://en.wikipedia.org/wiki/Corps_of_Army_Air_Defence.

20   Press Information Bureau. "Project AKASHTEER: First Batch of Control Centre for Air Defence Operations Delivered." Dec 26, 2024. https://pib.gov.in/PressReleasePage.aspx?PRID= 2088180#:~: text=Pr oject%20AKASHTEER:%20First%20batch%20of%20Control%20Cen tre,development%20for%20automation%20of%20Air%20Defence%20Operations.

21   Aerospace Newsletter, Centre for Air Power Studies, Vol IV, No 05, 05 May, 2024 https:// capsindia.org/wp-content/uploads/2024/05/Aerospce-Newsletter-Vol-IV-Issue-05.pdf

22   Ankit Kumar, IAD News. "Military Communication Satellites of India." 2023. https:// iadnews.in/military-communication-satellites-of-india/#:~:text=GSAT%207A%20(Angry%20 Bird),important%20role%20in%20today's%20IAF.

23   Lt Col Narendra Tripathi (retd), IAD Bulletin. "Harmonizing Military Space Ambitions with India's National Space Strategy: A Comprehensive Analysis." April 14, 2024. https://www.iad b.in/2024/04/14/harmonizing-military-space-ambitions-with-indias-national-space-strategy-a-comprehensive-analysis/#:~:text=(a)%20The%20establishment%20of%20the%20Defence% 20Space,in%202018%20con   solidated%20armed%20forces'%20space%2Drelated %20capabilities.

24   Yatish Mahajan, DefenceXP. "Role of Military Satellites in India's Security." October 4, 2020. https://www.defencexp.com/role-of-military-satellites-in-indias-security/ #:~:text=GSAT%207A,IAF%20network%2Dcentric%20warfare%20capability.

25   Ian Livingston. "Technology and the Third Offset: Foster Innovation for the Force of the Future." Brookings, December 9, 2016. https://www.brookings.edu/articles/technology-and-the-third-offset-foster-innovation-for-the-force-of-the-future/.

26   Ranjan Brothers, Indian Defense Analysis. "Himshakti: Indian Army's Most Lethal Electronic Warfare System." Last modified March 25, 2023. https://indiandefenseanalysis.wordpress.com/ 2023/03/25/himshakti-indian-armys-most-lethal-electronic-warfare-system/#:~:text=Samyu kta%2 0%E2%80%93%20largest%20electronic%20warfare%20system,CMC%20and%20 Tata%20Power%20SED.

27   Defence Update. "Himshakti EW: India's Indigenous Electronic Warfare System." Last viewed

April 07, 2025. https://defenceupdate.in/himshakti-ew-india-indigenous-electronic-warfare-system/#:~:text=Himshakti%20electronic%20warfare%20system%20seems,will%20to%20fight%20the%20war.

28 Arindrajit Basu, The United Nations Institute for Disarmament Research (UNIDIR), India's International Cyber Operations: Tracing National Doctrine and Capabilities, International Cyber Operations Research Paper Series Paper 5, https://unidir.org/files/2022-12/UNIDIR_India_International_Cyber_Operations.pdf

29 Data Security Council of India. "FAQs on IT Act 2000 and Amendments 2008. 01 December, 2010. https://www.dsci.in/resource/content/faqs-it-act-2000-amendments-2008.

30 Press Information Bureau, Enhancement of Capabilities of AI Technology, PIB Delhi, 01 AUG 2022 https://www.pib.gov.in/PressReleasePage.aspx?PRID=1846937

31 Press Information Bureau. "Defence Minister Launches New Initiatives to Strengthen Defence Production." 28 March 2022. https://pib.gov.in/PressReleasePage.aspx?PRID=1810442.

32 Press Information Bureau, Indian Army Establishes Quantum Laboratory at Mhow (MP), 29 Dec 2021 https://pib.gov.in/PressReleasePage.aspx?PRID=1786012

33 General Anil Chauhan, Chief of Defence Staff, Tactics Led Force Modernisation, 22 May 2025 https://bharatshakti.in/tactics-led-force-modernisation/

34 Manjeet Negi, Army restructures training of technical entry scheme officers, Apr 27, 2023 https://www.indiatoday.in/india/story/indian-army-brings-31-model-for-young-officers-training-2365592-2023-04-27

35 The Economic Times, Indian Army marks 76th Army Day with Chief Manoj Pande's call for 'Year of Technology Absorption', Jan 15, 2024 https://economictimes.indiatimes.com/news/defence/indian-army-marks-76th-army-day-with-chief-manoj-pandes-call-for-year-of-technology-absorption/articleshow/106853661.cms?from=mdr

36 Forty-Third Report, Standing Committee On Defence(2023-24), (Seventeenth Lok Sabha), Ministry Of Defence https://eparlib.nic.in/bitstream/123456789/2963533/1/17_Defence_43.pdf

37 Government of Canada. Integrated soldier system project. National Defence. Viewed April 11, 2025, from https://www.canada.ca/en/department-national-defence/services/procurement/integrated-soldier-system-project.html

38 Huma Siddiqui, Financial Express. Meet Lt Col Prasad Bansod, infantry school officer behind India's first indigenously developed 9mm machine pistol. 14 Jan 2021, https://www.financialexpress.com/business/defence-meet-lt-col-prasad-bansod-infantry-school-officer-behind-indias-first-indigenously-developed-9mm-machine-pistol-2171014/

39 Sharath S. Srivatsa, The Hindu. Products developed in-house by service innovators catch eyeballs. April 17, 2025, from https://www.thehindu.com/news/national/karnataka/products-developed-in-house-by-service-innovators-catch-eyeballs/article69206902.ece

40 Food Products and Technologies Developed by DRDO, Bimonthly S&T Magazine of DRDO, Vol. 25, No. 5, September-October 2017, https://www.drdo.gov.in/drdo/sites/default/files/technology-focus-documrnt/TF_Oct_2017_WEB.pdf#:~:text=It%20provides%2020.49%20per%20cent%20protein%2C%2013.29,which%20can%20be%20easily%20carried%20or%20handled.

41 Sharath S. Srivatsa, The Hindu. Products developed in-house by service innovators catch eyeballs. April 17, 2025, from https://www.thehindu.com/news/national/karnataka/products-developed-in-house-by-service-innovators-catch-eyeballs/article69206902.ece

42 Anuj Tiwari Indiatimes. "BrahMos to Akash: Here Is Indian Missile List." 21 Dec 2022. https://www.indiatimes.com/trending/social-relevance/brahmos-to-akash-here-is-indian-missile-list-588110.html.

43 Vajiram & Ravi. "Types of Missiles in India." April 17, 2025. https://vajiramandravi.com/upsc-exam/types-of-missiles-in-india/.

44   Brahmos Aerospace, Integrated Guided Missile Development Programme (IGMDP) https://www.brahmos.com/content.php?id=10&sid=25

45   Indian Space Research Organisation (ISRO). "Satellite Navigation." Accessed April 13, 2025. https://www.isro.gov.in/satellitenavign.html#:~:text=The%20system%20will%20be%20interoperable,basically%20two%20types%20of%20services.

46   Vajiram & Ravi. "RISAT-2 Satellite Makes Re-entry into Earth's Atmosphere." 26 Aug 2023. https://vajiramandravi.com/upsc-daily-current-affairs/mains-articles/risat-2-satellite-makes-re-entry-into-earths-atmosphere/.

47   Shantanu K Bansal, IAD News. "Addressing India's Space-Based Surveillance and Reconnaissance Needs." 2022. https://iadnews.in/addressing-indias-space-based-surveillance-and-reconnaissance-needs/.

48   Defense Innovation Unit. "About." Accessed April 17, 2025. https://www.diu.mil/about.

49   Government of India, Department of Defence Production https://www.ddpmod.gov.in/offerings/schemes-and-services/idex

50   Yuvraj Tyagi, HAL's CATS Warrior Quietly Achieves Major Milestone with Successful Engine Ground Run, January 22, 2025 https://www.republicworld.com/defence/global-defence-news/hals-cats-warrior-quietly-achieves-major-milestone-with-successful-engine-ground-run

51   Nilesh Christopher, Startup Tonbo Imaging keeps watch over India's borders, Sep 21, 2016, https://economictimes.indiatimes.com/small-biz/startups/startup-tonbo-imaging-keeps-watch-over-indiasborders/articleshow/54435138.cms?utm_source=contentofinterest&utm_medium=text&utm_campaign=cppst

52   Raghav Patel BBBS Unveils "Vajra Strike": An AI-Powered Direct Energy Weapon for Defense Against Drones Oct 8, 2024 https://defence.in/threads/bbbs-unveils-vajra-strike-an-ai-powered-direct-energy-weapon-for-defense-against-drones.10476/#google_vignette

53   Press Information Bureau, TDF scheme playing a crucial role in promoting 'Aatmanirbharta' in defence; Start-ups & MSMEs being encouraged to enhance capabilities in cutting-edge technology, 03 Jul 2024 https://pib.gov.in/PressReleaseIframePage.aspx?PRID=2030445#:~:text=Till%20date%2C%20a%20to tal%20of%20 77%20projects%2 C,have%20been%20suc cess fully%20realised%20under%20the%20scheme.&text=Combat%20Robotics%2C% 20the%20Pu ne%2Dbased%20start%2Dup%2C %20has%20successfully%20developed%20an %20Innovative%20Simulator%20for%20Unmanned%20Vehicles.

54   Ajai Shukla, Tata Advanced Systems to make wheeled armoured vehicles in Morrocco https://www.business-standard.com/external-affairs-defence-security/news/morocco-signs-with-tatas-for-building-wheeled-armoured-vehicle-in-morocco-124093001120_1.html

55   General Anil Chauhan, Chief of Defence Staff, Atmanirbharta In Defence, p. 110, Ready, Relevant and Resurgent A Blueprint for the Transformation of India's Military

56   TOI News Desk, Ministry of defence signs five major capital acquisition contracts worth Rs 39,125 crore, Mar 1, 2024, https://timesofindia.indiatimes.com/india/ministry-of-defence-signs-five-major-capital-acquisition-contracts-worth-rs-39125-crore/articleshow/108135355.cms

57   Government Of India, Ministry Of Science And Technology, Budget 2020 announces Rs 8000 cr National Mission on Quantum Technologies & Applications https://dst.gov.in/budget-2020-announces-rs-8000-cr-national-mission-quantum-technologies-applications

58   Sushim Mukul, Why C295 aircraft project is a game changer for India. Know in 5 points, Oct 28, 2024 https://www.indiatoday.in/india/story/c295-transport-aircraft-vadodra-facility-game-changer-india-explained-five-points-atmanirbhar-bharat-tata-airbus-2624262-2024-10-28

59   Indian Army Sambhav- A Leap in Defence Capabilities https://www.startupsprouts.in/indian-army-sambhav/#:~:text=In%20a%20groundbreaking%20development%2C%20the%20Ind ian %20Army,instant %20connectivity%2C%20even%20while%20 on%20the%20move.&text= The%20SAMBHAV%20mobile%20ecosystem%20employs%205G%2Drea dy%20handsets, robust %20security%20against%20eavesdropping%20and%20potential%20compromises.

60   Shivani Sharma, Army to boost firepower with 70,000 new AK-203 rifles as part of Russia deal Feb 6, 2025, https://www.indiatoday.in/india/story/india-russia-arms-deal-ak-203-kalashnikov-rifles-indian-army-combat-weapons-2675461-2025-02-05

61   Kathika Roy, P M Naik & P M Kurulkar, DRDO DECK: Remotely Operated Vehicle Based Engineering Solution For Hazardous Environment For Armed Forces, August 31, 2022 https://www.iadb.in/2022/08/31/drdo-deck-remotely-operated-vehicle-based-engineering-solution-for-hazardous-environment-for-armed-forces/

62   Press Information Bureau, A record over Rs 6.81 lakh crore allocated in Union Budget 2025-26 for MoD, an increase of 9.53% from current Financial Year, Feb 01, 2025, https://pib.gov.in/PressReleasePage.aspx?PRID=2098485#:~:text=This%20allocation%20is%209.53%25%20more,Capital%20Outlay%20on%20Defence%20Services.

63   Press Information Bureau, Raksha Mantri graces 'Shaurya Sandhya' organised as part of 75th Army Day celebrations in Bengaluru; Lauds Armed Forces for ensuring the country's territorial integrity & upholding rich tradition with unmatched bravery & sacrifice, Jan 15, 2023. https://www.pib.gov.in/PressReleaseDetailm.aspx?PRID=1891420®=3&lang=1

64   Press Information Bureau, Make in India Powers Defence Growth, Production hit 1.27 lakh crore in FY 2023-24, Exports cross 21,000 crore, Mar 24, 2025, https://pib.gov.in/PressReleasePage.aspx?PRID=2114546

65   General Anil Chauhan, Chief of Defence Staff, Atmanirbharta In Defence, p. 115, Ready, Relevant and Resurgent A Blueprint for the Transformation of India's Military

66   Press Information Bureau, Make in India Powers Defence Growth, Production hit 1.27 lakh crore in FY 2023-24, Exports cross 21,000 crore, Mar 24, 2025, https://pib.gov.in/PressReleasePage.aspx?PRID=2114546

67   What is the Information Technology Act, 2000 (IT Act)?, Apr 04, 2025 https://www.geeksforgeeks.org/information-technology-act-2000-india/

68   DD News, India's Defence Production Surges to Rs. 1.27 Lakh Crore in FY 2023-24, Exports Cross Rs. 21,000 Crore Mark, March 25, 2025, https://ddnews.gov.in/en/indias-defence-production-surges-to-rs-1-27-lakh-crore-in-fy-2023-24-exports-cross-rs-21000-crore-mark/

Afterword

"The real problem is not whether machines think but whether men do."
                                                            —B.F. Skinner

As I bring this exploration of *The Algorithm of War: Code, Compute, Conquer –
Insights for the Indian Army's Next Leap* to a close, I find myself reflecting on the
immense responsibility and opportunity that lie before the Indian Army. The pages
of this book have traced the arc of warfare through the lens of technology, from the
ancient forges that shaped steel to the algorithms that now shape battlefields. Yet,
this is not merely a story of machines or systems; it is a story of people, of soldiers,
scientists and leaders who will determine how these tools are wielded in service of
India's security.

Writing this book has been a journey of discovery and conviction. The deeper
I delved into the potential of emerging disruptive technologies, the more convinced
I became that the Indian Army stands at a pivotal moment. The technologies such
as artificial intelligence, quantum computing, biotechnology, hypersonics, and more,
are not distant dreams but tangible forces already reshaping the global strategic
landscape. For India, a nation with unique geopolitical challenges and aspirations,
the adoption of these advancements is not optional; it is existential.

The chapters of this book have sought to illuminate a path forward,
understanding the historical interplay of technology and warfare, dissecting the
capabilities of emerging innovations and applying them to the specific needs of the
Indian Army's combat arms and Space. The recommendations in the final chapter
are not prescriptive mandates but a framework for action, urging collaboration
with DRDO, ISRO, startups and the private sector, while fostering a culture where
soldiers and scientists learn from one another. The concept of "soldier as scientist
and scientist as soldier" is not a mere slogan but a call to reimagine the ethos of
military service in an age of rapid technological change.

As I conclude, I am acutely aware that the future is not a passive destination
but a reality we shape through our choices today. The Indian Army, with its storied
legacy of resilience and adaptability, is uniquely positioned to lead this
transformation. However, this will require boldness, to embrace new ideas, to invest
in human capital and to forge partnerships that transcend traditional boundaries.

It will also demand humility to learn from global leaders, to iterate through failures and to recognize that no single technology holds all the answers.

To the readers of this book, whether you are a soldier on the frontlines, a policymaker shaping strategy, a technologist pushing boundaries, or a citizen invested in India's future, I hope these pages have sparked both curiosity and resolve. The algorithm of war is not a cold equation; it is a human endeavour, driven by ingenuity, courage, and purpose. Let us commit to writing its next chapter together, ensuring that the Indian Army not only adapts to the future but defines it.

With gratitude for the opportunity to contribute to this vital conversation and with unwavering faith in the Indian Army's potential to conquer the challenges ahead,

June 2025

Maneesh Parthsarthy